Founders of
American
Industrial Design

AF776285

ALSO BY CARROLL GANTZ
AND FROM MCFARLAND

The Vacuum Cleaner: A History (2012)

The Industrialization of Design: A History from the Steam Age to Today (2011)

Founders of American Industrial Design

CARROLL GANTZ

McFarland & Company, Inc., Publishers
Jefferson, North Carolina

Library of Congress Cataloguing-in-Publication Data

Gantz, Carroll, 1931–
Founders of American industrial design / Carroll Gantz.
p. cm.
Includes bibliographical references and index.

ISBN 978-0-7864-7686-2 (softcover : acid free paper) ♾
ISBN 978-1-4766-1650-6 (ebook)

1. Industrial designers—United States—Biography.
2. Industrial design—United States—History. I. Title.
TS139.G36 2014 745.2092—dc23 [B] 2014021287

British Library cataloguing data are available

© 2014 Carroll Gantz. All rights reserved

No part of this book may be reproduced or transmitted in any form or by any means, electronic or mechanical, including photocopying or recording, or by any information storage and retrieval system, without permission in writing from the publisher.

On the cover: Sewing machine sketch
© Spencer Nugent/spencernugent.com

Printed in the United States of America

McFarland & Company, Inc., Publishers
Box 611, Jefferson, North Carolina 28640
www.mcfarlandpub.com

This book is dedicated to the few dozen pioneers of
industrial design who, in the 1920s and 1930s, and
in spite of the Great Depression that gripped America,
embraced the Machine Age and transformed millions of
mass-produced products from clumsy, ugly, mechanical devices into
svelte, attractive, modern designs appropriate to the 20th century.

Table of Contents

Acknowledgments

I am indebted to so many for their generous assistance in the development of this book. Among these are my good friend and design historian colleague, Victoria Matranga, H/IDSA (honorary member of Industrial Designers Society of America), who shares my passion in recognizing pioneers of the industrial design profession, and documenting their work and contributions to the field. She has been design programs coordinator for the International Housewares Association (IHA) for many years, organizing juries for annual IHA student design competitions, and serving as IHA's historian, publicist, and researcher. She is also an independent consultant, curator and researcher of houseware exhibits for museums, design firms, and private organizations, as well as an accomplished writer of design history. For this book, she shared numerous photographs and vital facts about pioneer designers from her vast collection of design history information.

My late friend, Tucker Madawick, before his death in 2006, graciously provided me with extensive literature and photos concerning the history of automotive design and many of its pioneers, for which I am eternally grateful. His career only started in 1939, so he was not one of the original pioneers of the early 1930s, but his artistic talent, business sense, leadership skills, and personality exceeded many of his worthy predecessors. He was a tower of strength to the profession, and is sorely missed.

Another valuable source was the Industrial Designers Society of America (IDSA), from whose files, newsletters, and publications over the years came a number of images for this book. Thanks to senior creative director Karen Berube and executive director Daniel Martinage, both of IDSA. Also thanks to Rosemary Morrow at Redux Pictures for image of Henry Dreyfuss, and to Larry Sommers at the Chicago Athenaeum and Sara Kitchen at Skyhorse Publishing.

Descendants of pioneer designers featured in the book were most helpful with photos and information about their deceased family members. Among these were William Waltman, grandson of Chauncey "Chick" Waltman; John Northup, great nephew of Amos Northup; Linda Braming, the daughter of John McLeod Little; Barbra Reinecke Sedory, daughter of Jean Reinecke; Merrily and Erik Holladay, the daughter and grandson, respectively, of Robert Davol Budlong; Alyce Gordon Rideout, the daughter of John Gordon Rideout; Harry Teague, son of W. Dorwin Teague, Jr.; and Tom Tjaarda, son of John Tjaarda.

Surviving organizations founded by pioneers were generous in providing photos and information about their founders. These include Mariela Cedeno of Nessen Lighting, founded by Walter and Greta von Nessen in 1927; Customer Service at Movado Group, Inc., for the image of Nathan George Horwitt; Curt Bailey of Sundberg-Ferar, founded by Carl W. Sundberg and Montgomery Ferar in 1934; Doug Studer at Deskey, founded in the 1940s by Donald

Deskey; Sarah Matheny of Teague, founded in 1926 by Walter Dorwin Teague; Carol Hwang of CMG Worldwide, representative for Loewy Design, LLC, founded by Raymond Loewy in 1929, for image of Raymond Loewy; Judy Crookes of Henry Dreyfuss Associates, founded in 1929 by Henry Dreyfuss; and Larry Sliker, a former employee of Henry Dreyfuss Associates.

Educational institutions, corporations, galleries, and museums were very helpful, as well. They include David Jameson, Iannelli Collection, ArchiTech Gallery in Chicago, for the image of Iannelli; Stephanie Coleman at Ryerson & Burnham Archives, the Art Institute of Chicago, for the image of David Chapman; Kathy Adelson, researcher at the General Motors Media Archive, for the image of Harley Earl; Janice Woo, director of libraries, California College of the Arts, for the image of Joseph Sinel; Lydelle Abbott at the Milwaukee Art Museum for images of Brook Stevens; Kay Menick at Art Resource and Antonia Moser at the Smithsonian Institution for the image of Donald Deskey; Chris Hunter at the Museum of Innovation and Science for the image of the Hotpoint refrigerator; Robert Dirig, archivist at the Art Center College of Design for information regarding Kem Weber; and Linda Baron at Herman Miller for the image of Gilbert Rohde.

I'd also like to thank the many individuals who volunteered information about the pioneers in this book, including Michael Lamm for information concerning Amos Northup; Christopher Geoffrey McPherson for information regarding Helen Dryden; Chip Rowe for information about Egmont Arens; David Nowacek and Gene Schreckengost for images of Viktor Schreckengost; and my good friends Gifford Jackson for information about Jo Sinel, Budd Steinhilber for personal information about Raymond Loewy, Paul Specht of Goldsmith, Yamasaki & Specht for an image of the Playmate cooler, Hampton Wayt for little-known information about Donald Dohner, and Gifford Jackson for information about many founders.

Preface

In January 1930, just three months into the Great Depression, my father bought a used 1927 Studebaker and took it on a two-week shakedown drive (at 30 miles per hour) from Reading, Pennsylvania, to Miami, Florida, and back, before he reported to his new job (he had just lost his old one). During the trip, he wrote to his 18-year-old fiancée (my mother-to-be) describing his trip, as well as how anxiously he looked forward to their planned marriage in June.

When my mother passed away in 1983, I found these letters that she had saved (he had died 30 years before her). They included trivial remarks and jokes from radio comedians of the time (Amos and Andy), as well as private phrases of endearment, descriptions of the weather, sights, accommodations and landmarks—all the ordinary things anyone might write to a loved one back home. What caught my attention, however, was the unusual way he referred to his car. Phrases such as, "Tell Pop [her father] the machine is just getting warmed up," or in a later letter, "I took the machine to Lansdale and I guess I will go over Friday night and get it."

I was puzzled: *machine*? Why didn't he call it the automobile? The auto? The car? The Studebaker? Why "the machine"? I now realize that like many in his generation who bought their first cars, he was incredibly proud to possess his very own "machine"! Not the machines exclusive to industry or corporations, such as steam engines, airplanes, zeppelins, skyscrapers, or dynamos that had enthralled the public for decades, but finally, his very own, personal, mechanical "machine," providing him the freedom to travel for fun, or for jobs in other places. The automobile was the magic carpet of the 20th century. People were no longer destined, like their parents and ancestors, to remain near the place they were born. My father was now emotionally part of, and committed to, the future: the "Machine Age," as were so many others of his day.

The Machine Age was a time when public fascination with machines and the future inspired a new art form, "modern design," which transformed ordinary consumer products from mechanical curiosities into futuristic fantasies. The creators of these designs were called industrial designers, and their work inspired me to become one of them. As I pursued my life-long career of industrial design, I began to realize that it hardly existed before my birth. I wondered, Where did it come from? How and why did it begin? Who or what started it? As my career progressed, I met and interviewed many founding designers who had in fact invented it, and I began to appreciate the incredible change in industry and popular culture that they had accomplished. So, after my retirement, I began to research these remarkable men and women, realizing that when they died, their role in this extraordinary period of history would be lost.

As a lifelong practitioner of industrial design, I already had a library of books, magazines, professional society newsletters, and articles concerning the early profession, but about 20 years ago, I began writing obituaries of deceased designers for my professional organization by referring to my files or by contacting their descendants. I also interviewed or corresponded with those who had retired in order to learn more about their early careers. Some provided informal autobiographies or documents they had written, while others related incredible stories about their early experiences. In some instances, they had donated their papers to various museum collections.

Most books about individual industrial designers focus on the handful of pioneer designers who became public celebrities in the 1930s, such as Raymond Loewy, Walter Dorwin Teague, Harold Van Doren, Norman Bel Geddes, or Henry Dreyfuss, all of whom wrote books about their careers and dominated the field well into the 1960s. Some may imagine that these few alone created the incredibly rapid and complete transformation of our physical environment to modern design in the 1930s, but this is not the case. There were dozens of others who shared in the creation of this new profession, some of whom preceded the big names by years, but who are little known or remembered today.

This book is about the origins of industrial design in that time of economic trauma, its evolution over the years, and in particular the remarkable, relatively small group of individuals who conceived and created it. Their work includes many of the most popular design icons of the 20th century. This is the first time their careers and biographies have been merged into a single, chronological narrative of the rise of the profession. You will meet them, one by one, as they enter the business arena as young men and women, and follow their careers, some to fame and fortune, others to anonymity.

Their unusual backgrounds, accomplishments, and careers, as can best be known at this time, are followed until their demise. Most were born between 1895 and 1910, and a few even survived into the 21st century. Of course, after these pioneers created the profession in the early 1930s, many younger designers followed them into the rapidly expanding field, including myself, until there are now more than 50,000 industrial designers. While a number of these designers are mentioned in the book as their ranks grew through the 1940s, 1950s, 1960s, and later, it is those few pioneers who started it all who are the stars and focus of this book.

Introduction

For those who are unfamiliar with the general subject matter of this book, it may be helpful to summarize the overall context of the time.

The "Machine Age" is a generic term that refers not just to machines, which had been around in industry for a century before, but to the new way in which the public perceived them. In the middle of the 1920s, people suddenly realized that mechanical machines were not just for large-scale industrial purposes, but that they actually could purchase machines such as cars, washing machines, refrigerators and vacuum cleaners at affordable prices for personal use around the home. They began to regard machines as personal possessions, rather than abstract technology.

At the same time, people also became aware of style, a quality previously considered only in decorative products for display in the home, but of little significance in functional, mechanical products, such as their new household appliances. Now that they were able to afford such products, and regarded them as personal possessions, however, they were extremely proud of them and wanted them to be seen by friends and neighbors as evidence of their increasing financial success. So they wanted them in the latest styles, just like some of the recent European designs being promoted by upscale department stores as fashionable.

In addition to these new perceptions, people began to realize that they could influence the way these products looked and functioned by expressing their preference of competing products on the market through purchase choice, or through customer surveys, commonly used by advertising agencies. Such public demand soon caused some manufacturers of functional, mechanical products to pay attention to the appearance of their products by engaging artists to redesign them. Most, however, saw no such need, since business was booming, and changing tooling was costly.

In the midst of these interrelated developments, the Great Depression hit in 1929, and manufacturers not only lost half of their sales volume, but faced going out of business completely, which many in fact did. They became desperate, and when they learned of industrial designers in 1934, they rushed to locate a design consultant, only to find that the few available were already over their heads with work. Manufacturers' demand for such designers was so intense that it attracted designers from numerous other fields to fill the need, and the industrywide transformation of industrial products into modern styles had begun. It has not stopped since. They were, and are, called industrial designers.

Today there are more than 50,000 industrial designers in the United States. By the time the profession of industrial design first became well known, in 1935, there were only about 36 pioneers who were practicing in the field. They came from all walks of life and businesses, mostly in the creative arts, to address a national problem: the economy was as bad as it had ever been, unemployment was at 25 percent, and industry was on life support because consumers were not spending. America was far behind Europe in design, particularly in mass-produced electromechanical consumer products, because manufacturers of these products relied solely on engineers for their design. Mechanical products were devoid of humanistic attributes such as attractive appearance, convenience of use, or the latest styles.

The pioneer "machine designers," primarily commercial artists, stage designers, and craftsmen, changed all this within a decade, by working with engineers, salesmen, and the public to transform the man-made landscape into what we now call the modern world by redesigning mechanical products to become sleek, stylish, shiny, and attractive. Their creative work attracted consumers, who began spending the money they had been hoarding because of bad economic times. Increased sales reinvigorated industry, brought America out of the Depression, and led to the highest standard of living in the world after World War II.

As will be described later, it was the first successful marriage of art and industry, a visionary objective sought by many European design educators since the beginning of the 20th century to provide well-designed products to the masses. Before that, 19th-century craftsmen had dreamed of providing ordinary people with handcrafted designs previously only affordable by the wealthy. Now, in the midst of the Great Depression, the machine designers accomplished this dream by founding a new profession, which created affordable and artistic designs available in mass-produced quantities.

The original pioneer designers continued to work diligently, and by the end of the 1930s they dominated the rapidly growing field. After World War II, some headed huge consulting firms or corporate design departments to meet the enormous postwar consumer demand for new and better products. As the leaders of the profession, the pioneers confirmed the status of their activity as a legal profession, established professional organizations, recommended educational programs, implemented codes of ethics, and initiated national annual design awards programs. A few became international celebrities and led the formation of international design organizations, or assisted in the economic and design development of war-torn nations after World War II as design consultants abroad. By the time some of the pioneers approached retirement, they had established large, successful organizations that continue to this day.

This is the story of those few dozen pioneers who founded modern industrial design in America and changed the landscape of consumer and industrial products within a decade, one product at a time.

CHAPTER 1

The "Artless" and "Art" Industries

The Machine Age (1918–1941), as it was known even at the time, not only included the Great Depression, but also the preceding decade after World War I when America led the world in industrialization, science, technology, engineering, and consumerism. Except for a brief postwar depression from 1920 to 1921, as President Warren Harding took office, the 1920s in America were a period of economic boom and prosperity, as American industry returned to peacetime production to meet consumer demand for the many new mechanical and electrical products then becoming available on the market for the first time in high quantities and at affordable prices.

Among these were automobiles, led by Ford's Model T, which by 1918 comprised 50 percent of the auto market and which epitomized American mass production, also known then as Fordism. Many home appliances used electricity, which, although only half the country was electrified, mostly in urban areas, was becoming more widespread. Machines, large and small, were the leading edge of technology. Refrigerators were replacing iceboxes, ranges were replacing woodstoves, and vacuum cleaners were replacing carpet sweepers. There were electrical appliances such as mixers, fry pans, coffee pots, waffle irons, juicers, irons, roasters, and toasters. To many consumers, many of whom were women, these industrial products for the kitchen were perceived to be as magical a technology as personal home robots would seem to us today.

Prices sound very reasonable today ($260 for the Model T and $52 for a vacuum cleaner in 1918), but in 2013 dollars, the Ford would cost $4,020 and the vacuum cleaner $804. Still, many households had good wages ($2,000 per year in 1918, or $30,700 in 2013 dollars), and since almost 50 percent of working-class wives worked for wages, many were two-income families. Credit buying was also available. Can you imagine paying 40 percent of your annual salary for a vacuum cleaner? Still, to many families, and with installment plans becoming quite common, it seemed like the American Dream coming true. After 50 years of invention and development of a massive industrial infrastructure, including gigantic steam engines, hydroelectric plants, electrified cities, factories, railroads, bridges, and paved roads, average citizens were finally seeing the products of mass production and industrialization in their own homes. Technology had finally reached their personal, daily lives.

The production system had been refined, improved, and made more efficient by Philadelphia engineer Frederick Winslow Taylor (1856–1915), who in 1911 published *The Principles of Scientific Management,* a system of efficiency experts using stopwatches to measure every step of production in order to reduce time and labor. This process of industrial organization became known as Taylorism, and was known everywhere as the companion to Fordism, the twin jewels in the crown of mass production.

All mass-produced, new, electrical, and mechanical consumer products were invented, designed, and built by engineers, the backbone and champions of 20th-century industrialization. Many were famous and known worldwide. Herbert Hoover, who would become U.S. president in 1928, was himself a wealthy mining engineer. Thomas Edison (1847–1931), Henry Ford (1863–1947), and Orville Wright (1871–1948) were still alive and being celebrated as American heroes as late as 1929.

They had brought electric light, the phonograph, the automobile, and the airplane to the world. All engineers were highly respected and admired as heroes of the age. The number of engineers had increased exponentially since 1870, when there were only a hundred. By 1900 there were 45,000, and by 1930, half a million. They were building modern bridges, dams, roads, skyscrapers, airplanes, ships and trains, as well as consumer products such as automobiles, cash registers, telephones, appliances, industrial machinery, clocks, tractors, shovels and pitchforks.

Engineers were trained in hundreds of technical schools and were familiar with a range of materials, stresses, structures, manufacturing processes, costs, and the many technologies, such as electricity, steam, and combustion engines, that were required to power the manufacturing equipment or the mechanical devices. The overriding objective of all engineers was to create a physical structure which, whether a suspension bridge, an airplane, a vacuum cleaner, or a toaster, was strictly functional, economical, and efficient. These were, after all, machines intended to provide humans with transportation, power, and labor saving, and these were the functions expected, highly admired and acclaimed by the public.

Most consumers were delighted with any of the new mass-produced electromechanical goods being made available, despite the fact that they were generally rather mechanical looking and ungainly in appearance, and although for many they were still rather expensive. But their mechanistic look was initially a positive image of technology. The main thing was that they performed their functional objective, were available at reasonable prices, and offered a higher standard of everyday living. For those many new consumers who had never had a vacuum cleaner before, its function, not its appearance, was of primary interest and the greatest source of user satisfaction. Functional performance, power, and labor saving were the coins of the realm, and all the new appliance manufacturers competed successfully on that basis.

Therefore, there were few "artists" needed or desired in the new mechanical product industries, other than those that applied a few surface decorations to the finished product, often just pinstripes or flourishes suggesting a token "human" or "artistic" touch. It is no surprise then that these products, which included autos, locomotives, cash registers, clocks, optical goods, sewing machines, washing machines, fountain pens, refrigerators, and electrical/mechanical machinery, were generally categorized at the time as the "artless industries," because they were generally designed and produced by engineers, not artists.

When we talk about the Machine Age, it is this specific category of mass-produced products, the "artless industries," that is referred to, not handcrafted artistic products such as decorative objects or crafts with limited production. Not only did machines of mass production manufacture these mechanical products in the millions, but in individual function and character, the products themselves were in fact mechanical, or machinelike devices, often with multiple moving components, such as motors, mechanisms, or engines, and were often operated by electricity.

When we talk about "design," we mean the formal definition: "a plan or drawing produced to show the look and function of an object before it is built or made." A designer, of course, is the person producing such a plan or drawing. Designers differ from artists in that they create work that has a purpose other than that of individual self-expression or esthetic beauty. Often the purpose is to meet the needs of an individual, a company, an audience, a functional operation, or all combined.

When we talk about "machine designers," the primary subjects of this book, we are talking about a small, innovative group of artistic individuals in the 1930s, who accomplished what no other artists in history had done: they abandoned the artist's traditional personal freedom of self-expression and applied their professional, artistic, creative, and analytical skills to mass-produced products that previously were designed solely by engineers, in order to make them more attractive to consumers, to make them more convenient and comfortable for users, to use more appropriate and economical materials, and yet to accommodate the normal business interests of cost, schedules, and manufacturing ease.

To achieve this, these artists had not only needed to create attractive, hand drawn, concept sketches of their ideas to communicate them to others, but to produce mechanical drawings to instruct engineers in the specific dimensions and final shape to be manufactured by production tools. They had to work, not as an individual, but as a team member with other specialized industry disciplines, such as engineering, marketing, manufacturing, and sales. In short, they became multitalented artists-designers-engineers-businessmen-salesmen, inventing a brand new profession that as yet had no specific name.

The term "artless industries" had evolved to distinguish certain mechanical products from the traditional "art industries," as defined by the Twelfth Exhibition of Contemporary American Industrial Art at the Metropolitan Museum of Art, New York, October 12 to November 22, 1931.[1] The "art industries" were the only industrial "art" supported, promoted, and exhibited by museums. These were industries producing decorative arts, which included ceramics, glassware, enamelwork, furniture, leatherwork, printing, jewelry, art metalwork, lighting fixtures, carpets, drapes, silverware, textiles, and wallpaper, or any product intended as a decorative object for the home. A number of these were in fact functional as well as decorative, so their designers were necessarily involved with form and function as well as surface decoration.

By and large, these "decorative arts" industries grew out of the Arts and Crafts movement of the 19th century, where skilled craftspeople originally selected appropriate materials and made both decorative and functional objects individually by hand, shaping and/or decorating the surfaces manually or by hand-operated machines of the era. The craftsman was in control of all the physical and functional aspects of a product, as well as its final appearance. Even after machines were developed to produce such products in quantity, many were still designed before manufacture by craftspeople, who therefore still controlled both the functional form and its external decorative treatments.

This personal involvement of a single individual craftsman in the total design of a particular product was common historically, and resulted in famous craftsmen names such as Paul Revere (silverware), Thomas Chippendale (furniture), Eli Terry (clocks), and Duncan Phyfe (furniture). Ever since Leonardo DaVinci (1452–1519), art and engineering had gone hand in hand. From the beginning of industrialization, art and industry were closely related. In the early 19th century, the "mechanic arts" or "useful arts,"[2] as they were called, included a wide range of technical skills, such as mechanical design, decorative design, mechanical drawing, materials analysis, crafts, machine design, production planning, art, factory design, physics, mathematics, etc. Over the century, as industrialization advanced and as educational institutions tried to teach such comprehensive technical skills, these all became individually named specialties and were separated from one other. Thus did art and engineering become separate disciplines, with specialized skills and differing motives.

As 20th-century manufacturing produced more and more consumer products in greater and greater quantities, and as factories become more complex technologically, more impersonal, and more industrial in character, fewer and fewer American craftspeople or artists were interested in the artless industries dominated by engineers, dirty machinery, time clocks, and crass commercialism. Even the art industries held limited appeal to artists because such applied art generally went unrecognized and required working nine to five in a factory environment that was less than clean, quiet, or dignified.

So even though many educational institutions offered programs in "industrial art" or "applied art," most aspiring artists tended to prefer the "fine arts," such as painting or sculpture, where they could receive personal recognition and prestige as exhibitors in galleries or museums and could work in their own personal studios. This was the essential conclusion of a $120,000 comprehensive study of industrial arts education from 1920 to 1922 by Charles R. Richards of the Cooper Union, funded by the General Education Board of the federal government.[3]

At the time, the terms "industrial art" and "applied art" were the most commonly used terms for art careers in industry, and as noted above, they were not popular choices for aspiring artists. But there was another, more obscure term used occasionally in the fine arts field since 1839: "industrial design," also referring to the use of artists for industrial purposes. This term had more recently been specifically defined for industrial purposes in 1913 by the U.S. commissioner of patents, Edward B. Moore, who urged modifications of existing regulations to protect the property of "industrial design, the distinguishing *form* of products that have marketable value" (italics in original).

The existing regulations he referred to were design patents, which protected the exterior appearance of an object against copying since 1842, when they were first initiated by the patent office for "any new and original shape or configuration of any article of manufacture not known or used by others before."[4] Many museums encouraged the copying of European products by American manufacturers because they were not protected outside of the country of origin. Commissioner Moore was concerned that even U.S. design patents were not being enforced or upheld aggressively in America, thus discouraging innovative designs by Americans.

America found itself in a conflicted situation in the 1920s. It commanded the most productive manufacturing industry in the world, as well as the most lucrative market for consumer products. Factories were making as many products as possible, and a stable economy enabled a mass market of middle-class consumers to buy as many products as industry could produce. Nevertheless, a few art museums and organizations, who were familiar with the European potential to dominate world markets, were warning that America was far behind Europe in applied arts and was in desperate need of more future designers in industry to compete globally. In 1918, Walter Sargent (1868–1927), director of the Department of Fine and Industrial Art at the University of Chicago, had written,

> The situation regarding industrial design in the United States is improving, but so far as we can estimate ... we shall need after the war about *fifty thousand more industrial designers* [author's italics] than are now available or in training, and probably few can be imported. Each country will need its own ... we shall have to depend on ourselves more than in the past, not only for designers, but also for styles of design.[5]

Meanwhile, during the same time, the situation in Europe was quite different. While America enjoyed the postwar benefits of mass production and a strong economy, Europeans were suffering from wartime devastation and inflation, with virtually no industrial production to restore economic health and recovery, and practically no consumer markets. America was fortunate in that it had no foreign competition.

Europeans had seen the American industrial model as the key to the future for some years before the war. In 1909, Italian poet Filippo Tommaso Marinetti (1876–1944) founded a fine arts movement called Futurism, which promoted speed, technology, youth, automobiles, and airplanes, but also, sadly, fascism and war. Followers believed that the power of art and technology could change the social structures of the world and produce a dynamic, aggressive, and strong Italy, but political upheaval, scarce national resources, and World War II would prevent European countries from competing effectively with U.S. manufacturing for another 30 years. In Germany, the march of Nationalist Socialism would within ten years lead inexorably to Hitler and German rearmament, and in Russia, the Bolsheviks would transform the country into communism, where nationalized industry would support the government, not individuals.

However, Europeans still led the world in modern art. In fact, the modernist movement had begun in the last part of the 19th century. France, for example, had introduced Impressionism and led Art Nouveau at the turn of the century. Prominent artists such as Pablo Picasso, Henri Matisse, Fernand Léger, Georges Braque, Marcel Duchamp, Francis Picabia, and André Derain led a parade of new interpretations of art. From 1907 to 1911, Picasso, Léger, and Braque introduced Cubism, an art movement in which objects are analyzed, broken up, and reassembled in an abstract form, often depicting objects from a multitude of viewpoints. Cubism has been considered the most influential art movement of the 20th century affecting art, sculpture, architecture, and music.

The works of many of these European artists and a few Americans were first shown in America at the International Exhibition of Modern Art in New York in 1913, the so-called Armory Show, since it was held at the city's 69th Infantry Regiment Armory, at Lexington and 26th Street. Most American viewers, including President Theodore Roosevelt, rejected the new styles as obscene or primitive, certainly not art, and most American artists continued to paint or sculpt in highly traditional styles. America was not yet ready for modern art.

In 1914, as World War I began, modern artist Duchamp (1887–1968) took a step that challenged the very definition of art. He took a common manufactured wire bottle rack, signed his name to it, and declared that it was art because he *said* it was art. In 1915, he and Francis Picabia (1879–1953) both visited New York and were publicized as prominent French artists who for the first time were, as they said, "seeking America in matters of art." Picabia praised what he saw:

> I adore New York. There is much about it, which is like the Paris of the old days.... I have been profoundly impressed by the vast mechanical development in America. The machine has become more than a mere adjunct of human life.... I intend to work on until I attain the pinnacle of mechanical symbolism.[6]

Americans were flattered to be recognized and admired by famous French artists, and Picabia was true to his word. He soon produced a number of drawings inspired by spark plugs, meshing gears, pistons, and cylinders that were published by Alfred Steiglitz's art journal, *291.* Picabia's work clearly assumed that mechanical objects were worthy as works of art, a concept that at the time most Americans failed to share.

The unconventional work of European artists had no effect whatsoever upon the 1915 Panama-Pacific International Exposition in San Francisco, which had been planned since 1911, primarily to celebrate the completion of the Panama Canal, but also to herald the recovery of San Francisco from the disastrous effects of its 1906 earthquake.

The architecture, sculpture, and murals featured were clearly in the traditions of the 19th century, with highly realistic, classical, but allegorical statues and murals, mostly nudes, suggesting abstract ideas of abundance, progress, conquest, creation, power, etc., and historical depictions of conquistadors, pirates, and miners. The art at the exposition was a public display of America's imitation of European art as it was in the 1870s, not an appropriate expression of

America's 20th-century leadership in technology, science, or manufacturing. It was art of the past. A more effective global advertisement of America's obsolescence in art and design could not be imagined.

Yet despite criticism from traditionalists, modern art marched forward. In 1917 Duchamp submitted a mass-produced ceramic urinal, titled Fountain, to the Independent Artists Exhibition in New York as "art," which aroused outraged protests but nevertheless established a fundamental principle of modern art: art is whatever the artist chooses to define as art. In this case, it also suggested that common utilitarian objects could be viewed as art, which may have been a surprise to the manufacturer of the urinal, or any company making similar utilitarian products.

Other European countries evolved independent styles of modern art. In 1917, Piet Mondrian (1872–1944) and Theo van Doesburg (1893–1931) initiated the Dutch movement of de Stijl (the Style), which featured primary colors arranged asymmetrically in rectangular paintings, or abstract furniture. In Russia after the 1917 Revolution, Alexandr Rodchenko (1891–1956) and Varvara Stepanova (1894–1958) founded an artistic movement called Constructivism in 1921, and in 1922 published their manifesto to "set itself the task of finding the communistic expression of material structures."[7]

Leon Trotsky (1879–1940) inspired Communist Party encouragement of avant-garde art. He rejected fine art but promoted the more "productive" and "constructive" fields of architecture and mass production. So artists were asked to "go into the factories, where the real body of life is made."[8] Rodchenko and Stepanova inspired the founding of an avant-garde art and design school called Vkhutemas (an abbreviation of its formal title: Higher State Artistic and Technical Workshops).

After the war, Europe also led in architecture and design, which was often inspired by the American model of mass production. In the particular category of consumer product design, the primary focus of this book, the English Arts and Crafts movement dominated the latter 19th century. As the 20th century began, Arts and Crafts leadership broadened to other European countries, as France dominated Art Nouveau. In Austria in 1903, the Wiener Werkstätte (Viennese Workshops) were formed by architect Josef Hoffmann (1870–1956) and Koloman Moser (1868–1918) who, like most Arts and Crafts leaders, were opposed to mass production and rejected the evils of what they saw as cheap and shoddy products produced under abhorrent conditions. They wanted to counteract this machine-made trend by making handcrafted products that were both innovative in design and superior in quality and workmanship.

Arts and Crafts workshops in America followed this same philosophy, which drew a sharp line between handmade crafts and industrial mass production, segregating art from industry on the basis of morality and artistic superiority. The core belief of the Arts and Crafts movement was that production by cold machines and unskilled labor (often under poor working conditions), even though inexpensive, had no humanistic value, and in fact was a form of slavery.

However, acceptance of mass production was growing. In 1906, a member of the Dresden (Germany) Workshops, Richard Riemerschmid (1868–1957), designed a line of Machinenmöbel (machine furniture), where individual pieces were precut and assembled in a production process, rather than in the one-at-a-time tradition of craftsmen. Still, most Arts and Crafts workshops continued to oppose mass production as a threat to their livelihood, while innovative German applied arts educators radically sought a final solution (sorry—not the best term) to the problem by literally marrying art and industry.

In 1907, German applied arts educator Hermann Muthesius (1861–1927) formed the Deutscher Werkbund (German Union of Work), a trade association. It consisted of 12 industrialists and 12 artists/architects, dedicated to dominating global product markets for Germany.

That same year, to consummate this intended marriage of art and industry, Werkbund architect member Peter Behrens (1868–1940) was appointed as "artistic consultant" to Allgemeine Elekricitäts Gesellschaft (AEG), the German equivalent of America's General Electric, where he designed corporate graphics, products, and buildings, the first such collaboration of art and industry in practice. His designs were well in advance, designwise, of comparable products in the United States. It would be 50 years before an American corporation engaged a similar "artistic consultant" to integrate and control a single corporation's entire visual "identity" (IBM, 1956).

Muthesius in 1907 also proposed the development of a new Machinenstil (Machine Style) esthetic of product forms dictated by the purpose they serve, and as the "visible expression of inner driving force of the age,"[9] that is, to generate new types of forms that visually communicated their intended function, as well as their manufacture by machine processes. The significant purpose of this proposal was to restore the craftsman's role in shaping mass-produced consumer products by an artistic hand and eye, a role that had been lost at the beginning of the century when engineers took over the entire process of mass production. This had ended the Arts and Crafts era and relegated artistic contribution in manufacturing to mere surface decoration then known as applied art. The innovative concept of a Machine Style would soon be advanced by one of Germany's leading applied arts schools. In 1908, Bruno Paul (1874–1968), one of the 12 artist/architects who founded the German Werkbund, developed a Typenmöbel (standard furniture), which permitted an unlimited number of combinations of standardized, machine-made elements.

After the war in 1919, Walter Gropius (1883–1969), an architect, member of the Deutscher Werkbund, and former employee and protégé of Peter Behrens in Berlin, returned from military service to take over as director of the Weimar Kunstgewerbeschule (Weimar Arts and Crafts School), founded in 1902 by Belgian architect Henry Van de Velde (1863–1957), an Arts and Crafts leader who had become a member of the Werkbund, but as a foreign national was forced to leave the country and abandon the school during the war.

Gropius renamed the school Das Staatliches Bauhaus (State School of Building), commonly known only as the Bauhaus. Gropius was an avid admirer of American Fordism and Taylorism, and of American grain silos for their architectural simplicity and monumentality. He shared the belief of many in postwar Germany that setting new standards of industrial art would secure German global design leadership in the economic recovery. There at the Bauhaus, over the next decade, an avant-garde art and design faculty and their students developed principles for the design of modern architecture and products that featured simple, undecorated, geometrical modern forms, which were exquisitely crafted in quality materials.

Although staffed by many of Europe's most renowned modern artists, the school in Weimar was patterned after medieval craft guild workshops. Professors were known as masters, and students progressed from Lehrlinge (apprentices) to Gesellen (journeymen) to Jungenmeister (young masters). Gropius wanted to elevate Arts and Crafts to the level of fine art. In 1923, Làszló Moholy-Nagy (1895–1946) joined the faculty as master of the metal workshop. He was born in Hungary, studied law in Budapest, and later served in World War I. Convalescing from a war wound, he took up painting and worked his way across Germany as a letterer and sign painter, arriving in Berlin in 1921, where he took up painting and experimental photography. In 1922 he participated in the Dada and Constructivist conference organized by Theo van Doesburg in Weimar.

In general, many of these student designs met the idealistic standard of Muthesius's Machinenstil, and a few would be manufactured in limited quantities for the commercial market. Although the school was plagued by financial and political difficulties, it would flourish until

1924, when political demonstrations and Weimar budget cuts forced Gropius to relocate the school to Dessau, where it would come under increasing pressure by the Nazis, who were determined to rid the country of forward-looking (avant-garde) intellectuals.

Another former employee and protégé of Peter Behrens, Swiss-born architect Charles Edouard Jenneret Gris (1887–1965), was also intrigued by American Fordism and Taylorism, only he used it in reverse. Designing a slaughterhouse in Paris in 1916, he included a conveyor system to "disassemble" the animals. This practice, used in the United States since 1867, was often claimed by Ford to be the inspiration of his assembly line for automobiles. In 1917 Gris established a small brick-making factory, and by 1920 he considered himself an industrialist as well as an architect. It was then that he adopted his pseudonym, "Le Corbusier" (the Raven), under which he began publishing *L'Esprit Nouveau* (The New Spirit), a periodical with themes of order, rationality, engineering, and technology. Twenty-eight issues were published over the next five years.

As an architect, Corbusier dreamed of producing houses the way Ford produced automobiles. In 1921 he published sketches of a house called Maison Citrohan (Citrohan House), using the analogy of France's mass-produced car based on Ford's assembly line, the Citroën. His concept, he declared, could complete houses in three days made of artificial materials such as concrete. Over the next few years, Corbusier designed a number of private homes in Paris based on his Citrohan House concept. They featured stark interior white walls, furniture made of tubular steel, and single bare bulbs in light fixtures. The effect was pure, simple, modernist design that followed the German Bauhaus style and which he called "a machine for living in." In 1922, Corbusier developed plans for a "Contemporary City" for three million inhabitants that he hoped would reorganize society along lines of American industrial models. In 1923, he published *Vers une Architecture* (Toward an Architecture), a book summarizing his philosophy of a future world functioning like a machine.

Most Americans, enjoying the first fruits of mass production in a new consumer culture, had no idea of the extensive European design or art theories that had evolved on the continent either before or after the war. Modern art was simply not part of American culture. Art, to most Americans, was pretty landscape or portrait paintings on museum walls; Greek sculptures in gardens; or expensive, handcrafted, decorative home furnishings in traditional styles imported from Europe. The concept of combining functional, mechanical products and art was literally unimaginable.

By 1925, Europe had not yet recovered economically, but Paris was not only still at the leading edge of modern art, attracting scores of aspiring artists from around the world, including young Americans, but had also become the center of fashion, both in style and entertainment. African American entertainer Josephine Baker (1906–1975) became a stage sensation for her exotic dancing at the Folies Bergére cabaret and would later popularize a dance step called the Charleston. Jazz was the most popular music of the time, but the term "jazz" also meant anything exciting and fun. George Gershwin's composition of 1928, "An American in Paris," evoked the sights and sounds of Paris in the 1920s, and constituted most Americans' sole impression of Paris.

Gabrielle "Coco" Chanel (1883–1971) had opened her boutique in Paris in 1919, setting global fashion trends in women's clothing with her "little black dress" and borrowed men's clothing accessories, such as suits, ties, and shirt collars, for feminine fashions in the mid–1920s. She introduced Chanel No. 5 in 1924, the first perfume to feature a designer's name. She epitomized Paris of the 1920s, where women tasted new freedoms to wear what they liked, to cut ("bob") their hair short, to wear short skirts (above the knees), to wear straight dresses without waistlines, to wear makeup and perfume, to expose bare legs, to wear high heels, to wear "cloche" (helmet-type) felt hats, to drive automobiles, or to smoke in public. Whatever happened in Paris soon became fashionable around the world.

Young American women of the "Roaring Twenties," who had only recently been granted the right to vote (19th Amendment to the Constitution, May 19, 1919), shortly after being deprived of the legal right to buy alcohol (Prohibition: 18th Amendment to the Constitution, January 19, 1919), were influenced by these new Paris fashions. Many who followed the new styles were called "flappers," and their behavior was perceived pejoratively by many during Prohibition as outlandish, irresponsible, frivolous, sexually provocative, and in general as flouting social and sexual "norms." Retrospectively, of course, women were merely expressing their freedom from residual Victorian restraints against women, and at the same time establishing "fashion" as a driving social force of the 20th century, particularly for women as they became consumers of the new household appliances now becoming available.

Such appliances were featured in France in 1923 at the Salon des Arts Ménages (Salon of Domestic Science). French manufacturers collaborated with domestic designers by paying them to develop new styles for presentation, and also by paying them royalties for each reproduction of their design, thus sharing risks as well as rewards. Many of the new designs were based on the basic geometric shapes of Cubist modern art. By and large, such products were not produced in large quantities, as manufacturing in Europe had not yet recovered from the war, nor had it evolved to the level required for significant competition with America.

But Europe still retained its historical reputation as the center of art and design, and in fact Europeans have a number of legitimate historical claims for initiating what we now call industrial design because of the many Arts and Crafts groups that advanced the quality of design esthetically in the late 19th and early 20th centuries. However, what we call industrial design today is more broadly defined than as simply designing consumer products in a modern style, but as the process of designing products for mass production by machines in collaboration with engineering, marketing, and manufacturing, for the purpose of consumer acceptance and affordability, as well as for corporate profit, as it is practiced today. It is the Machine Age in America, in the 1920s and 1930s, that defined, pioneered, and implemented this modern process in reality, rather than in European theoretical prototypes that were designed without any manufacturing, cost, or sales considerations, and were therefore only of academic, or esthetic value.

Before that, as we have seen, the only major industries in the United States involving artist/designers were the "art industries" in general, and of course there were many designers in these fields for decades before. Some of these designers made the transition to the new field of industrial design before others and could claim, as some did, to be the first "industrial designers" in the United States, long before it became well known as a specific profession. A number of them were veterans of World War I.

One of the art industries was the manufacture of decorative lamps, and in about 1920, 24-year-old **Chauncey Eugene (Chick) Waltman** (1896–1962) was already well known for his lamp designs for the Art Lamp Company on Wabash Avenue in Chicago, where he served as designer, salesman, and advertising manager. The most successful lamp he designed at about this time was the Almco lamp. Waltman was born near Columbus, Indiana, the son of a minister. The family moved to Cincinnati, Ohio, and lived near the famed Rookwood Pottery, where Waltman, as a boy, was inspired by artists he saw working there. The family moved several times around Michigan, ending up in Grand Rapids, where Waltman was at the top of his art class in high school. While still in school, he worked in the decorating department of the Berkery and Guy Furniture Company, and by 1912 he started a decorating department at the Imperial Furniture Company.

After high school he took a job in Chicago with John A. Colby and studied decorating and furniture design at night school, but after the start of World War I in 1917, he joined the Air Force and was trained at Eastman Kodak, in Rochester, New York, to prepare maps with photos and hand sketching from airplanes. In 1918, during the flu epidemic, he was sent to Cornell University for further training and to earn a commission as second lieutenant. But suddenly the war was over and he returned to Chicago and resumed night school.

He soon got a job as a lamp designer for Valentine and Seaver and later with the Art Lamp Company, where, as advertising manager, he used ads with borders by Walter Dorwin Teague (more about him later). In 1920, he and his first wife built a home in Oak Park, the upscale Chicago suburb, near where Frank Lloyd Wright's home and studio were located. But the marriage foundered and he took a new position with the Kaplan Lamp and Shade Company.

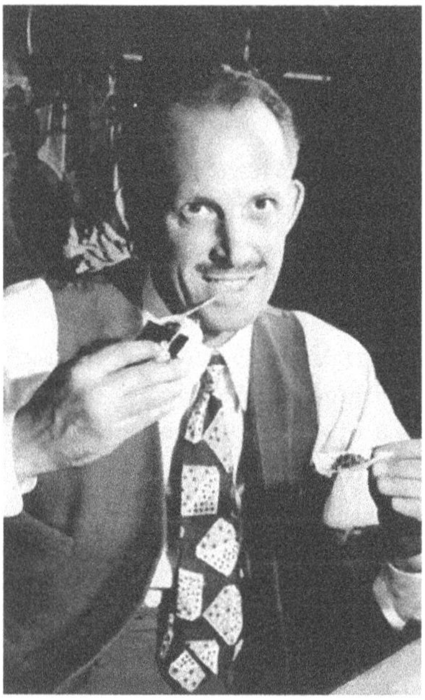

Chauncey "Chick" Waltman, ca. 1930s (courtesy William Waltman).

In 1924, Waltman made a lucrative business deal with Kaplan. For half his normal weekly salary of $150, he would go to Paris to study art, and spend two days a week drawing lamp designs in the new styles and send them to Kaplan for manufacture. For about a year, he studied at the Grande Chamber des Arts, biked through England and France, and traveled to Italy and North Africa. When he returned to Chicago in 1925, he set up his own freelance office, designing in wood, cast metal, wrought iron, and glass.[10] After industrial design became well known in the 1930s, his office would participate as Waltman Associates, at 165 Chicago Avenue.

Also among the art industries was the furniture industry, which almost exclusively produced traditional and historical styles (many still do). Among designers in this field was **Kem Weber** (1889–1963) who began in 1921 at age 32 to work in the design of furniture for the Barker Brothers store in Los Angeles, then the largest furniture retailer in the United States. Weber had been born in Germany as Karl Emanuel Martin (K.E.M.) Weber, apprenticed as a cabinetmaker and had graduated from the Berlin Kunstgewerbemuseum (Berlin Academy of Fine Arts) in 1912, studying under the well-known Bruno Paul (1874–1968) and after graduation working for Paul.

In 1914 Paul sent him as a designer for Germany's exhibit at the 1915 Panama-Pacific Exposition in San Francisco, but he was stranded there by the start of World War I. At first he opened a modern design studio in Berkeley. But after the sinking of the *Lusitania* in May 1915, American sentiment turned against Germany, and as a German, he had trouble finding clients, so he worked as a lumberjack in the High Sierra. Toward the end of the war he operated a chicken farm in Los Gatos with his young wife Erika. In 1918 they settled in Santa Barbara, California, where Weber taught art classes and opened a small factory in an old church to produce modern and period furniture, but he met with only modest success. When Barker Brothers offered him a position in its design department in 1921, he jumped at the chance and moved to Los Angeles.[11]

Other artists were in the advertising business, which promoted mass-produced artless industry consumer products such as automobiles, appliances, furniture, machinery, tools, and other functional products designed by engineers. Gradually, as the demand for more attractive products grew, these artists suggested ideas for product improvement to their advertising clients.

One such individual was New Zealander Joseph "Jo" Sinel, who in 1923 at age 34 was working for the Lennon Mitchell advertising agency in New York, and who designed graphics and products for clients, although specific product designs by him are not known until 1929. Sinel became a prominent industrial designer and educator, and we will later describe his career in more detail.

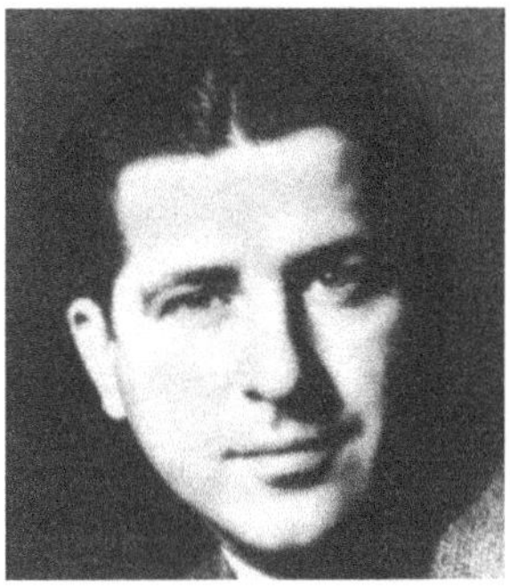

John Vassos, ca. 1934 (February 1934 *Fortune* magazine).

Another artist who would become legendary in industrial design, **John Vassos** (1893–1985), designed a face lotion bottle in 1924 that became popular (with the lotion removed) when used as a hip flask during Prohibition, and sales increased dramatically by 700 percent. Vassos was born John Plato Vassacopoulos in Romania. His family moved to Constantinople where his father became a Greek publisher and diplomat. As a teenager, young Vassos drew a cartoon in his father's newspaper that offended the Turkish senate, and he was forced to leave the country in 1909 at age 16. He fled on a British ship to Palestine. During World War I, he served as a cabin steward on a Belgian steamer, which was torpedoed. He was rescued by the U.S. Navy and landed in London in 1915. He obtained papers and traveled to Boston in 1919 where his older brother Alex lived.

There, he studied at the Fenway Art School; studied art and illustration with John Singer Sargent (1856–1925), the leading portrait artist of his generation; and assisted with stage designs for the Boston Opera Company. He came to New York in 1924, became an American citizen, changed his name to Vassos, opened a commercial studio called the New York Display Company, studied at the Art Students League, worked as an illustrator for *Harper's* and *New Yorker* magazines, and illustrated and published books.

Furniture designers became prominent in the emerging field of industrial design as more modern designs were demanded. Among these was **Leo Jiranek** (1900–1990), who in 1924 took over his father's furniture design business at age 24 and became one of the industry's first freelancers. In fact, he was considered by many to be the "Dean of Furniture Designers." Jiranek, known as Jerry, was born in Grand Rapids, Michigan, and after prep school at Western Military Academy, he spent two years at Norwich University, then transferred to Princeton University in 1920, where he graduated as an engineer in 1922. Over the years, he would contribute to more furniture companies than any other designer.

But in 1925, there was one furniture designer who abandoned all tradition by introducing a style intended to represent the vitality of modern America, inspired by the new steel towers rising all over Manhattan. Generally, his pieces were bookcase-cabinet combinations or desks, which featured unornamented block forms that rose in terraced levels and provided various functional storage boxes and shelves, often lacquered red, blue, or green. Typical were his Skyscraper cabinet and Skyscraper bookcase. Pieces were tall, but with a narrow footprint, perfect for small Manhattan apartments. The designer was 38-year-old **Paul T. Frankl** (1886–1958), whose new Skyscraper furniture immediately became popular and was probably the first evidence of a modernistic style originating in America.

Frankl was a furniture designer and maker, architect, painter, and writer born in Vienna, Austria, the son of a wealthy real estate dealer. He had been a member of the Wiener Sezession (Vienna Secession) that had early in the 20th century sought to create a new style that owed nothing to historical influence. After completing his architectural studies at the Berlin Polytechnic, he spent time in Berlin and Copenhagen before arriving in the United States in 1914 and settling in New York. He brought an outsider's perspective and a desire to create a uniquely American design esthetic.

By 1922 he had established Frankl Galleries on 48th Street in New York, and by 1924 he had a showroom on Madison Avenue, where he sold furniture of his own design as well as imported modern textiles and wallpapers, calling his company Skyscraper Furniture. He described his attraction to the Skyscraper style:

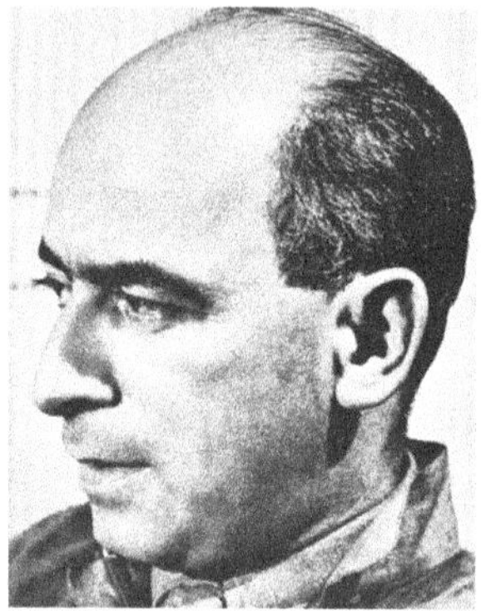

Paul T. Frankl, ca. 1938.

> Modernity expresses itself in the energy of the vertical thrust of towering buildings that push themselves through the crumbling red husks of the city of the past.[12]

In fact, Frankl knew that designing a furniture line inspired by these structures would have a mass appeal to Americans, and he was right. But he was a rare practitioner of modern design in America. Europe was where the action really was.

Chapter 2

The Arrival of Art Deco

France was the main attraction in the art world of the 1920s and naturally wanted to promote and reinforce its prewar reputation and culture of fashion, style, decorative arts and fine arts. By the early 1920s, many aspiring young American artists were being influenced by the dramatic impact of French modern art and considered a visit to, or study in, Paris as an essential part of their artistic education in modern art and culture, as well as a carefree, youthful experience in the good life of Europe.

So from April to October of 1925, an Exposition International des Arts Décoratifs et Industriels Modernes (International Exposition of Modern Decorative and Industrial Arts) was held in Paris, the first "World's Fair" since the one in San Francisco ten years before. The lengthy French title would much later be shortened to Art Deco to describe the many European varieties of new designs and their subsequent impact around the world, but at the time, the designs were generally described as "modernistic." The exposition was unique in one key respect: former similar expositions traditionally celebrated past national accomplishments. In this one the intention of the organizers was to establish new standards. Exhibitors were required to display only "those exhibits that were not dependent of the art of the past."[1] Invitations went out to all countries except Germany, which was still regarded as the villain of the Great War. Twenty-six countries accepted, except for China and the United States.

China did not accept because of disruptions in political matters, and the United States ostensibly because the State Department did not recommend appropriations to do so, due to lack of support from commercial interests. As Herbert Hoover, then secretary of commerce and in charge of the Coolidge administration's industrial policy, explained to the Fourth Annual Exposition of Women's Arts and Industries:

> The advice I received from manufacturers was that while we produced a vast volume of goods of much artistic value, they did not consider that we could contribute sufficiently varied design of unique character or of special expression in American artistry to warrant such a participation.[2]

Obviously, U.S. manufacturers saw no significant consumer market potential in Europe, but more to the point, they had nothing to display that met the exposition's intent of new standards or its specific criteria of products that were "not dependent of the art of the past." Most U.S. products of the time were, in fact, copies of European designs. Hoover's decision did not set well with the cultural establishment, where many were already quite aware of how advanced European design and fashion had become. When Charles Richards, the same art educator who conducted the 1922 disappointing study of industrial arts education, asked

Hoover to delegate a commission of 100 representatives of trade associations, along with some artists and designers, to visit and report on the exposition, Hoover gracefully complied.

The six-month exposition opened on April 28 with 4,000 guests at the inauguration, and thousands of visitors arrived on following days. It was located between the esplanade of Les Invalides and the entrances of the Grand Palais and Petite Palais, a 57-acre site, two-thirds of which contained French displays. Exhibits included work by famous glass and jewelry designer René Lalique (1860–1945), who also designed the exposition's crystal tower fountain. There were displays by a number of Parisian department stores such as Bon Marché, Printemps, and Magazins du Louvre, which created entire room "ensembles" in modernist styles. The work of French furniture and interior designer Émile-Jacques Ruhlmann (1879–1933) was displayed in a special, classical "Pavilion of the Collector" designed by French architect Pierre Patout (1879–1965). In a far corner of the exposition, generally ignored by the press, was Le Corbusier's Pavilion de L'Esprit Nouveau, a spare modern home interior and furnishings along the lines of his Maison Citrohan concept, that attracted both admiration and criticism for its lack of decoration.

Russia had a pavilion displaying the work of Alexandr Rodchenko, Varvara Stepanova, and students of the art and design school, Vkhutemas, which they had established. The students won several prizes. The popular Swedish pavilion was simple in form and devoid of decoration, featuring modern crafts in typical Scandinavian styles inspired by Gregor Paulsson's (1889–1977) 1919 book, *More Beautiful Things for Everyday Life*, which challenged the past and encouraged Arts and Crafts workers to use new forms of the modern age. Danish architect and designer Arne Jacobsen (1902–1971), still a student, won a silver medal for his chair design. Austria contributed Frederick Kiesler's (1890–1965) City in Space exhibit to house displays of the work of Josef Hoffman (1870–1956), architect, furniture designer, and one of the founders of the Vienna Secession.

With 16 million people visiting the exposition, it was destined to have a lasting impact on applied arts around the world. Certainly Le Corbusier influenced architecture in what would become known as International Style in 1932. The decorative arts were influenced by the unusual new styles that were generally called modernistic but which included an entire range of imagery and motifs that have been variously described as streamlined, geometric, symmetric, or as a Machine Age look. Some imagery was clearly derived from French Cubism, German Bauhaus, Italian Futurism and Russian Constructivism, all described earlier.

American critic Helen Appleton Read (1897–1974) reported on several important principles of modern design she observed at the exposition:

> It is characteristic of the new décor that the nature of the material is at all times respected and is allowed to dictate its treatment. Wood is wood, iron is iron; the bad taste that invariably ensues when the attempt is made to give the quality of one material to another, to make wood look [like] iron or marble like lace, is avoided. This universal tendency toward simplicity of outline should be of the greatest interest to the American designer if he will recognize that this is the fundamental note. All design in this country is governed by the factor of whether or not it can be reproduced in mass production. It costs no more to get out a good design than it does a bad one, and the fact that the best designs of the new décor are the simplest to the point of being geometric makes them so much the more easy to put on the market.[3]

Read's assumption that simpler forms were easier to produce may seem logical, but in fact most manufacturing processes could produce complex forms as easily as simple ones. Few design forms are as complex as the Model T or a 1922 kitchen stove, both popular designs of the time. Many manufacturers would argue that copying an existing design was much easier than investing in a new design, which required extensive design work as well expensive new tooling. They would argue that the cheapest of all was to keep producing the old design as long as possible. Read went on to explain her theory of the reasoning behind the new modern forms:

> The whole trend of modern design toward the geometric and the unadorned had its inception through the omnipresence of the machine. Because we had absorbed the shapes of the machine into our consciousness, we ended up by creating designs which reflected it, and which in turn are the most suitable for machine made processes.[4]

Read may have had a point here. Designers over a generation may have been so preoccupied with the impact of machines and mass production on society that they unconsciously felt compelled to invent new forms that suggested machines. The simplest explanation, however, may have been that as artists and designers often do, they simply made things different than the past to demonstrate progress in the creative arts, or to be more inventive than those who went before. It is always speculative to precisely pinpoint the intent of creative people. They just love to create new things!

The American Commission submitted a report to the Department of Commerce, warning that the modern movement would inevitably reach the United States, and that Americans needed to

> initiate a parallel effort of our own upon lines calculated to appeal to the American consumer.... As a nation we now live artistically on warmed-over dishes. In a number of lines of manufacture we are little more than producing antiquarians. We copy, modify, and adapt the older styles with few suggestions of a new idea. It is true that this practice of reproducing older forms has been an invaluable education to our people. It is also true that the adaptation of old motives when performed with intelligence and skill continues and probably will continue to give us a large proportion of the decorative manifestations acceptable to American taste. It would seem equally true, on the other hand, that the richness and complexity of American life call for excursions into new fields that may yield not only innovations but examples well suited to the living conditions of our times.[5]

The ambiguity of this report clearly reflected conflicting opinions among commission members as to what to do about the obvious differences between the new European design direction and the traditional practices of design in the United States. Copying of older styles was justified as educational and acceptable, but innovation may yet be of some value. Some members doubted that the new modern styles would appeal to Americans. The report gave the Department of Commerce no clear recommendations for action, and even if it had, it is doubtful that Commerce could have done anything to change the current traditions and practices of manufacturers. Any shift in direction would have to come from the manufacturers themselves based on consumer demand, and at this moment in history, the general public was blissfully unaware of what was happening in Paris. This would soon change.

Two and a half million Americans now had radios and were becoming more aware of world events. In England, John Logie Baird was even demonstrating a mechanical television system. Reports from other Americans who attended the Paris exposition quickly circulated within the art and design communities and the press. The Art-in-Trades Club issued a manifesto challenging American artists to compete in the design of modern furnishings. In January 1926, an editorial about the Paris exposition in the *New York Times* urged industry to change because the public would soon catch up with Europe in recognizing the competitive value of new designs:

> American captains of industry must soon change their traditional characters. True, the public may still cling to its conception of them as iron-jawed, level headed giants of efficiency ... but if the

> present international competition keeps up, we may find them, like the Germans, openly cooperating with museums and art schools; or, like the French, proclaiming themselves merchants of beauty.[6]

One of the attendees at the Paris Exposition had been Kem Weber, mentioned earlier, who had been working at the Barker Brothers furniture store in Los Angeles and represented their interest abroad. He had become an American citizen in 1924 and had Americanized his German name to only its initials, K.E.M. In 1926, he had become art director for Barker Brothers. Up to that time, Barker had exclusively featured period revival furniture in its five stores. Inspired by what he had seen in Paris, Weber in the fall of 1926 opened a new store-within-a-store, christened "Modes and Manners," on the fourth floor of Barker's eleven-story building. Weber designed the wall treatments, display cases and furniture in a space with jagged shapes, bright colors, and wild patterns. Modern furniture by Weber's Austrian friend in New York, Paul Frankl, was also displayed. It was not only the first outlet for modern furniture and accessories in Southern California, but also one of the first in the country. Over the next decade, Weber would work to promote modernism with the furniture he designed there in his store studio, the Silver Tree. The store had its own production facilities to manufacture the furniture and soon opened a new special Hollywood branch, which Weber also designed.[7]

Museums were anxious to promote the new designs, and in February the American Association of Museums imported a collection of 400 French and Swedish objects from Paris, called Selected Collection of Objects from the International Exposition of Modern Decorative and Industrial Art, which was exhibited at the Metropolitan Museum of Art and subsequently sent to tour other American member museums. The exhibit explained that it was

> not to stimulate demand for European products, nor to encourage copying of European creations, but to bring about an understanding of this modern movement in design in the hope that a parallel movement may be initiated in our country ... to bring us to new forms appropriate to the living conditions of the twentieth century.[8]

In *Vogue* magazine, patterns of textiles and paper goods were illustrated and described as "modern designs" or of "modernistic influence." Advertisements in national magazines began to adapt Art Deco design formats and motifs to give their ads and products a sense of modernity. Even ads for the most modest, utilitarian objects could be glorified when presented in a modern artistic format. Ready-to-wear fashions quickly adopted Art Deco elements such as straight-line forms, squared necklines, triangular and diagonal accents, and geometric patterns. Such styles soon appeared in every category of fashion accessories, particularly jewelry, compacts, handbags, hats and scarves.

Younger generations today probably wonder why there was such a fuss about these European designs, when fashionably designed products are available today from all over the world. It is a bit hard to relate, I know. But depending on your age, you only need to remember what it was like in the 1950s, when home television sets became available, or in the 1980s, when personal computers were introduced, or in the 2000s when iPhones hit the market.

The public is always excited when a new technology, previously available but only affordable by large institutions, finally becomes inexpensive enough and available to ordinary people, or when a fashion trend, such as short skirts in the 1960s, casual wear in the 1980s, or sport shoes in the 1990s, suddenly erupts everywhere in a short period of time. In the 1920s, the latest technology was, in fact, mechanical devices, such as new electric appliances or any new household decorative accessories, furniture, or bathroom and kitchen equipment. People simply wanted to buy the most modern-looking things, and at that time, "modern" was dramatically different from the past. Or as flappers would have said in the 1920s, "Modern design is swell!

The bee's knees! The cat's meow!" (Translation: Cool! Awesome! Mind-blowing!) They were no different from young consumers today.

While the new modern design styles were being promoted aggressively everywhere, there were additional incentives encouraging consumers to buy. Increasing efficiencies in manufacturing over the last ten years had enabled a reduction in production and labor costs, thus offering lower consumer prices. The vacuum cleaner that cost $52 in 1918, in 1926 was priced at $29.95, 49 percent lower (in 2013 dollars, that's a reduction from $804 to $394). So the new mass-produced products were much more affordable, for many more people. In addition, more jobs were available at higher salaries. The economy was better than it had ever been since the war. People were primed to buy, and they wanted the latest styles.

Now it was up to the manufacturers, many of whom had neither the inclination nor the specialized skills to provide such new styles.

The first of the so-called artless industries of mass production in the late 1920s to recognize the public desire for new design styles was in fact the automobile industry, in which 43 firms produced four million new cars each year. Travel by car was increasing rapidly with the opening of the Lincoln Highway, the first coast-to-coast roadway, connecting San Francisco and New York (now Route 40). In 1926, the Model T, which in 1923 had commanded 52 percent of the U.S. market but was virtually unchanged since its 1913 introduction, was now challenged by Chevrolet's new 1926 model with bold new styling and colors.

General Motors (GM) head Alfred P. Sloan had initiated annual styling changes because he was convinced that new styling was the key to success. Chevy sales doubled, and Ford dropped by 25 percent. Within a year, Chevrolet would win 45 percent of the market, and by 1927 Ford's share would drop to 15 percent. Ford was forced to drop the Model T and produce its redesigned Model A in 1928, designed by Edsel Ford, Henry's son. Some manufacturers figured that if Henry Ford himself, the king of Fordism, needed styling, they had better follow suit.

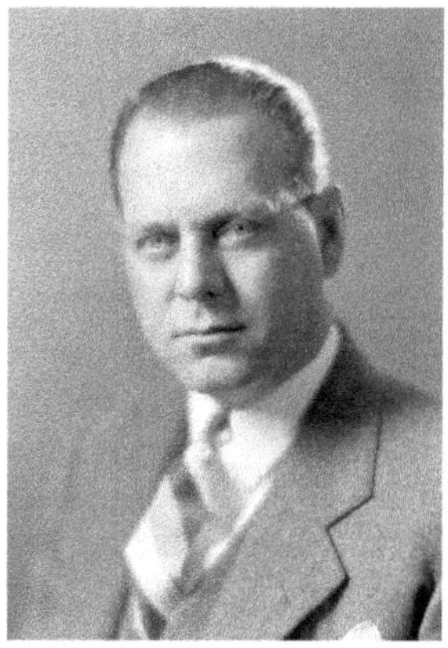

Harley J. Earl, ca. 1930s (General Motors).

Now, in early 1927, about when Charles Lindbergh flew solo across the Atlantic, GM's Sloan charged **Harley J. Earl** (1893–1969) with designing the 1927 LaSalle, a cheaper, production-model version of GM's luxury Cadillac. It was to be the first mass-produced car to have its appearance completely styled from bumper to bumper by a single individual, and the first to use chrome plating on the trim. Earl used full-scale clay models to preplan every detail and blend all components into an integrated form.

Earl had been born in Hollywood, California, had attended the University of Southern California in 1912 and studied engineering at Stanford University from 1914 to 1917. He began his career with the firm his father had founded in 1889 as the Earl Carriage Works, and after 1908 called the Earl Automobile Works, which built custom car bodies for film stars starting in 1911. Clients included Roscoe "Fatty" Arbuckle, Tom Mix, Mary Pickford, Cecille B. DeMille, and Al Jolson, among others. Earl designed the first complete car produced by the firm in 1918.

That year, Don Lee, a West Coast Cadillac distributor, head of Don Lee Coach and Body Works, bought the Earl Company and kept Harley on as designer and general manager of its custom body shop in 1919. In 1921 Earl patented a design that suggested the wraparound windshield he later would pioneer. In 1926, the Fisher Brothers of General Motors, impressed with

Earl's technique of using modeling clay to develop the forms of his designs, hired him as a consultant to Cadillac, and he relocated to Detroit. His first assignment was to design the LaSalle, a low-cost Cadillac. The 1927 LaSalle would become an instant success and would convince Sloan to establish the first internal corporate design department at GM, called the Art and Colour Section, and to name Earl as its first director. Within a year Earl would hire a staff of 50 automotive designers.

In 1927, promotion of modern design to the public continued on many fronts. The Metropolitan Museum of Art held an exhibit of Swedish industrial arts, introducing Scandinavian design to the U.S. public for the first time. In May, Jane Heap (1887–1964), editor of a literary journal called *The Little Review,* organized a Machine Age Exposition in Steinway Hall in Midtown Manhattan. Inspired by Le Corbusier's Pavilion de L'Esprit Nouveau, and by Soviet Constructivism, Heap displayed actual machines: a Studebaker crankshaft, a Curtis airplane engine, and a Hyde Windlass propeller, along with photographs of skyscrapers, the Bauhaus in Dessau, Soviet industrial buildings, and architecture by Walter Gropius. E. B. White (1899–1985) wrote in the *New Yorker* that the exposition demonstrated "that the machine is the tutelary symbol of the universal dynamism."[9] Department stores, such as Macy's and Franklin Simon of New York, Wanamaker's in Philadelphia, and Marshall Field in Chicago, inspired by the success of Paris department store ensembles during the 1925 exposition, staged exhibitions in May 1927 featuring the new Art Deco designs. Typical was Macy's Art and Trade exhibition, which was attended by 40,000 people and was hyped aggressively by radio announcements and talks by notable participants.

Consumer goods manufacturers were being advised to incorporate modern design into their products by advertising agencies. In the August 1927 issue of *Atlantic Monthly,* Ernest Elmo Calkins (1868–1964), head of the Calkins & Holden advertising agency, who was called by many the "Dean of Advertising Men," wrote an influential article, "Beauty, the New Business Tool," in which he stated the case for more attractive products:

> We passed from the hand to the machine, we enjoyed our era of triumph of the machine, we acquired wealth, and with wealth, education, travel, sophistication, a sense of beauty; and then we began to miss something in our cheap but ugly products. Efficiency was not enough. The machine did not satisfy the soul. Man could not live by bread alone. And thus it came about that beauty, or what one conceived as beauty, became a factor in the production and marketing of goods.[10]

Calkins continued to define beauty:

> The only art that can survive and grow is art that is related to our life and our needs, and that has a sound economic foundation. It is to be hoped that manufacturers in the search for design to beautify their products will start with a clear conception of what beauty is, especially beauty in an article of use ... the surest guide in divining new beauty in machine-made things is to grasp and interpret the beauty they naturally and intrinsically possess.[11]

In other words, beauty in useful products is not extraneous artistic decoration or embellishment, but the honest and simple visual expression of its natural function through form and character. Whether Calkins knew of the Bauhaus or not, he was expressing the exact same philosophy of design, but in business terms. Calkins asked his advertising colleagues, "How could ads work, if the styling of products did not express the qualities claimed for it?"[12]

Many manufacturers by this time were paying attention to their advertising agencies, which were already incorporating the modern look into their ads. In many cases, such manufacturers asked their agencies for help in designing products, requests that were honored by nameless artists on the agency's staff and given no credit. Other manufacturers sought such design assistance by advertising for employees or consultants. One can imagine what such a help wanted ad for an industrial designer might have looked like in 1927:

> **Wanted**: Artistically trained industrial artist to create new modern designs for products made by company X. Applicants need to be familiar with manufacturing processes such as casting, stamping, grinding, milling, welding, fastening, and assembling, and willing to work collaboratively under chief engineers and efficiency experts in actual factories. Designs must be able to be reproduced easily and inexpensively by machines, and must be attractive to consumers in the modern style. Applicants must provide examples of mass-produced products they have designed recently. Compensation for such work is to be negotiated.

An advertisement like that would have attracted few applicants, because at that time, practically no one existed with such extensive qualifications that required both art and engineering training, and who would be willing to work in mass-production facilities. As we have seen, few art students would even consider such a position, even if they had been trained in engineering or manufacturing matters, and of course none were. As noted earlier, most graduates of art schools wanted nothing to do with industry.

Engineers who *were* trained in manufacturing processes, and may have designed mass-produced products, had absolutely no art education, nor any inclination to acquire it. The industrial art position described in the ad would have been, as some would say, "neither fish nor fowl," that is, neither true artist or true engineer. Traditional separation of art and industry had, at least in America, left industry without an essential, competitive, business tool.

Jane Heap, who had organized the Machine Age Exposition in New York, wrote that an effective artist "must affiliate with the creative artist in other arts and with the constructive men of his epoch: engineers, scientists, etc."[13] In 1927 there were a few professionals who recognized the increasing demand for industrial design, were trained in art and design, and responded enthusiastically to the call by manufacturers, knowing full well that this would require their willing collaboration with industry, engineers, and scientists. They were the far-seeing individuals who almost single-handedly became pioneers in what was to become a brand new profession. They were interesting and unusual men and women. They were the machine designers.

Here's where things began to get interesting. As the demand for modern design increased both by manufacturers and by the public, professional people in a number of art-related, advertising, or commercial fields responded with interest, or were sought out and targeted by industry. Some, of course, were already in the art industries, including furniture and lighting products.

One of these was **Donald Deskey** (1894–1989), then 33, who had just opened an office, Deskey-Vollmer, in New York in 1927, with his partner, Phillip Vollmer, specializing in the design of furniture and lighting. Deskey was born in Blue Earth, Minnesota, and had studied architecture at the University of California, Berkeley, and painting at the California School of Design, the School of the Art Institute of Chicago, and the Art Students League of New York. After military service in World War I, he started working with advertising agencies in Chicago and New York, but soon left to study painting in Paris from 1920 to 1922, where he also married. He returned to the United States to begin

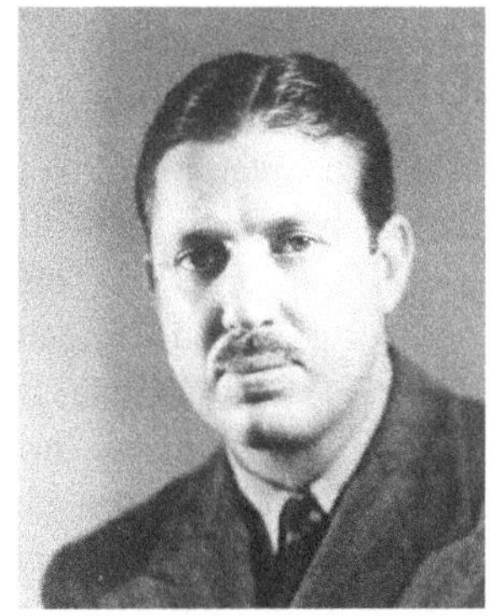

Donald Deskey, ca. 1930s (Cooper-Hewitt, National Design Museum, Smithsonian Institution/Art Resource, NY. Photograph by Matt Flynn).

a career in painting; to teach art at Juniata College in Huntington, Pennsylvania; and to exhibit his work in Pittsburgh. In 1925 he returned to France to attend the Paris exposition and visited the Bauhaus in Dessau, both of which greatly influenced his approach to design, and upon his return in 1926, he settled in New York to work for furniture designer Paul Frankl, mentioned earlier. He designed furniture and lamps for Frankl, but also designed an apartment interior for Adam Gimbel, the founder of Gimbel's department store, who had bought out rival Saks & Co. and opened an uptown branch called Saks Fifth Avenue in 1923.

Meanwhile, in California, Kem Weber left the Barker Brothers store in 1927 to open an independent design studio in Los Angeles, but continued to work as a consultant for Barker, shaping its products well into the 1930s. His design of an armchair for the John Bissinger residence in San Francisco in 1928 was in the sleek and sophisticated modernism that would come to define the Hollywood of the 1930s. In his new office, he sought out collaboration with manufacturers and explored new materials.

Deskey; Kem Weber; set designer Lee Simonson (1888–1967); set designer, architect, and artist Frederick Kiesler (1890–1965); Paul Frankl; and a few other designers were justifiably concerned with the dishonest but common practice of manufacturers copying designs, a practice that museums historically encouraged by displaying traditional European designs as superior models of design quality. At the time, there was no reason to believe that the new innovative modern designs in America would not be subjected to the same practice. Now, as early as 1927, the new modern designers realized that, for the first time, American designs could be more innovative than European ones and felt that such borrowing, as Frankl put it, "degrades the producers and corrupts the taste of the public."[14] They formed an organization called the American Union of Decorative Artists and Craftsmen (AUDAC), which idealistically aimed to "cooperate with manufacturers and the public in the placing of American arts and crafts on a basis of honesty, dignity, and merit."

Another lighting designer was **Walter von Nessen** (1889–1943), then 38, who opened Nessen Studios in New York's Murray Hill section in 1927 with his wife, **Greta von Nessen** (c. 1900–1975), then 27, to design and fabricate architectural lighting. His designs, such as his innovative swing-arm lamps for table, floor, and wall applications, were simple, polished metal, modern, and highly functional in allowing the lamp to be easily pushed into a range of horizontal positions without moving the floor or desk lamp base. The Nessens also designed furniture and accessories for the Chase Brass and Copper Company, including a tubular steel chair. Von Nessen was born and trained in Germany, and his wife studied at the School for Industrial Arts in Stockholm, Sweden. They both emigrated to the United States in 1925.

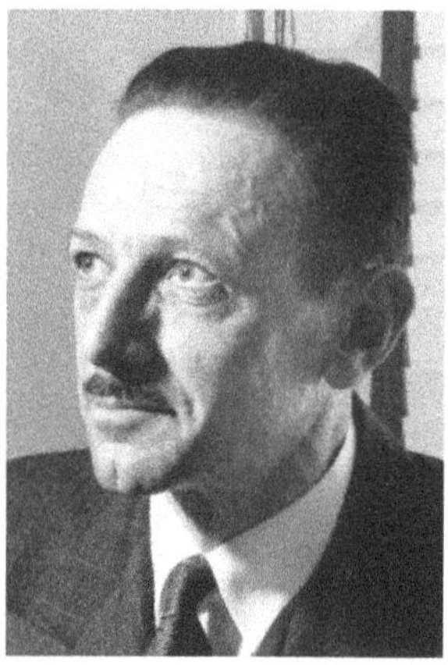

Gilbert Rohde, ca. 1940 (courtesy Herman Miller, Inc.).

There was also **Gilbert Rohde** (1894–1944), then 33, who in 1927 traveled to France and Germany and was inspired by modernism there to design and produce custom furniture in a modern style. That same year, he opened his own office in New York, designing furniture and showrooms and consulting with manufacturers such as General Electric and the Hudson Motor Car Company. Rohde was born in New York and became familiar with cabinetmaking in his German father's shop. He studied at the Art Students League and the Grand Central School of Art, both in New York. His early work was as a drama and music critic, cartoonist, reporter, and furniture illustrator.

In an environment where women led the fashion industry,

it seemed appropriate that some of these women would be attracted to the design of modern products. One of these was **Helen Dryden** (1887–1981) who, inspired by the flurry of modern design activity after the Paris Art Deco Exposition of 1925, turned her career in costume design to industrial design in 1927 at about age 40. She designed tableware and lamps for the Revere Corporation and as art director with the Dura Company in Ohio, where she designed chrome decorative objects, textiles, and glassware and was soon acclaimed as "the highest-paid woman artist."

Dryden was born in Baltimore with unusual artistic ability, designing and selling clothes for paper dolls that appeared in newspaper fashion sections. She trained in landscape painting under Hugh Breckenridge and studied briefly at the Philadelphia Academy of Fine Arts before moving to New York to begin her career in fashion design illustration at *Vogue* magazine in 1909. From 1913 to 1922, she designed dramatic and outstanding *Vogue* covers, as well as covers for *Charm*, *Harper's Bazaar*, *House & Garden*, *Vanity Fair*, and *Harper's Weekly*. From 1914 well into the 1920s, she designed costumes for Broadway shows and fabrics for Stehli Silks and Aberfoyle Fabrics. In 1925, she lived at 9 East 10th Street in New York with her 25-year-old Philippine-born cook and butler, Ricardo Lampitok.

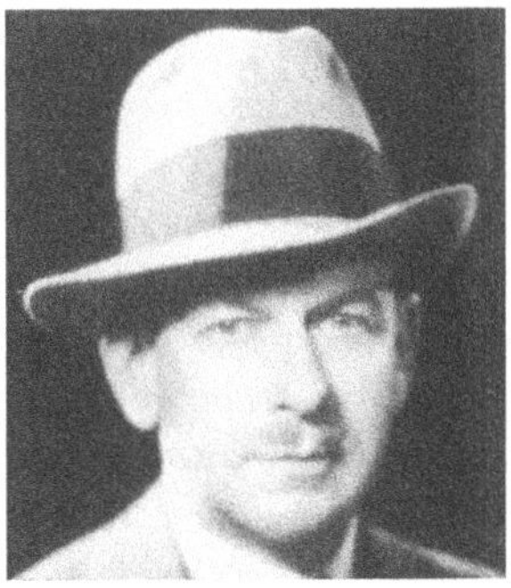

Walter Dorwin Teague, ca. 1934 (February 1934 *Fortune* magazine. Photograph by Lazernick).

Other entrants to the new profession were in the advertising business. In 1927, **Walter Dorwin Teague** (1883–1960), then 44, was awarded a contract with Eastman Kodak to design cameras, starting on January 1, 1928. Teague immediately added "Industrial Design" to his firm's letterhead, since he was already a highly successful advertising artist, who had opened his own typographic studio in 1912 and by 1920 had been involved in commercial packaging.

Born in Decatur, Indiana, as one of six siblings, Teague's grandfather had moved from North Carolina to Pendleton, Indiana, where one of America's largest Quaker communities existed. Teague's father became a circuit-riding Methodist minister and later a tailor in Pendleton. At age 16, while still in school, Teague worked as a handyman at the local newspaper, and later as a reporter. He was inspired to become an artist by books on architecture in his high school library. At age 19, he left Indiana for New York City, where he studied painting from 1903 to 1907 at the Art Students League of New York, an art school founded in 1875 for both amateurs and professionals that offered reasonably priced classes. Students could study part or full time, but there were no degree programs or grades awarded. The school is still operating today in the same manner.

Teague earned money by checking hats at the YMCA in downtown Manhattan, and also by painting signs. His skill in lettering soon evolved into drawing illustrations for mail-order catalogs, such as for neckties and shoes. This led to more elaborate advertising illustrations, which caught the attention of Walter Whitehead, an advertising executive at the Ben Hampton Advertising Agency he had met at the YMCA, and who promptly hired him.

When Whitehead left the Ben Hampton Agency in 1908 and joined the more prestigious New York agency of Calkins & Holden, Teague went with him. In his four years there, Teague became known for his distinctive frames, or borders, for advertising art, which blended Baroque and Renaissance influence with a unique simplicity. After a year of freelancing, Teague left Calkins & Holden in 1912 to work from his own typographic studio, creating graphic designs for magazine ads. In 1920, when he designed frames for the famous Arrow Collar ads, he was already recognized widely for what were known as "Teague borders."

By the mid–1920s, the demand for border designs weakened, and Teague became involved with commercial packaging. He was intrigued by the Paris exposition of 1925, and in June 1926 he traveled to Europe to view Bauhaus work in Italy and was inspired by the work and writings of Le Corbusier. Upon his return, he reestablished his office devoted to package design.

In 1927, Richard F. Bach (1887–1968), curator of the Metropolitan Museum of Art, recommended Teague to Adolph Stuber, a top manager of the Rochester, New York–based Eastman Kodak Company, which was considering engaging an artist to design cameras. Teague took Stuber on a tour of the modern art shops in New York, including the contemporary shop of Rena Rosenthal, in order to show him the clean lines of home furnishings that were coming into style.[15]

Teague had no knowledge of cameras or how they were made, but he proposed working on-site with Kodak's engineers, claiming that "designing according to engineering necessities ultimately leads to greater beauty and heavier sales."[16] Stuber apparently agreed. Teague's relationship with Eastman would result in dozens of camera designs over many years, and he would continue to consult with Eastman for the remainder of his life.

Teague's first designs for Kodak would include its 1930 Art Deco gift camera and packaging, and its Beau Brownie, both of which featured a colorful geometric pattern of rectangles and squares reminiscent of Dutch artist Piet Mondrian (1872–1944) and De Stijl art. By February 1928, Teague was already promoting the new profession of industrial design by writing an article, "Modern Design Needs Modern Merchandising," in *Forbes,* a business magazine. He advised, "The designer who gets results for the manufacturer plans with all departments of a business before he ever lays a pencil to drawing board."[17]

That same year, 1928, manufacturers were trying to make design changes in their products. But rather than changing the form of their old products, which would require expensive tooling costs, it was much easier for them to change the color of their product. In 1930, *Fortune* magazine recalled that in the early 1920s, "kitchen and bathroom were a porcelain and surgical white. Furniture was in natural wood tints, surfaced with a colorless finish. The stove was black. Faucets were metallic." Now, in a typical home in 1928,

> in the cellar stood a Redflash Boiler, more crimson than the flames of the adjacent furnace. On the door of the orange refrigerator was perched a green parrot. Not only the dishes, but the pots and pans were decked out in bright gay tints and tones. A green-tiled bathroom was set off with green tub, green towels, and, monstrously enough, green toilet paper. For afternoon tea there were blue glasses and a yellow cloth. A yellow alarm clock—a cherry garbage can, a scarlet typewriter—surfaces had turned suddenly into palettes.

Fortune concluded, "The acceptance of color was undoubtedly significant from the standpoint of old traditions being abandoned and old suppressions being released."[18]

The word was getting around the advertising world that designing products for industry was financially lucrative. An early example of this was the ticket scale designed for International in 1929 by **Joseph "Jo" Claude Sinel** (1889–1975), a 40-year-old artist working for the Lennon Mitchell advertising agency in New York. Sinel in a later interview recalled,

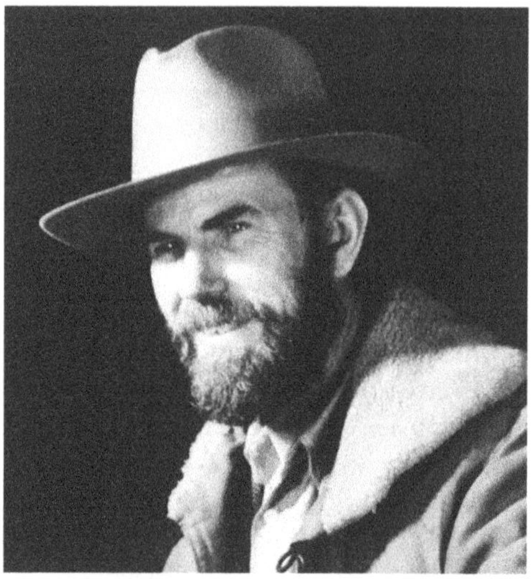

Joseph "Jo" Sinel, 1928 (courtesy California College of the Arts Libraries).

> One of the first designs I made was the International Ticket Scale, and I made a model from a Del Monte (wooden) fruit case.... I took the model in to show the executives of the company and they paid me a fee of $10,000 [$136,000 in 2013 dollars] and royalties for 25 years.[19]

This design is well documented and is in a style that many would call "skyscraper style," with vertical and graduated geometrical elements, typical of the period.

Sinel led a most unconventional life for his era. In later times, he might have been called a wandering "hippie," but his earnings in commercial art demonstrated outstanding skill and diligence in his career. He was born in Auckland, New Zealand, where his father, Thomas, ran a stevedoring operation and sired ten children. Jo's primary school's headmaster recognized his talent for drawing and urged him to attend the Elam School of Art. From 1904 to 1909, he served an apprenticeship as a lithographic artist for the *New Zealand Herald* newspaper. He did freelance work for printing firms in Wellington and Christchurch, then gained valuable commercial art studio experience in Australia as well as working as a harvest hand and sheep shearer.

Impulsively, he left for England and arrived in Liverpool in the middle of World War I, but he had no trouble landing a job at Hudson-Scotts, England's largest lithographer, in Carlisle. Next was Carlton Studios, England's most prominent art studio, and after that, the C. F. Hyme advertising agency in London. The war became increasingly dangerous to Londoners due to Zeppelin air raids, so Jo shipped out as a merchant seaman during wartime to avoid military conscription. After unloading guns off a British ship in Newport News, Virginia, he landed once again in Australia where he easily obtained a high-salary art job. When Australia passed a statute saying that men of military age could not leave the country after a certain date, Sinel decided it was time to leave.

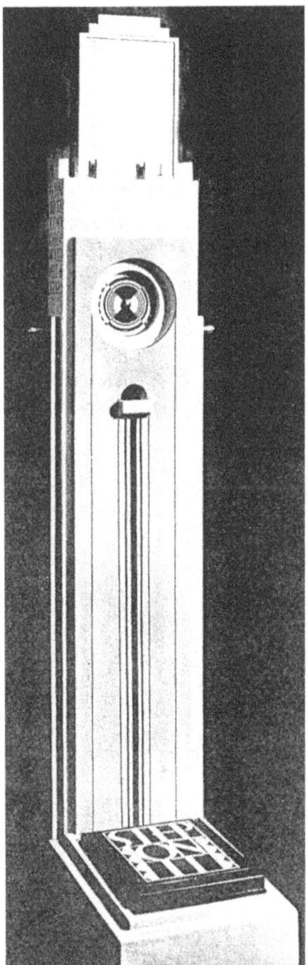

1929 International Ticket Scale by Sinel (*Designscape* magazine, February 1977).

In 1918 he left for the United States and landed in San Francisco at the time of the influenza epidemic, when people on the street were wearing surgical masks. Again, he landed an art position at Foster & Kleiser advertising agency but soon quit his job and set out for the high Sierras to build a cabin with an artist friend on the shores of Lake Susie. When snow arrived, he headed back to San Francisco to find several job offers waiting.

The one he chose was in New York with First National Pictures, a film company in which Charlie Chaplin and Douglas Fairbanks were associated. For First National, he designed promotional graphics for film distribution, but soon, driven by wanderlust and fascinated with automobiles, bought one in 1920 and drove across America, which was virtually unpaved at the time, to see its natural wonders and cities. He had advertisements for First National Pictures painted on the side of his car that attracted photographers at many cities, fueling Jo's natural flair for publicity. Returning to San Francisco, he worked simultaneously for Foster & Kleiser and the western branch of the McCann-Erikson advertising firm. When Jo refused to do a certain job at Foster & Kleiser, they fired him, but he and three others, including Harold von Schmidt, set up an independent business to design posters for various San Francisco advertising agencies, including Foster & Kleiser.

In 1921 and 1922, Sinel taught design and lettering at the California School of Arts and Crafts (CCAC, but now, California College of Arts, CCA) under its founder, Frederick Meyer. Jo would continue his association with CCAC, off and on, until his death. But in 1922 he took a position as art director for C. C. Ronalds, who was starting a new advertising agency in Montreal, Canada.

After five weeks there, Sinel went to New York to do freelance work with von Schmidt in a studio in Greenwich Village, where they were members of the Society of Illustrators and the Artists' Guild. Through these contacts, Sinel landed a job with the Lennon Mitchell advertising agency in 1923. In 1924, he published a book, *A Book of American Trademarks and Devices*. During this period, Sinel claimed, so he probably did, work on a number of product designs, as was not uncommon for advertising agencies of those days, but documented examples have not yet been found. After his International scale design in 1929, he would design many products for industry throughout the 1930s.[20]

CHAPTER 3

Modern Design Meets the Depression

Another profession attracted to the new field of modern design was the creative theatrical and stage world. The theatrical set design profession was suffering from the increasing popularity of motion pictures, which in 1927 had just introduced sound in *The Jazz Singer,* starring Al Jolson. Films were drawing crowds away from Broadway, and many theaters that in the past had staged live performances were now converting to movie theaters.

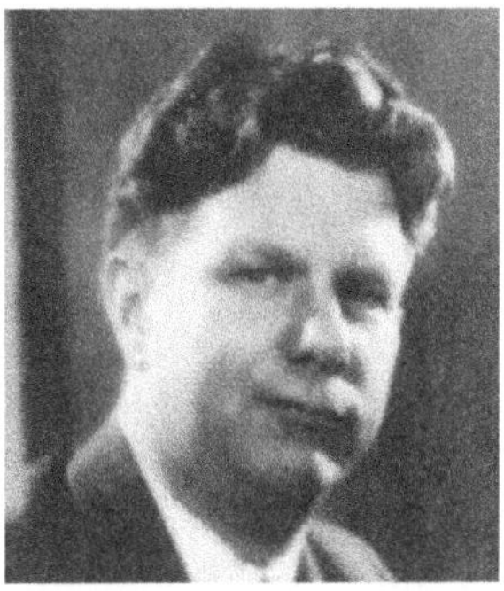

Norman Bel Geddes, ca. 1934 (February 1934 *Fortune* magazine. Photograph by Maurice Goldberg).

Norman Bel Geddes (1893–1958), a nationally recognized set designer, saw the potential for his talents in automotive design when urged by Ray Graham, of the Graham-Paige Motor Corporation in 1927. Then 34, Bel Geddes proceeded to design a series of five brass concept car models for Graham, each design representing a 20 percent increase in modernization over the previous one, demonstrating progressive annual model changes, each of which was not to be out of style for five years. Car No. 1 was designed in 1928, intended to be prototyped in fall of 1929, but it was never built due to the stock market crash and the onset of the Depression. Despite this setback, he was to become one of the most flamboyant and imaginative pioneers of the new profession.

Bel Geddes was born in Adrian, Michigan, as Norman Melancton Geddes. Raised in New Philadelphia, Ohio, he was a high school dropout and tried unsuccessfully to acquire formal art training at the Cleveland Institute of Art and the School of the Art Institute of Chicago. In 1913 he began work as advertising illustrator and art director for the Peninsular Engraving Company in Chicago, and then for the Barnes-Crosby Advertising Agency in Detroit. He also worked in 1916–1917 for Aline Barnsdall's Little Theatre in Los Angeles as a set designer, where he was attracted to the theater. In 1916, when he married Helen Belle Schneider, they incorporated their names to form the surname Bel Geddes. In 1918, he became scene designer for the Metropolitan Opera and designed and directed a number of very successful theatrical works, as well as an ice show starring worldwide skating star Sonja Henie. Bel Geddes's daughter, Barbara Bel Geddes, was born in 1922 and would become a well-known Broadway and film actress starting in 1947. The essence of Bel Geddes's theatrical set design career was to create environments and costumes to entertain and take people into magical worlds of imagination.

In the mid–1920s, Bel Geddes, for his own amusement and that of his friends, began creating elaborate table games. In the previous ten years, a public fascination with scale models had been developing. Construction of model airplanes that actually flew became popular with young boys, and electric model railroads at family Christmastime evolved into hobbyists constructing elaborate and extensive basement layouts with realistic miniature trains, bridges, lakes, mountains, trees, people and buildings. The idea of realistic miniature replications of the real world apparently also fascinated Bel Geddes.

He started with a five-by-three-foot football field, with players kicking, running, and tackling. That was followed by a mechanical baseball game and an indoor golf game, built with wooden blocks, wire screen, and paste-soaked newspaper. He added green velvet and "shrubbery." The game took over the dining room table and disrupted family life.

In 1925, Bel Geddes was creating film set designs for D. W. Griffith and the Cecil B. DeMille film *Feet of Clay*. He also designed costumes for Max Reinhardt, the Austrian-born stage and film actor and director. During this period, his wife of ten years left him and took his two children. Depressed, Bel Geddes was taken to Belmont Stakes racetrack for distraction, and he was inspired by the pageantry there to invent a new game, which he called the Nutshell Jockey Club. It was a 20-inch-wide racetrack built on a 28-foot-long table.

Running times were identical to real track times. There was green velvet turf, white fencing, bleachers, a judges' stand, and boxes for stable owners. The game was electrical and featured 800 cast-iron horses, each three inches long and realistically painted. Twenty horses lined up behind the electric starting gate and were pulled by invisible silk threads connected to unseen pulleys. A "chance machine" varied the speed of individual horses to simulate an actual race. The setup took over the whole basement at his Manhattan brownstone on East 37th Street that doubled as his office, and was estimated to have cost Bel Geddes $4,000, plus the labor of his staff. Every Saturday evening, eight races were run, announced over an amplifier, with a recording of racetrack bedlam and the sound of horses' hooves to lend atmosphere. Race betting, sale of horses and stables, and odds were conducted through a racing secretary. It was a live horse race in every respect except the weather.

Within two months, nearly 100 people owned stables in the Nutshell Jockey Club. The game was a magnet for visiting Hollywood notables, since racing in California would not be legalized until 1933. They included Charlie Chaplin; Douglas Fairbanks, Jr.; Leslie Howard; Ethyl Barrymore; Adolph Menjou; and Basil Rathbone. It was so successful that New York mayor Jimmy Walker, one of the game's enthusiasts, assigned a patrolman to keep "the wrong people" out.[1] Bel Geddes was already famous.

After he opened his industrial design office in 1927, Bel Geddes left the theatrical field, except for creating sets for the 1935 Broadway production of *Dead End*. Despite offers by gambling syndicates to operate the Nutshell Jockey Club (one offered "the blond little Leonardo ... with habitually tousled hair" a thousand dollars per night to operate the Nutshell at Saratoga Springs, New York, for a season), Bel Geddes declined and put it in storage.

Department stores continued to enthusiastically promote modern design. In 1928, the B. Altman department store in New York held an exhibition called Twentieth Century Taste. Lord and Taylor announced an international competition in design and trade symbols. Franklin Simon engaged none other than Norman Bel Geddes to develop a modern approach for window displays. His windows featured dramatically lit, very modern-looking, Cubistic pyramids and metallic mannequins. When revealed to the public, they created massive traffic jams and crowded sidewalks. Within 90 days, every store on Fifth Avenue had changed their windows to the new modern style. Also in 1928, Bel Geddes was commissioned by the Simmons Company to design steel beds that would go to market in 1932.

Macy's staged a dramatic International Exposition of Art in Industry in 1928, organized by the new head of its department of design, Austin Purvis. The exposition included 15 rooms, most designed by well-known European designers, such as a powder room with products from the Wiener Werkstätte (Vienna Workshops) by Josef Hoffmann, director of the Austrian Academy of Fine Art; a dining room by French jewelry designer Maurice Dufrêne; a man's study and dining room by Bruno Paul, head of the German Vereinigte Staatsschulen für Freie und Angewandete Kunst (United State Schools for Fine and Applied Arts); and a butcher shop and living room by Italian designer Gio Ponti (1891–1979), who had just founded *Domus* (Home), an Italian architectural and design magazine that promoted the Wiener Werkstätte as well as the Italian Novocento (Nine Hundred) group to which Ponti belonged since its founding in 1926.

European émigré designer Kem Weber, who had just designed a skyscraper side table/vanity; architect William Lescaze (1896–1969); and architect Eugene Schoen (1880–1957) designed the American rooms, which included work by German silversmith émigré Peter Müller-Munk, who later would became a leading industrial designer. European design was still the main draw.

From February 12 to March 24, 1928, Richard Bach, of the Metropolitan Museum of Art, decided to emulate the 1925 Paris exposition with French room ensembles in an exhibition, Contemporary American Design, the eleventh in the series, which was intended to conclude in March, but by then, 100,000 had attended and he extended the exhibition to September 2. The subtitle, The Architect and the Industrial Arts, revealed that Bach favored prominent American architects such as John W. Root, Ralph Walker, Raymond Hood, Ely Kahn, and European architects such as Eliel Saarinen, Joseph Urban, and Eugene Schoen, to lead the new profession of industrial design, just as architects had led in Europe, and that was in fact the very purpose of the exhibition. As Bach said in the catalog,

> [The Museum] has secured the enthusiastic cooperation of nine architects of distinction in this and other cities, who have made the exhibition practically their own by planning every detail of it themselves, from the general design of the gallery as a whole to the minutiae of their individual exhibits. Hence the appropriateness of the special title of the exhibition, "The Architect and the Industrial Arts," as it is a fine and instructive illustration of what the architect can do outside of the field which we commonly associate with his profession, and it was gratifying to us to learn that they welcomed the opportunity of showing this very thing.[2]

American designers and AUDAC members were understandably appalled that they were excluded from these shows in favor of European and American architects, particularly in the industrial arts arena, where they had demonstrated their abilities in product design far more than architects had. In retaliation, they began planning their own show for the following year.

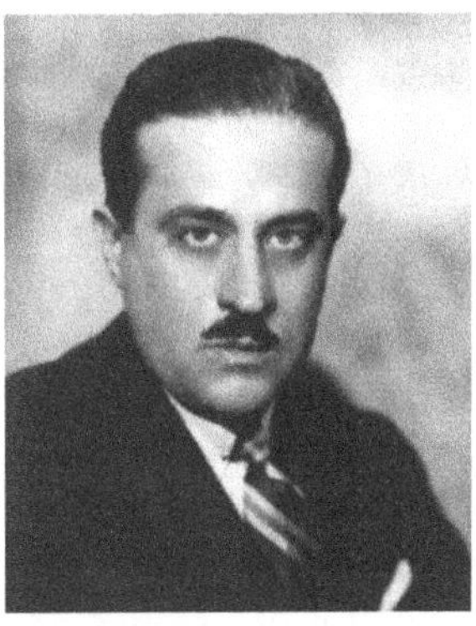

Raymond Loewy, ca. 1934 (used with permission of Loewy Design LLC, www.RaymondLoewy.com. Photograph by Hal Phyfe).

Another designer, 36-year-old **Raymond Loewy** (1893–1986), opened his New York office in 1929. His first product client was Sigmund Gestetner, head of the British Gestetner Company, for which he redesigned a Model 66 duplicator (mimeograph) machine from a hodgepodge of shapes into a smooth single form after building a clay model. Gestetner liked it so well he paid

Loewy $2,000. It would be introduced to the market in 1933, and the same design would be used for 15 years afterward.

Raymond Fernand Loewy was born in Paris, and as a 15-year-old in 1908, he saw the Brazilian inventor Alberto Santos Dumont (1873–1932) fly his handmade airplane about a hundred yards over a polo field in the Bois de Boulogne in Paris. Inspired, Loewy designed a rubber-band-powered scale-model airplane and named it the Monoplan Ayrel. It won the James Gordon Bennett Cup for model airplanes. Bennett was the American editor of the Paris edition of the *New York Herald* and the *Paris Herald.* Teenage boys offered to buy Loewy's plane in such numbers that he patented it and formed a company, the Ayrel Corporation, to produce quantities in rented space. It was advertised and merchandized all over France, and it sold so well he had an offer to purchase his company, and promptly sold it so he could continue his studies.[3]

He graduated from Paris University in 1910 and from 1914 to 1918 served in the French Army as a captain in the Engineering Corps during World War I, where at the front he decorated his dugout with flowered wallpaper, draperies, and tufted pillows. During four years of war, during which he was burned by mustard gas, he rose from private to captain and received the French Legion of Honor. In 1918 he graduated from the Ecole de Lanneau in Paris.

Loewy immigrated to America in 1919 to join his brother Georges, a doctor in Manhattan, with only $40 in his pocket and still wearing his captain's uniform (the only clothing he had). His sketching on the ship so impressed Sir Harry Gloster Armstrong, then British consul general in New York, that he gave Loewy a note of introduction to the publisher Condé Nast, where he got his first job. He also worked as a fashion illustrator for *Vogue* magazine, *Harper's Bazaar* magazine, and Wanamaker's, Saks Fifth Avenue, and Bonwit Teller department stores. He designed costumes for Florenz Ziegfeld before establishing his office. Later, when asked his first impressions upon arriving in America, Loewy replied,

> To put it briefly, I was amazed at the chasm between the excellent quality of much American production and its gross appearance, clumsiness, bulk, and noise. Could this be the leading nation in the world, the America of my dreams? I could not imagine how such brilliant manufacturers, scientists, and businessmen could put up with it for so long.[4]

As mentioned earlier, the acclaimed "Dean of Advertising," Ernest Elmo Calkins, had been a vocal proponent of more stylish and attractive products for several years. Now he put his money where his mouth was. In 1929, he appointed 40-year-old **Egmont Arens** (1889–1966) as art director for Calkins & Holden, where Arens initiated and headed an internal industrial design department. Arens was born in Cleveland, Ohio, and at a young age was sent by his family to New Mexico to recover from tuberculosis. From 1911 to 1914 he attended the University of New Mexico, and from 1915 to 1916, the University of Chicago. He began his career as sports editor of the *Albuquerque Tribune-Citizen* in 1916, and from 1917 to 1923 operated his own Washington Square Bookstore on West Eighth Street in New York, where he established the Flying Stag Press, printing magazines with an upstairs press.

His interest in the arts led to editing and printing *Playboy* magazine (no, not that one) in 1919, the first American magazine devoted to modern art, named after Irish playwright John Millington Synge's 1907 play, *The Playboy of the Western World.* In 1922, while still publishing and running the bookstore, Arens was art editor of *Vanity Fair,* one of the leading style and fashion magazines in the country, and later served as managing editor of *Creative Arts* magazine.

A prolific writer, Arens's interest in modern design led him into products, because by

1927 he was designing and seeing to the manufacture of a line of lamps[5] and other consumer products, which had caught the eye of Calkins & Holden.

At an American Management Association (AMA) annual meeting in 1929, an officer of the KitchenAid Manufacturing Company of Dayton, Ohio, E.B. French, reported that Egmont Arens's 1928 redesign of his company's KitchenAid mixer (possibly KitchenAid Model G) had cut its weight in half and improved its appearance. As a result, sales had increased 100 percent. Another attendee reported that while a simple radio costs only $50, "in an artistic cabinet it brings $150." A representative of the L.C. Smith typewriter company said that his company introduced color in 1926 after years of black only, and soon only 2 percent of sales were black.

Business leaders were beginning to accept the reality of modern design as something they needed to deal with. Earlier in 1929, the AMA had conducted seminars for businessmen on whether "intelligent and profitable use [can] be made of the modernistic style of design." Seminar titles included "How the Retailer Merchandizes Present-Day Fashion, Style, and Art"; "How the Manufacturer Copes with Fashion, Style, and Art"; and "The Renaissance of Art in American Business."

At the AMA annual meeting cited in the paragraph above, keynote address speaker E. Grosvenor Plowman, advisor on merchandising problems to the Associated Industries of Massachusetts, stated that the modernistic style was the "offspring" of the Jazz Age, and that "the modern style will remain as long as the machine in its present form is the characteristic of world production, [changing] its outward form from year to year, just as women's dress styles change," and that manufacturers would have to learn to balance faddism and permanence.[6] Plowman clearly thought that modern design was "faddish" and that "permanence" (the way products always looked) was quite acceptable. The date of this particular AMA annual meeting was coincidentally on October 29, 1929, the exact day of the stock market crash that initiated the Great Depression.

Investors had been buying stocks on margin, that is, on credit, for years. Brokerage firms would lend $9 for every $1 an investor deposited, which vastly overpriced the stock market. When the market fell, brokers called in these loans, which could not be paid back, so many investors were wiped out. RCA stock, at $549 per share in 1929, dropped to almost $0 by 1932. Even before the stock market crashed, many businesses were making more products than they could sell and were overstocked. But now, many Americans drastically reduced their spending on consumer goods and services. Manufacturing and sales were soon down by 50 percent, and unemployment would rise to 25 percent.

Many people lost their jobs, and among these was Helen Dryden, the "highest-paid woman artist." She held her art director job at the Dura Company until the stock market crash of 1929, when she was let go and replaced by **George W. Walker** (1896–1993), who had also just lost his job at Graham Paige. Walker was born in Chicago and attended the Cleveland School of Art and the Otis Art Institute, Los Angeles, in 1916. In 1920 he began his career as an advertising illustrator in a Los Angeles department store. He returned to Cleveland as a freelance illustrator and established his own firm, working for department stores and Premier Press.

He entered the automotive business in 1928 by working for the Peerless Auto Company, and then for the Baker-Raulang body company, both in Cleveland. **Amos Northup** (1889–1937), then chief designer for the Murray Corporation of America in Detroit, visited Baker-Raulang and urged Walker to come to Detroit, where he interviewed and worked briefly for Harley Earl and John Tjaarda at General Motors, and then for Graham Paige, working with Amos Northup until the company failed in 1929 after the crash. Now he was with Dura, which was also an automotive parts supplier.

Northup, an automotive designer born in Bellevue, Ohio, had attended Cleveland Polytechnic Institute, where he studied design, and began his career as a furniture and interior designer. In 1918, he joined Pierce Arrow as the company's truck designer, but soon he was designing interior colors and trim for passenger cars. In 1921 he left Pierce Arrow and opened his own private studio, doing interiors for Pierce Arrow, Leon Rubay, and the 1924 Wills Sainte Claire. In 1927, he was hired by the Murray Corporation, the second-largest independent body producer, and shared in the design of the 1928 Hupmobile. In 1928, he would leave Murray to join Willys-Overland in Toledo, Ohio.

Norman Bel Geddes was now concerned with other enterprises. In 1929, Bel Geddes was designing products and buildings for the Toledo Scale Company, a commission enabled by his uncle, Frederick Geddes, a director of the company and a trustee of the Toledo Museum of Art, within which Bel Geddes would design a theater stage. But because of the stock market crash, his product and factory designs for Toledo never saw fruition.

One of Bel Geddes's former apprentices, also a successful set designer, ventured into the new profession. In 1929, 25-year-old **Henry Dreyfuss** (1904–1972) opened his own office at 580 Fifth Avenue, New York, for stage and industrial design activities. He had just won a competition for the design of a "Phone of the Future" sponsored by Bell Laboratories and would begin work on it in 1930, although it would not be produced until 1937. Dreyfuss was born in Brooklyn, New York, to a family in the theatrical materials supply business. He completed studies as an apprentice to Norman Bel Geddes in 1924 and produced 250 stage sets for a number of theaters before 1928, when, as we know, Bel Geddes closed his career in the theater and entered the industrial design field. Dreyfuss was now on his own.

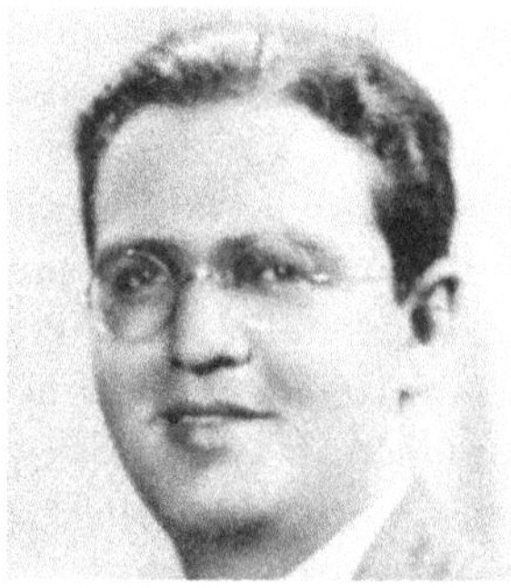

Henry Dreyfuss, ca. 1934 (February 1934 *Fortune* magazine. New York Times/ Redux).

Over the year before the 1929 stock market crash, several art/design schools had initiated programs to educate engineers in art and design, at the request of major manufacturing corporations. In Toledo, Ohio, the Owens Bottle Glass Company convinced the Toledo Museum School of Design to initiate courses in industrial design for its engineering employees in 1928.

In Pittsburgh, Pennsylvania, that same year, Westinghouse Electric requested an evening course at Carnegie Institute of Technology (CIT) "in the fine art of design as applied to electrical machinery" for engineers, which began in 1929 and was called Design and Appearance. Wilfred Readio, head of CIT's Industrial Arts program, Department of Painting and Decoration, which taught commercial illustration, costume design, textile design, interior decoration, stage design, and printing, headed the course.[7]

There were three sections in the course, and Donald Dohner, who taught design, and Alexander Kostellow, who taught drawing, taught the other two. They were both on the faculty under Readio.

Westinghouse in 1930 hired **Donald Roscoe Dohner** (1892–1943), one of the CIT instructors, to become director of art in the engineering department at Westinghouse, a position in which he controlled the appearance of all products manufactured in the company's 25 plants. Over the next four years, with a staff of eight, most recently graduated students from CIT, Dohner would design many Westinghouse products from locomotives to ashtrays, including meters, control panels, and switchboards. He worked with toolmakers, die makers, shop foremen, and superintendents. He studied markets, materials, and manufacturing operations and made a point of working in close contact with engineers, not only on appearance and sales appeal, but on cost, ease of manufacture, convenience, and economy of maintenance.

Dohner had grown up in Indiana and starting in about 1912 studied at the John Herron Art Institute, Indianapolis; the Academy of Fine Arts in Chicago; and the Chicago Art Institute. He devoted two years to work in advertising, one year in merchandising, and in 1916 he studied the sales influence of product design. In 1918 he went to Pittsburgh to study set design at the Carnegie Institute of Technology (CIT), but while waiting for classes to begin, he took a job as industrial arts instructor in the Pittsburgh public school system. After his graduation from CIT in set design, in about 1925 he began teaching at CIT as an associate professor of design. Now, at age 38, he became one of the first corporate heads of U.S. industrial design departments. Robert Lepper, a 1927 graduate of CIT's industrial arts program in painting, had just returned from Europe and was hired as an associate professor to teach Dohner's night course. More on him later.

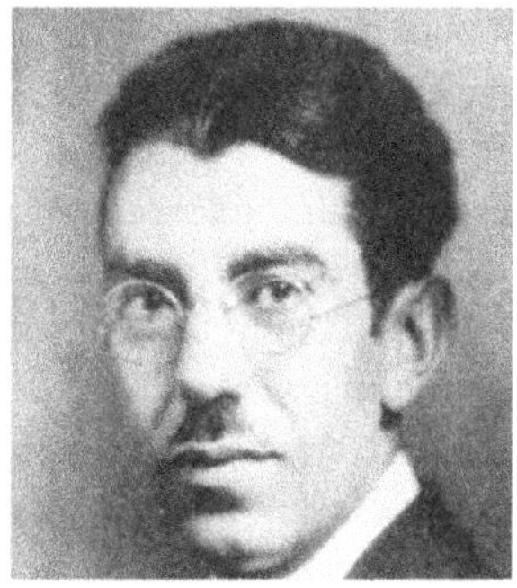

Donald R. Dohner, ca. 1934 (February 1934 *Fortune* magazine. Photograph by Bachrach).

Another major manufacturer, the Owens-Illinois Glass Company, in 1930 hired **Edwin W. Fuerst** (1903–1988) as a designer and stylist of its Toledo design studio, which specialized in glass containers and proprietary bottles. Born in Ottawa, Ohio, Fuerst left high school during his junior year after the death of his father and worked at the Detroit Steel Range Company. He then worked at Willys-Overland Motors, Inc., for a short time and returned to Detroit Steel Range, where he learned tool and die designing while working as a night superintendent and paymaster. While there, in 1927, he started attending classes at the Toledo Museum School of Design, mentioned earlier, until 1932. But in 1929, at age 26, he established his own design office in Toledo. One of his first clients was Owens-Illinois, who hired him as designer and stylist of its design studio.

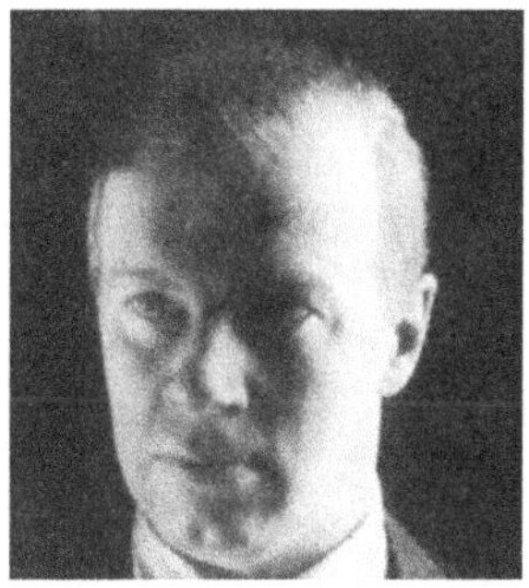

Gustav Jensen, ca. 1934 (February 1934 *Fortune* magazine. Photograph by Ruth Bernhard).

Other manufacturers were beginning to see the light. In 1930, **Gustav Jensen** (1898–1954) was hired by the International Nickel Company to design a new sink using Monel metal (a copper-nickel alloy) to popularize this new material. He also designed a telephone for AT&T (as one of ten entries in an AT&T "Telephone of the Future" competition won by Henry Dreyfuss) and a water heater for L.O. Koven & Brothers. Jensen was born in Copenhagen, Denmark, where his father was a banker and lawyer, and his mother came from a long line of ministers. He studied philosophy at the University of Copenhagen but wanted to become an opera singer. Nevertheless, he developed an interest in architecture, which led him to art and typography.

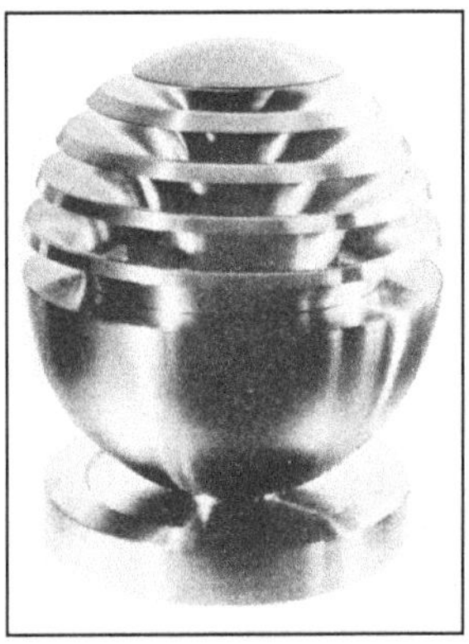

Radio by Jensen (*Pencil Points* magazine, 1937).

He emigrated to the United States in 1918, settling in New York City at the age of 20. He soon had clients including Colophon Quarterly, Covici-Friede, United Drug Company, and DuPont, for whom he designed book jackets, bindings, and cosmetics. He had begun work in industrial design in 1928 and was soon designing letters, borders, perfume bottles, cosmetic boxes, telephones, radios, silverware, toasters, and, as noted above, kitchen sinks. He called himself a "designer to industry" and would be productive up to World War II.

Forty-two-year-old famed sculptor **Alfonso Iannelli** (1898–1965) in 1930 turned to industrial design by beginning to work as a designer for Sunbeam Appliance Company. He designed logos and appliances with modernist shapes, including a Cubist style, chrome-plated faucet for the Mueller Mixing Valve Company. Born in Italy to a poor family, he apprenticed with a decorator before immigrating to America with his family in 1898. He studied at the Art Students League in New York and in 1900 moved to California to work in architecture, architectural decoration, and stained glass. From 1910 to 1915 he designed hundreds of posters and show cards for the Orpheum Theater Chain. In 1914 he came to Chicago to design sculptural ornamental work for Midway Gardens, a huge entertainment center on Lake Michigan, designed by Frank Lloyd Wright, but Wright claimed all of Iannelli's work as his own. Iannelli refused to ever work for Wright again. In 1915, he established Iannelli Studios and began teaching at the Art Institute of Chicago, where in 1922 he was honored with a one-man exhibition of his work, which included modernist lamps and furnishings.

In 1930, **Nathan George Horwitt** (1898–1990) designed the Beta chair, using steel, chrome-plated tubing, for the Howell Manufacturing Company. Horwitt was born in Russia and as a child immigrated to the United States, where he attended the City College of New York and New York University. He later attended the Art Students League of New York and served in the U.S. Army during World War I. He was hired as an advertising copywriter and was eventually promoted to director of advertising. He established the firm of Design Engineers in Manhattan in the late 1920s, and after that he developed patents to be sold to manufacturers.

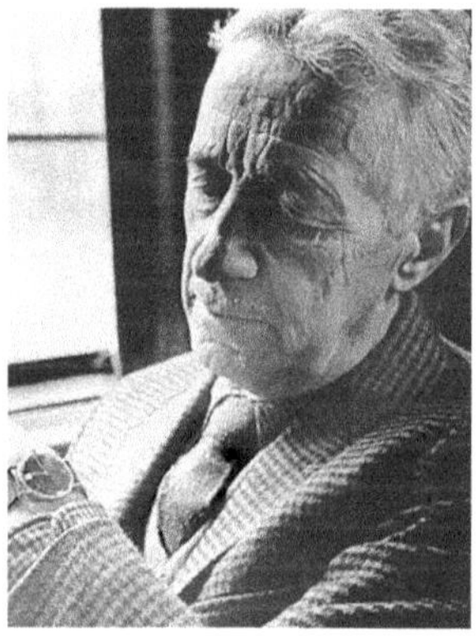

Nathan George Horwitt, ca. 1960 (courtesy Movado Group).

Norman Bel Geddes was becoming interested in the future. He wrote an editorial in 1930 entitled "Ten Years from Now" for the *Ladies' Home Journal,* predicting 50 industrial innovations, including the following:

> Aluminum will replace steel in railroad car construction ... synthetic materials will replace products of nature in structural materials ... synthetic materials will replace wool, cotton, and silk in textiles ... aircraft will attain a speed of six miles per minute [360 mph] ... a net of airlines will encircle the world ... [and most prophetically] artists will be thinking in terms of industrial problems of their age.[8]

His vision of the future would within two years inspire him to predict and design fantastic futuristic trains, planes, and ships in a trendsetting book. Bel Geddes was actively promoting his work during 1931–32 with similar magazine articles and mounting dozens of exhibitions of his work across the country and abroad, including Toledo, New York, and Vienna.

Designers were beginning to work together to promote their work. Members of the American Union of Decorative Artists and Craftsmen (AUDAC), formed in 1927 but ignored by museums and department stores in 1929, quickly reorganized and finally got a chance to display what they could do. They mounted their own exhibition in early 1930 at the Grand Central Galleries in New York to counter museum promotions of European designs. The exhibition featured five modern interiors filled with American products already on the market to demonstrate that human comfort, practicality, and utility alone were appropriate American design criteria, rather than the typical European artistic designs favored by museums.

Well over 120 exhibitors were represented, including 27 industrial designers. The October 1932 issue of *Popular Mechanics* reviewed the exhibition and reported that the exhibition was "proof that art and the machine have formed a partnership" and that "the best manufacturing enterprises ... were seeking counsel from the small group of men who are equipped to design beauty into the products of factory machines." Among the many exhibits of utilitarian objects were Gustav Jensen's Monel metal sink, Walter Dorwin Teague's scale model of his Marmon car design, Jo Sinel's International ticket scale, and Russel Wright's cocktail shaker. Wright also designed an elegant set of sterling silver flatware with simple, geometric shapes that was totally modern and would much later be reproduced by the Museum of Modern Art. Wright's designs were so successful that he began a business, Russel Wright Incorporated, with his wife, Mary, designing and producing spun aluminum accessories for the home.

Russel Wright (1904–1976) was born into a historic old American family in Lebanon, Ohio. His mother had direct lineage with two signers of the Declaration of Independence, and his father and grandfather were local judges. Wright trained in art under Frank Duveneck, American figure and portrait painter, at the Art Academy of Cincinnati while still in high school. In 1921 and 1924, he attended the Art Students League in New York and the Columbia University of Architecture. Despite his interest in art, he dutifully followed his family tradition by studying for a legal career at Princeton University. While there, he was a member of the Princeton Triangle Club and won several Tiffany & Company prizes for outstanding World War I memorial sculptures. When he left Princeton, he abandoned law as a career and immediately became a theatrical set designer for Norman Bel Geddes in New York, and later for other film and theater directors. In 1928, however, he had started his own design firm, making theatrical props and small decorative cast metal objects.

In 1931, AUDAC would publish its *Annual of American Design* to promote its members and to record its exhibition of 1930 with 176 pages, 300 photos of products, and essays about design. Frank Lloyd Wright (1857–1959) wrote about the "Principles of Design"; historian Lewis Mumford (1895–1990) wrote about "Culture and Machine Art"; Norman Bel Geddes wrote about "Business Interiors"; Kem Weber described the "Modern Shop"; Paul Frankl wrote about the "Home of Yesterday, Today, and Tomorrow"; and Richard Bach, of the Metropolitan Museum of Art, wrote about the "Resurgence of Quality," among other contributors. Sections of the book included "The Shop," "The Home," "Textiles," "New Materials," "Industrial Design," "Architecture Halls, Walls, and Fittings," "Graphic Arts in Advertising," and "Commercial Photography." Bel Geddes praised the successful collaboration between art and industry.[9]

The publication, a 9¼ × 12¼ inch hardcover book, was a masterpiece of Art Deco design and probably the earliest documentation of American industrial design, as it entered the public scene. Soon there would be much more visibility for industrial designers, with the help of museums, business magazines, and national merchandizing organizations.

From May 2 through July 1, 1931, the Brooklyn Museum held an exhibition of Modern Decoration and Design by members of the American Union of Decorative Artists and Craftsmen (AUDAC), the organization's first large general museum exhibition, which was billed by the museum as "about the first opportunity the public has had to see a representative indication of the trend and importance of modern design among American craftsmen." Well-known exhibitors included Frank Lloyd Wright, Rockwell Kent, Lee Simonson, Joseph Urban, Paul T. Frankl, Egmont Arens, Donald Deskey, Helen Dryden, Norman Bel Geddes, Nathan George Horwitt, Gustav Jensen, Walter Dorwin Teague, Walter von Nessen, and Russel Wright, among others. Kem Weber, of Hollywood, California, designed a suite of rooms for the 'investor of moderate means."[10]

By 1931, several additional major organizations were initiating departments of industrial design. One was Montgomery Ward in Chicago, the mail-order giant, which hired 31-year-old **Anne Swainson** (1900–1955), a Swedish émigré and fashion coordinator, as head of its new Bureau of Design, which was described as "a service to buyers in evaluating the quality and appearance of products that Montgomery Ward ordered from thousands of outside suppliers." Swainson was educated in applied arts at Columbia University and had been working as a "fashion coordinator" at Chase Copper & Brass. The Bureau concept was her invention.[11]

She and her staff redesigned Ward's catalog, replacing old-fashioned woodcuts with photography, and insisting on the authority to approve or reject all projects. She was the first woman executive at Montgomery Ward and staffed her department with arts graduates from the Armour Institute of Technology (later to become the Illinois Institute of Technology, IIT) and the Art Institute of Chicago. Among them was **George A. Beck** (1908–1977), a recent graduate of the Chicago Academy of Fine Arts. Over the years, the Bureau would employ dozens of designers who would later become prominent in the field. With her staff, Swainson assisted in merchandising, product, or packaging design for many of the companies that produced soft or hard goods to be listed in the catalogs, as many often had no internal design capability. Products repackaged or designed by the Bureau of Design were adorned with the "MW" (for Montgomery Ward) logo.

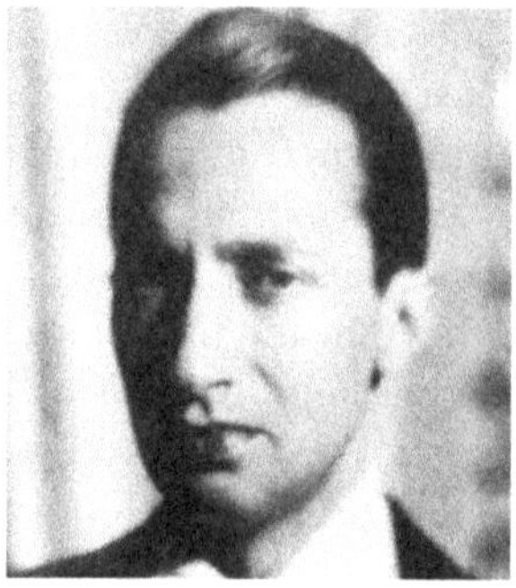

George Sakier, ca. 1934 (February 1934 *Fortune* magazine).

Another major manufacturing firm, the American Radiator & Standard Sanitary Corporation, the Kohler of its time, which produced bathroom and kitchen fixtures in porcelain enamel, as well as heating equipment, engaged a 34-year-old designer from one of the art industries, glassware. **George Sakier** (1897–1988) began in 1931 as a consultant and later became head of the company's Bureau of Design Development.

Sakier was a man with multidisciplinary talents. Born in the United States, he studied at the Engineering School of Columbia University and at Pratt Institute in Brooklyn. At age 19, in 1916, he composed the text *Machine Design and Descriptive Geometry*. During World War I, he worked as a camouflage technician, and after the war he continued to work as an artist, combining technology with classicism. In the 1920s, he worked in art direction for *Vogue*, *Harper's Bazaar*, and *Modes and Manners*. As a contributor to European art periodicals, he stimulated the rescue of a Mayan collection of artifacts from the cellar of the Palais du Trocadéro, a palace built in 1867 for the World's Fair in Paris, proving to France that America had an ancient art.

Sakier's fine art paintings were shown in the Julien Levy ultramodern gallery in New York, established in 1931, which also displayed the work of Pablo Picasso, Max Ernst, Marcel Duchamp, and Salvadore Dali. From 1929 until 1979, Sakier worked as primary consultant for the Fostoria Glass Company of Moundsville, West Virginia, and designed the company's Art Deco designs, including black Art Deco vases No. 2404 and No. 2428 in 1930.

Other companies were engaging design consultants to ensure that their products were of the latest modern design. The Toledo Scale Company in Toledo, Ohio, in December 1931 engaged a firm to design new public scales in various colors using a new plastic material, Plaskon, to complement different store interiors. The new design firm formed that year was the partnership of Harold Van Doren and John Gordon Rideout, forming the firm Van Doren and Rideout. Toledo Scale was their first client, and their first assignment was for a line of public health "person weigher" scales in colorful, Art Deco, skyscraper styles.

Harold Livingston Van Doren (1895–1957) was born in Chicago, finished high school in New Jersey, and graduated from Williams College in 1917. After serving in the Army during World War I, he studied at the Art Students League in New York from 1920 to 1921, and went to Paris from 1922 to 1924 with a fellowship in art history. He worked as a lecturer at the Louvre in Paris, as an artist for the *Chicago Tribune*'s Paris edition, as an actor in a Jean Renoir film, and as a translator of books on French artists Cézanne and Renoir. He would have had an opportunity to attend the 1925 Art Deco exhibition in Paris. In 1927, he became assistant director of the Minneapolis Institute of Arts while serving as an art director for Minneapolis newspapers and writing articles for the *Saturday Evening Post* and *Encyclopedia Britannica*.

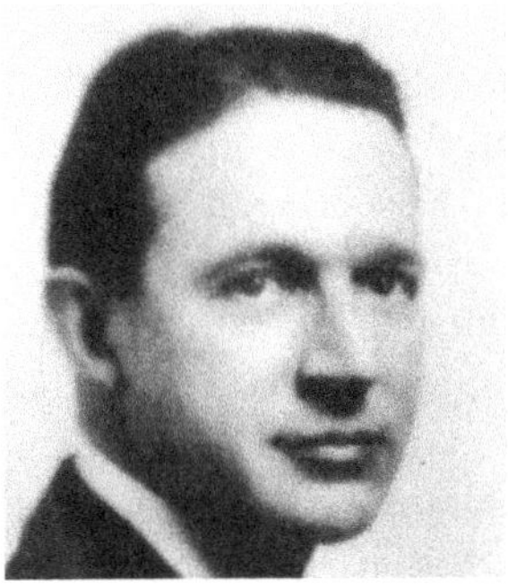

Harold Van Doren, ca. 1934 (February 1934 *Fortune* magazine. Photograph by Souren).

In 1930, his college classmate H.D. Bennett recruited him to Toledo as a color consultant for Bennett's new enterprise in the development of synthetic materials. Van Doren developed commercial applications for Plaskon, a plastic molding compound, and formed his design partnership with **John Gordon Rideout** (1898–1951), who was born in St. Paul, Minnesota. Rideout entered the University of Washington as an architecture student, but World War I interrupted his study and he served in the Navy in 1917. In the 1920s he worked as a graphic designer in Chicago and was one of the founders of the Chicago Society of Typographic Arts. In 1931 he came to Toledo to partner with Van Doren in industrial design.

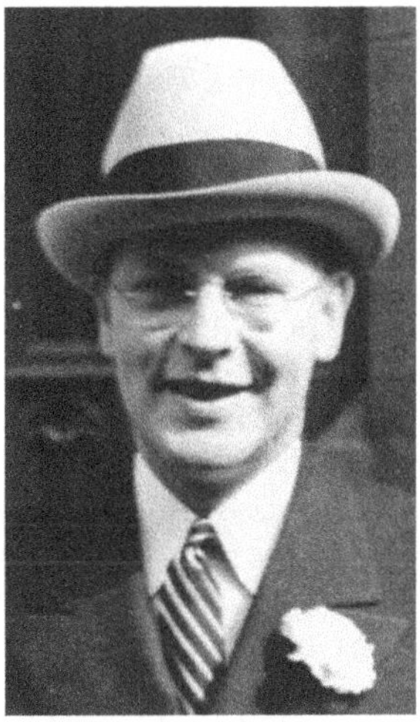

John Gordon Rideout, 1937, on his wedding day (courtesy Alyce Gordon Rideout).

With the Depression deepening and more manufacturers seeking modern design to boost sales, more and more designers were getting into this new profession of industrial design. **John McLeod Little** (1906–1996) opened an industrial design firm, J.M. Little and Associates, in Minneapolis in 1931. Born in Minneapolis, he studied at the Chicago Academy of Fine Arts, the Art League of Minneapolis, and the Minneapolis School of Art from 1927 to 1929, after which he worked for the Batten, Barton, Durstine & Osborn (BBDO) advertising agency. While managing his office in the 1930s, he accepted professorships in the field of design at the University of Michigan, Ann Arbor, and later at the University of California at Berkeley.

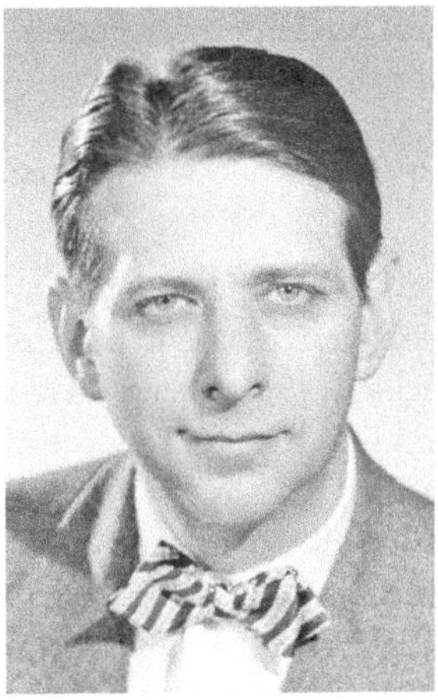

John M. Little, ca. 1949 (courtesy Linda Braming).

Museums were still promoting modern design. The Twelfth Exhibition of Contemporary American Industrial Art was held at the Metropolitan Museum of Art in New York, October 12 to November 22, 1931. Industrial designers among the exhibitors of furniture included Norman Bel Geddes, Donald Deskey, Paul T. Frankl, Walter von Nessen, Gilbert Rohde, Joseph Sinel, and Walter Dorwin Teague;

exhibitors of metalwork and lighting fixtures included Donald Deskey, Henry Dreyfuss, Paul T. Frankl, and Walter von Nessen; and an exhibitor of silverware was Peter Müller-Munk.[12]

The public was increasingly aware of modern design through popular motion pictures, which almost inevitably were filmed in sets that were completely modern in Art Deco décor, costumes, furnishings, and accessories. While most people suffered economically because of the Depression and lived quite frugally, they seemed to love escaping to theaters to see their favorite stars in modern, luxurious and extravagant settings. With all the bad economic news described in the Depression, it is easy to assume that many did not have enough money to spend on consumer goods. But the most prominent advertiser of the time disputed that.

In 1932, Elmo Calkins of the leading advertising agency, Calkins & Holden, wrote an introduction to a book, *What Consumer Engineering Really Is* (an unfortunate title for what would have been better called industrial design), written by Egmont Arens and Roy Shelton of Calkins' industrial styling division. Analyzing the Depression, Calkins wrote,

> Many stopped buying while still able financially to continue, and many are still able but restrained by fear and misplaced thrift. The sudden cessation of buying slowed up the entire industrial machine. Retail storekeepers curtailed orders to factories. Factories cut down production, reduced wages, laid off men, still further reducing the number of customers for goods. In a comparatively short time, with all the resources of the country still intact, we had Depression. There were no longer enough buyers for the large quantities of goods we had learned to make and distribute so abundantly.[13]

Not only did Calkins correctly analyze the problem, but he had a theory, "obsoletism," for solving it:

> Obsoletism is another device for stimulating consumption. The element of style is a consideration in buying many things. Clothes go out of style and are replaced long before they are worn out. That principle extends to other products—motorcars, bathrooms, radios, foods, refrigerators, furniture. People are persuaded to abandon the old and buy the new to be up-to-date, to have the right and correct thing. Does there seem to be a sad waste in the process? Not at all. Wearing things out does not increase prosperity, but buying things does. Thrift in the industrial society in which we now live consists of keeping all the factories busy. Any plan that increases consumption of goods is justifiable if we believe that prosperity is a desirable thing. If we do not, we can turn back the page to earlier and more primitive times when people got along with little and made everything last as long as possible. We have built up a complicated industrial machine and we must go on with it, or throw everything in reverse and go backward.[14]

Calkins expressed what all industrial designers sensed instinctively but many manufacturers and businessmen of the time did not. Although "planned obsolescence" would acquire a negative connotation 25 years later (*Hidden Persuaders,* by Vance Packard, 1957), nevertheless it was precisely what was needed to turn around the Depression, and is still a basic source of our material prosperity and economic health. Calkins could have added that by providing a "previously owned" market for goods, many lower economic classes were able to own products at far less cost than new ones, thus sharing in an ever-expanding standard of living.

The Museum of Modern Art (MoMA) had been founded in 1929 to promulgate and become an arbiter of modern art. Its first exhibit that year featured the "Big Four" of Postimpressionism: Cézanne, Gauguin, Seurat, and Van Gogh. In February and March, 1932, MoMA held its first architectural exhibition, The International Exhibition of Modern Architecture, and published a book, *The International Style: Architecture Since 1922.* It was MoMA's intention to establish architecture as a form of modern art to rival painting, and as the intellectual origin of international modern design. Organizers of the exhibition were Alfred H. Barr, Jr. (1902–1981), the founder and head of the museum; Henry Russell Hitchcock (1903–1987), an architectural historian; and the wealthy connoisseur/collector/architect Philip Johnson (1906–2005).

Hitchcock and Johnson were fascinated with European modernists such as Bauhaus architects Walter Gropius and Mies van der Rohe, as well as architects Le Corbusier and J.J.P. Oud, all of whom Hitchcock had called the "New Pioneers" in his 1929 book, *Modern Architecture.* Other architects represented included George Howe, Richard Neutra, Raymond Hood, and Rudolph Schindler. This new style featured spare, rectangular, steel and glass structures. The term "International Style" became the formal name of modern architecture and would dominate the architectural scene for the next 30 years.

Interestingly, Frank Lloyd Wright (1867–1959), America's greatest architect, was invited to participate but declined because he had misgivings about this new style, which he detested because it was counter to his own "organic" style of architecture. In fact, Wright had inspired many of the lauded European architects of the so-called International Style, including Gropius, Theo van Doesburg, Le Cobusier, and Richard Neutra, with his Wasmuth Portfolio published in Europe in 1910. Now, in 1932, Wright was regarded by many U.S. architects as an old-school, irascible, has-been. Philip Johnson had in fact disparagingly referred to Wright as "America's greatest 19th-century architect."[15]

The architect that Wright had inspired in Germany, Walter Gropius, was now, however, having difficult problems with the increasing power of Adolf Hitler.

Chapter 4

Streamlining and the Chicago Fair

The German Bauhaus, much admired by MoMA, was in serious trouble in Dessau, Germany. Faced with political and financial pressure, Gropius and Moholy-Nagy had resigned from the Bauhaus in 1928 and went to England, while architect Hannes Meyer, a committed Marxist who attempted to link the school to a vision of social and political transformation, assumed leadership. Some students were communists, others Nazis, both escalating political tensions. Architect Ludwig Mies van der Rohe (1886–1969) had become director of the Bauhaus in 1930, replacing Meyer, who had been forced by Nazis to resign at the same time that communist students were expelled.

In the month before the MoMA exhibition, the Dessau city council, pressured by the powerful National Socialist (Nazi) German Workers Party under Adolph Hitler, who was not yet in full power, dissolved the Bauhaus in that city, due to a "Jewish-Marxist art manifestation," a travesty that may have inspired MoMA to elevate the legacy of the Bauhaus to near martyrdom.

As a last-ditch effort to save the school, Mies van der Rohe, the last director of the Bauhaus, would move the school to Berlin, but within a year, Hitler would close the school in April 1933, and its faculty and students would scatter around the world, many coming to the United States later in the decade. There were other similar schools in Germany that were much larger in scope and number of students than the Bauhaus, such as the Industrial Art School in Berlin, with distinguished graduates such as Bruno Paul, Adolph Meyer, Kem Weber, and Mies van der Rohe himself, but such schools were not shut down by Hitler and thus were essentially forgotten in the United States.

In July 1932, **Belle Kogan** (1902–2000) opened her own design office, Belle Kogan Associates, at 185 Madison Avenue in New York City with a retainer from the Quaker Silver Company. The company had hired her in 1929 and sponsored her training in glass and silver manufacturing. Kogan was born in Ilyashevka, Russia, and had emigrated to the United States in 1906 with her parents, who settled in Allentown, Pennsylvania. She was the first of eight children, and she showed interest in art at an early age. Her Bethlehem High School art teacher encouraged her to study mechanical drawing, where she was the only girl in the class.

After graduation, she chose to attend the Pratt Institute in Brooklyn while teaching her high school class in mechanical drawing. But in 1920, she was forced to leave during her first semester at Pratt to manage her father's jewelry store. During this period, she attended the Art Students League of New York as time permitted. After Quaker hired her, the company sent her to train as a silver designer at the Rhode Island School of Design, and in 1930 at the German Art School in Pforzheim, Germany. The first female industrial designer was now in business for herself.

Other industrial designers were already making names for themselves. Both Raymond Loewy and Henry Dreyfuss appeared on major magazine covers. Dreyfuss was interviewed by *Popular Mechanics* magazine and pictured in the October issue working on his Toperator washing machine. In it, 28-year-old Dreyfuss described the process followed by industrial designers as more than simply adding beauty, but requiring research and incorporating practical considerations of cost, function, and convenience.

Gilbert Rohde had designed furniture for the Heywood Wakefield Company, including a side chair in 1931, 250,000 of which would be sold over the next eight years. He had established a relationship with D.J. DePree, president of the Herman Miller Furniture Company, in 1929. Rohde convinced DePree that the company should manufacture modern furniture. In 1932, DePree appointed him as director of design to shift the company's orientation from traditional to modern furniture, and instituted the policy of putting Rohde's name on each product. Rohde's designs of 1933 and 1934, including the clock he designed in 1933 for the Herman Miller Clock Company, would be equally strikingly modern, spare, and simple as many later designs of the 1950s.

A Design for Machine exhibit held in Philadelphia in 1932 featured furniture by Gilbert Rohde and designs by Russel Wright in a breakfast room that included lamps, tables, flatware, and accessories in aluminum, a material that had been used previously only in cookware. Since 1930, Wright and his wife had been designing and making decorative aluminum pieces in attractive, modern styles, some of which were cooking utensils attractive enough to be carried directly from the stove to the serving table. Wright had also designed an early table radio model along with his aluminum housewares.

More manufacturers were seeking designers. Jo Sinel in 1932 designed an Accousticon hearing aid for Dictaphone, a Radio-Victrola for RCA, Toastmaster tray accessories, and a Folmer-Graflex camera. **Joseph B. Platt** (1895–1968) in 1932 was engaged by the Parker Pen Company to design its Vacumatic fountain pen and its famous arrow-shaped

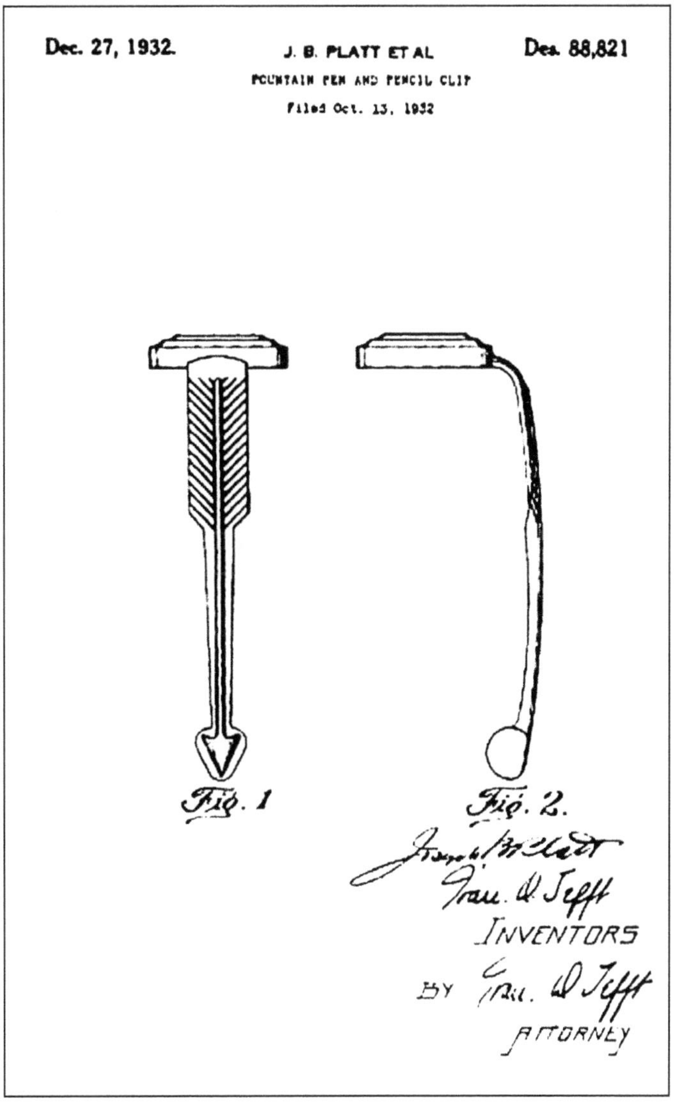

Parker Pen clip by Platt, 1932 (U.S. Patent Office, D88,821).

pocket clip, which became a Parker trademark on all its pens to this day. Platt in 1919 had married June Evans (1898–1977), the daughter of Rudolph Evans, the sculptor who would create the 1,000-pound bronze statue of Thomas Jefferson that would be added to the Jefferson Memorial in Washington in 1947.

In the 1920s, Platt and his new wife had relocated to the French commune of Senlis, an expatriate American community north of Paris, where Platt worked as a correspondent, editor, and illustrator for *Vanity Fair*, *Vogue*, and *House & Garden* magazines. He illustrated magazine covers from about 1923 to 1925, and probably attended the Paris Art Deco Exposition. He and his wife hoped to become community leaders at Senlis, but this did not work out. Returning to the United States in about 1929, the Platts lived in New York City and in Little Compton, Rhode Island, a beach resort.

Their apartment in New York was in the same building as that of Raymond Loewy and his new wife, Jean Thompson, and the couples became good friends. Loewy was probably responsible for getting Platt the job at Parker Pen. Platt's wife, June, was also, according to Loewy, "an incomparable and creative cook," and starting in 1936 with *June Platt's Party Cookbook*, she used Loewy's name as an endorsement on the book jacket. She would write a number of popular cookbooks well into the 1950s, including the *June Platt Cook Book* in 1958. During the Depression, Platt and his wife collaborated in painting murals, probably WPA projects, one of which wrapped the dining room of the Country Club of Detroit in Grosse Pointe Farms, Michigan.

In 1932, another major manufacturer, the B.F. Goodrich Company in Akron, Ohio, engaged **H. Creston Doner** (1902–1991), who had just opened his design office in Akron, Ohio, to design the company's rubber products. Doner was born in Nevada, Ohio, and was raised on farms in Nebraska and Pennsylvania. After high school he installed radios and worked on oil wells in Oklahoma. Doner was flamboyant in both appearance and behavior. In 1929 he had designed a stylish English sport jacket in Oklahoma and had it tailored to his own specifications and patented. He always dressed fashionably and regularly wore an ascot with a jacket. In 1954, he would be selected as one of the 25 men to become a charter member of the American Society of Beau Brummels (honoring those well-dressed men in the spirit of George Bryan Brummell, the leader of English fashion at the beginning of the 19th century).

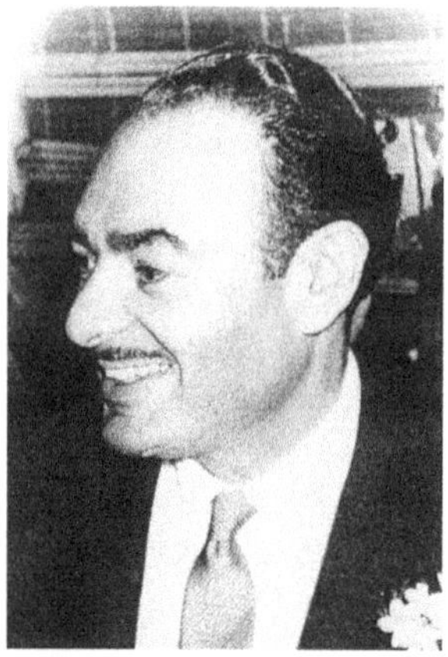

Alfons Bach, ca. 1970s (courtesy Anita S. Bach).

Alfons Bach (1904–1999) opened his own design firm in 1932, Alfons Bach Associates, in New York. Clients included Heywood-Wakefield, its subsidiary, Lloyd Manufacturing Company (Bach was a colleague of Kem Weber, a fellow German, who also worked for Lloyd); International Silver; Philco; and Trans World Airlines (TWA). Bach was born in Magdeburg, Germany, grew up in Munich; and studied in Berlin. He worked in set design and studied architecture at the Reimann School and the University of Berlin. He knew people at the Bauhaus but immigrated to the United States in 1926 and became a freelance designer.

Adolph Daniel "Al" Lambach, Sr. (1912–2008) arrived in New York in 1932 to work for the Progressive Table Company designing dinette furniture. Lambach was born in Germany and immigrated with his parents to Chicago in 1923. He attended Washburn Trade School there studying mechanical engineering drafting, then served an apprenticeship learning furniture design in Europe, and returned to Grand Rapids, Michigan, then the center of furniture design in the United States. He worked for an office furniture company there while attending Davenport Business College.

The Aluminum Cooking Utensil Company of New Kensington, Pennsylvania, a division of the Aluminum Company of America (Alcoa), bought the services of **Lurelle Van Arsdale Guild** (1898–1986) in November 1932 to design its line of Kensington Ware kitchen giftware utensils, which would be aggressively advertised and remain in production until World War II. Guild, born in Syracuse, New York, attended Syracuse University where he studied painting and 17th- and 18th-century design and crafting methods. He also did some freelance design work. After graduation, he and his wife relocated to New York City. Using his training in art and mechanical drawing, Guild worked in art industries, mainly home furnishings and the decorative arts.

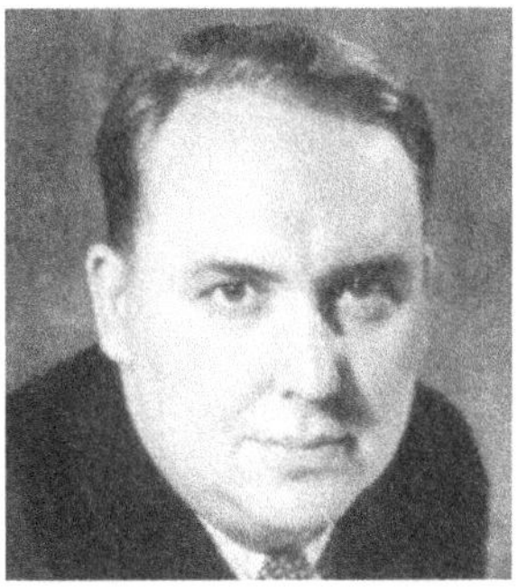

Lurelle Guild, ca. 1934 (February 1934 *Fortune* magazine. Photograph by Rudolph Hindemith).

In 1924, Guild started his own business, Lurelle Guild Associates, which, with a staff of eight, made models and worked on product development. He also started Dale Decorators, a door-to-door decorating service, which employed 100 women who marketed wallpapers, curtains, and carpeting. He scoured the country for antiques and sold them. He placed samples of his products in retail shops and surveyed potential customers. When he invented a new product, he patented it and then assigned the patent to the manufacturer, charging a fee plus royalties. His early interest in crafts led him to study Chauncey Jerome (1793–1868), a New Haven clockmaker who was one of the fathers of mass production. Fascinated with history, Guild moved an entire early American village from New Hampshire to Darien, Connecticut, where he lived.

Guild designed the permanent museum at Alcoa's New York offices, which displayed individual pieces as works of art, utilizing modern display tables of extruded aluminum and Z-shaped window treatments that eliminated glare and allowed daylight to enter the room at various angles. For Kensington, Guild also designed the Kensington showroom in Rockefeller Center in New York City. Indirect lighting was used from below display shelves. The walls were painted so that they graduated from dark blue-green to white, and the floors were color-matched to them.

What attracted these early designers to enter a strange world of machinery and mass production? Of course, as mentioned earlier, industrial design had suddenly become a financially lucrative field, as manufacturers scrambled to find talent to meet intense public demand. But aside from that, there is something very satisfying to a creative person to know that his or her design would not just be printed in a magazine, exhibited in an art gallery, win an award, or sold in small quantities as a decorative craft, but would be mass produced, sold by the thousands, and distributed throughout the country to a large audience of the general public who would use and interact with his or her design on a daily basis. It was a feeling, not only of public recognition and acceptance, but one of a personal contribution to society, making the world a better place to live. This was a powerful incentive.

New York, of course, was the American center of culture, business, and style, and many of the early leading designers were working there. John D. Rockefeller, Jr. (1974–1960), had developed Rockefeller Center, a 12-acre complex in Midtown Manhattan, beginning in 1929. In 1930, Donald Deskey had designed home interiors for Rockefeller, a client connection that helped him win the competition for the interior design of Radio City Music Hall in Rockefeller Center, an extravagant Art Deco landmark that opened in December 1932 with a lavish stage show featuring Ray Bolger and Martha Graham. Deskey would work on the interior until 1934. The Music Hall would be restored to its original condition in 1999.

But the biggest attractions in New York were its famous skyscrapers. New York skyscrapers were then capturing the imagination of the world. The Chrysler Building in New York City had been completed in May 1930, becoming the tallest building in the world (71 stories, 1,046 feet). Started in 1928, it was a classic Art Deco design by architect William Van Allen (1883–1954) for Walter P. Chrysler (1875–1940), president of the Chrysler Corporation he had formed in 1924 to manufacture Chrysler automobiles. Chrysler had been *Time* magazine's Man of the Year in 1928. Eleven months after the Chrysler building was completed, in April 1931, the 102-story Empire State Building was completed to capture the title of the world's tallest, at 1,454 feet.

These skyscrapers had been the inspiration of Paul Frankl for his stylish Skyscraper furniture in 1925, and he had written several books to publicize his business and to promote the new modern styles. In 1928 he had published *New Dimensions: The Decorative Arts of Today in Words and Pictures,* and in 1930, *Form and Re-Form: Practical Handbook of Modern Interiors.* Now, in 1932, he published *Machine Made Leisure,* capitalizing on the popular term "machine," which to the public was shorthand for the wonder of industrialization and mass production, just as "digital" or "computer" would be shorthand for the "Information Age" of the 1980s and 1990s.

But the skyscraper styles popularized by Frankl would soon be replaced in popularity by "streamlining," a style based on the public fascination with high speed, as celebrated in the new age of aircraft and racing cars, such as the Supermarine S.6B British seaplane which set the world speed record of 340 mph at the Schneider trophy air race of 1931 (and which would be developed into the famous British World War II fighter plane, the Spitfire), and Malcolm Campbell's Blue Bird racing car of 1931 (245.733 mph). The sleek, streamlined forms of shining metal embodied in these vehicles symbolized speed and were used by Norman Bel Geddes to envision the future in his 1932 book, *Horizons.* Bel Geddes based his designs on the concept of speed, as he said in his book, "speed is the cry of our era, and greater speed one of the goals of tomorrow."[1]

Bel Geddes did not invent the term "streamlining," although he popularized it in America. The term was originally used by English biologist/mathematician D'Arcy Wentworth Thompson (1860–1948) in his influential 1917 book, *On Growth and Form.* He used the term to describe organic structures that offer the least resistance when they are in motion, and how a fluid medium tends to impress its "stream lines" on a deformable body, like snowdrifts and sand dunes. He concluded that the same principle occurred in the evolution of the bodies of fish and birds. German Count Ferdinand von Zeppelin incorporated the same concept into the design of zeppelins, rigid airships named after him that he developed at the start of the 20th century, and which were popularized in the early 1930s by transatlantic flights from Germany to the United States.

Early aircraft and racing cars emulated similar streamlined forms. A pioneer was Austro-Hungarian Paul Jaray (1889–1974), an employee of the Zeppelin Company in Germany, who observed that automobiles roll along the ground, but they also move through the air. He developed scale-model cars to test in the Zeppelin wind tunnel to determine the best shape to lower air resistance. The ideal shape was a teardrop with rounded windscreen and enclosed bottom. In 1922, he had a full-scale prototype, the Ley T6, built by the German carmaker Rudolph Ley AG in Armstadt. Jaray was soon consulting with a number of European carmakers, and his principles were soon incorporated into racing cars by Maybach, Audi, and Mercedes-Benz, as well as aircraft designs. In 1922, Romanian engineer Aurel Persu (1890–1977) also applied aerodynamics to auto design, based on the ideal shape of a falling drop of water. In 1923, he built a prototype in Berlin, which reached high speeds, and patented it in the United States in 1927.

Detroit had picked up on the streamlining concept in 1931, when Amos Northup, then the chief body designer of the Murray Corporation, first used streamlining principles, although modestly, in the design of the Reo Royale Eight, the first mass-produced car to display the streamlined esthetic by smoothing the traditional sharp angles of car bodies into soft curves. A rounded, peaked shell surrounded a vertical V-shaped radiator, and the headlights were mounted in elongated pods. Front fenders swept rearward almost to the middle of the car. A curved panel provided a tapered rear end and covered the fuel tank. Murray claimed the curved fenders increased strength and that the shape reduced wind resistance, although this was negligible under 60 mph.

The design was an immediate public sensation, proving that streamlining could sell cars and encouraging Murray and Northup to design a much bolder streamlined body for the 1932 Graham Blue Streak, which had the first chromed, tilted back, V-shaped grille concealing the functional radiator, the first fender skirts to hide the "spatter zones" behind the wheels, and a trailing "beaver tail" that concealed the gas tank. This design made all previous automotive design obsolete and began a progression of automotive streamlining that continued well into the 1950s and beyond. By 1934, there would be no U.S. car without front, and sometimes rear, fender skirts.

A few early American industrial designers, besides Harley Earl of General Motors, were attracted to automotive design, which was, as it is today, a romantic and popular icon of lifestyle, speed, and freedom, for example, Walter Dorwin Teague and his son, **Walter Dorwin Teague, Jr.** (1910–2004, later called W. Dorwin Teague). Then a 19-year-old engineering student at MIT, the son had teamed up in 1929 with his father to design the Marmon 16 for the Marmon Motor Company, headed by Howard C. Marmon. Both father and son later claimed individual credit for the design. The car was introduced in 1932 and was a classic and impressive design, but it could not be described as streamlined. Four hundred would be produced before the Depression put the company out of business.

Raymond Loewy patented his first car design in 1930, and in 1932 designed a "cycle fender" Hupmobile for the Hupp Motor Car Company, which was in a fairly traditional style. He would design a more radical, more streamlined Hupmobile in 1934 called the Aerodynamic Hupmobile. Because executives were skeptical of his proposal, he decided the only way to get the job was to build a car by hand, using an existing chassis, to show them he was right. It took him several months, but when the executives saw the innovations in three dimensions, they agreed, and he got the job. The exercise cost Loewy over $20,000, in 1932! (In 2013 dollars: $341,000.) But he said it was worth it. He was to have been paid for the design as a percentage of all Aerodynamics sold over 10,000, but Hupp sold fewer than that, so he never received a dime. Later in his career, Loewy would become the most renowned non–Detroit auto designer. Fascinated with transportation in general, Loewy even ventured into ship design by designing the *Princess Anne* ferryboat in a streamlined style for the Pennsylvania Railroad Company in 1933. It was the year he took an office on the 54th floor of a skyscraper at 500 Fifth Avenue.

Bel Geddes, as we know, had entered the industrial design field in 1928 designing cars and was well aware of these streamlining developments in automotive design. He apparently concluded that transportation with streamlining was the wave of the future. His book, *Horizons*, not only astounded the design world with futuristic designs of transportation that bordered on science fiction, but jolted the transportation industry into action. The book was

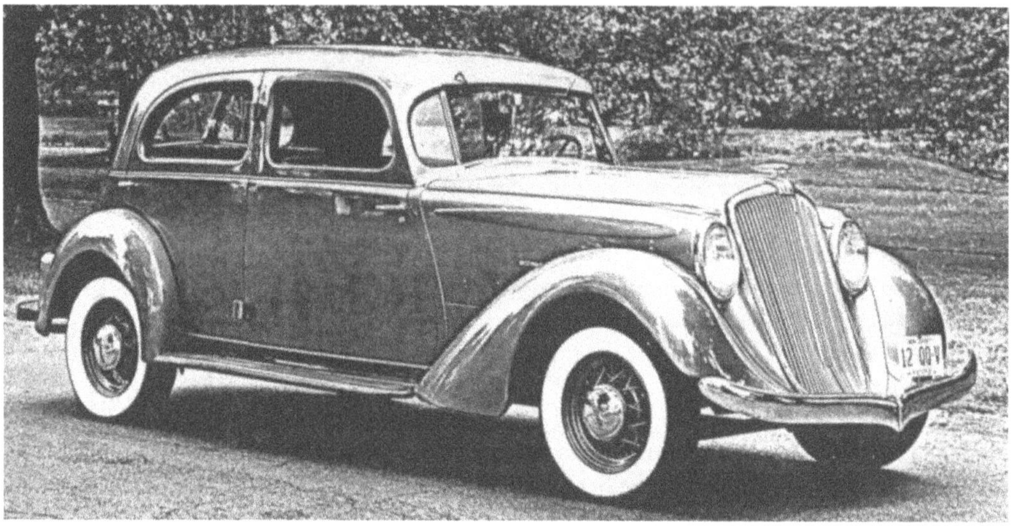

1934 Aerodynamic Hupmobile by Loewy.

filled with advanced concept designs that were never built, but seemed realistic, detailed, and easily achievable in the foreseeable future, because they included not only dramatic artist renderings of the external appearance, but detailed cross sections, mechanical specifications, and plans for interiors.

His Motor Car Number 5 sedan with a teardrop shape had a huge rear fin to stabilize it at high speed, which anticipated General Motors' car fins of the 1950s by several decades. His double-decker Motor Coach Number 2 had a similar teardrop shape and fin but seated 53 people and included a lavatory and buffet service. Bel Geddes's Locomotive Number 1 resembled and anticipated 1960s aluminum "bullet trains" with coaches and locomotive having near-cylindrical cross sections, as well as a rear lounge observation car, also bullet shaped, with generous glass windows.

Horizon concepts included designs for airports, such as a floating rotary airport at the tip of Manhattan, and an incredible Air Liner Number 4, a swept-wing, giant, amphibian, tailless, flying wing. The wingspan was 528 feet, four times that of the 1958 Boeing 707, and on top of the wing were mounted 20 engines, producing a maximum speed of 150 mph. The wing itself, which varied in thickness from 10 feet at the tip to 22 feet at the center, provided adequate space for 451 passengers (with sleeping facilities) and a crew of 155.

Bel Geddes's *Ocean Liner* in *Horizons* was a 1,088-foot-long behemoth of 70,000 tons that looked more like a torpedo than a ship, with streamlined funnels and 10 decks. The stern deck was enclosed in glass, which opened in good weather, and the stern itself tapered to a point, completing the teardrop appearance. The liner included an aircraft in a hanger, a tennis court, a sand beach, swimming pool, gymnasium, cinema, and three restaurants, with 2,000 first-class passengers and a crew of 900.

Also in the book were concepts and plans for private homes, theaters, auditoriums, open-air cabarets, and a 256-foot-diameter rotating aerial restaurant which seated 1,500 at the top of a 32-foot-diameter, 27-story-high central tower. Bel Geddes intended the design for the 1933–1934 exposition being planned in Chicago. Another was a design of a glass-walled factory and machine shop for the Toledo Scale Company, as well as product concepts for counter and island scales, but the latter were never produced.

The book also included a few more mundane commercial designs by Bel Geddes, including his Franklin Simon windows (1929), his car design for Graham-Paige (1928), his counter and island scales for Toledo Scale (1929), his steel beds for Simmons (1928–1929), his famed 1931 gas stove for the Standard Gas Equipment Corporation (SGE), and his radio for RCA (1931). But it was his incredible visions of the future of transportation in *Horizons,* and his groundbreaking, streamlining style that would earn him the title "the P.T. Barnum of Industrial Design" by some. Ironically, the noncredited renderings of many of the spectacular designs in Bel Geddes's book were actually drawn by a young recent graduate in architecture from the University of Pennsylvania, **C. Stowe Myers** (1906–1995). Myers was born in Altoona, Pennsylvania, and received his degree in architecture in 1930. He would go on to work for Walter Dorwin Teague in 1934.

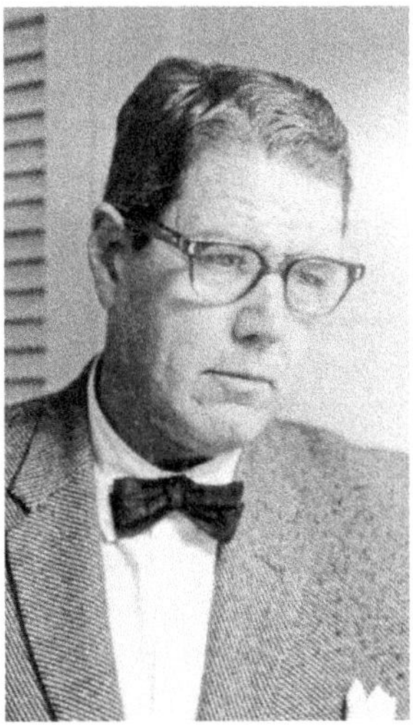

C. Stowe Myers, ca. 1950s (courtesy Industrial Designers Society of America).

But the future visions of *Horizons* were by Bel Geddes alone. Later in life, Raymond Loewy admitted that he and most other pioneers had little idea of what the future would bring. They were just making a living by improving products with good appearance:

> Looking back to the early days ... I can say that we proceeded with considerable naïveté. We didn't realize what the repercussions might be decades later ... that a new profession was beginning around us![2]

But Bel Geddes did. He recognized full well that he and others were transforming the visual landscape of America. On page 1 of his book, he states, quite presciently,

> In the perspective of fifty years hence [1982], the historian will detect in the decade of 1930–1940 a period of tremendous significance. He will see it as a period of criticism, unrest, and dissatisfaction to the point of disillusion—when new aims were being sought and new beginnings were astir. Doubtless he will ponder that in the midst of a world-wide melancholy owing to an economic depression, a new age dawned with invigorating conceptions and the horizon lifted.[3]

He was absolutely right. These were the men and women who pioneered in the new field of industrial design before it became well known by either the business community or by the public. Compared to many other artists practicing in the art industries, they were boldly entering the business and industrial world of mass-produced mechanical products, where artists normally feared to tread. They embraced the change that modern design offered to a nation

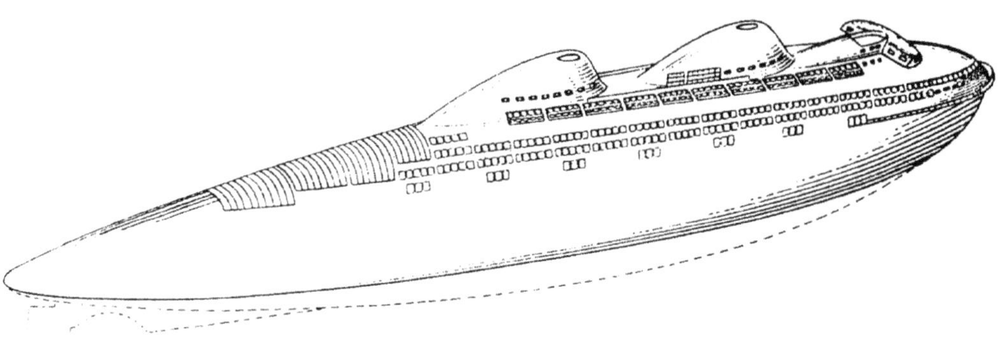

Ocean liner by Bel Geddes (Patent No. D91,579, U.S. Patent Office).

mired in traditional, historical, and antiquated design habits. They committed their careers to advancing that change and sought to make America a leader in international design, rather than a follower.

By January 1933, when President Franklin D. Roosevelt initiated his administration, America became focused on several burning issues. The Depression was affecting not only the many individuals who had lost their jobs (25 percent) and/or investments (millions). Farm commodities were only at 40 percent of what they had been in 1929 as a severe drought ravaged the agricultural heartland. Seven billion dollars in deposits had been frozen in failed banks after the March Emergency Banking Act (called the "Bank Holiday"), which closed all banks for four days and allowed only Federal Reserve-approved banks to operate. The day after the Bank Holiday ended, the stock market rose from 53.84 to 62.10 points, a gain of 15.34 percent, the largest single-day percentage gain to this day. When you are at the bottom, there's nowhere to go but up.

Nevertheless, American exports had declined from $5.2 billion in 1929 to $1.7 billion in 1933. Manufacturing was struggling to survive in an environment where consumers had little means or incentive to buy products. The lowest point of industrial production in the Great Depression had been reached, which was at 50 percent of what it had been in 1929 before the crash, as were retail sales. Auto production was at a quarter of its pre–Depression levels. The Depression had never been darker, despite the sudden rise in the stock market.

A boost in national morale was needed, and several events in 1933 contributed to raising spirits. First, Roosevelt repealed Prohibition, the policy that had made public use of alcohol against the law since 1920, but which had done little to curb the illegal activities of rum-running gangs or speakeasies that provided liquor to continued public demand. The second was the opening of the Century of Progress International Exposition, a World's Fair, held in Chicago to celebrate the city's centennial. The fair had been in the planning stages since January 1928 and had been slowed by the start of the Depression in October 1929. Now, on May 27, 1933, the fair opened on 427 acres along the Lake Michigan shoreline between 12th and 39th Streets in Near South Side, Chicago.

The Chicago Century of Progress World's Fair opening in May 1933 was a showcase of American technology, transportation, architecture, and modern design. The Fair's motto, "Science Finds, Industry Applies, Man Conforms," emphasized the belief that "man" needed to, as we say today, "get with the program." The program was the future. Buildings for the exposition were multicolored, creating a "Rainbow City" as opposed to the "White City" of the 1893 Chicago Fair (the World's Columbian Exposition) that introduced electric lighting. The 1933 fair opened when light from the star Arcturus was detected by photoelectric cells in a series of astronomical observatories and was transformed into electrical energy that was transmitted to Chicago. The star was chosen because its light had started its journey to earth 40 years before, at the same time as the previous Chicago Fair of 1893.

Unlike the 1893 fair, which featured neoclassical architecture and themes, this one generally followed "modern" architecture, and its symbol was the Sky Ride, a transporter suspension bridge perpendicular to the shore of the lake on which one could ride from one side of the fair to the other in "rocket cars," or ride elevators up the two 628-foot towers for a spectacular view. The midway was filled with nightclubs and the re-creation of important scenes from Chicago history. Future stars such as Judy Garland and the Andrews Sisters performed at the nightclubs, along with fan dancer Sally Rand. The first Major League Baseball All-Star Game was held at Comiskey Park in conjunction with the fair, and yes, Babe Ruth hit the first All-Star home run.

A 1933 Homes of Tomorrow Exhibition featured 12 modern homes demonstrating modern conveniences and practical new building materials. A 12-sided, glass-and-steel-framed House of Tomorrow designed by architect Fred Keck had an all-electric kitchen, plastic wall partitions, air conditioning, fixed window panes, and, incredibly, an airplane hangar. The Briggs Company exhibited a Kitchen of Tomorrow designed by John Tjaarda, which included a cylindrical electric range center island lit by a lamp that rose from the center like a periscope.

Entry to the exposition was through "Passimeters," modern turnstiles designed for the Perey Company by John Vassos, who, since his entry into the design field in 1924, had written, illustrated, and published a number of fiction books: *Contempo* in 1929, *Ultima* in 1930, and *Phobia* in 1931. Now, in 1933, he established an internal industrial design department for the Radio Corporation of America (RCA) and would remain as RCA design consultant until 1964, while continuing to practice independently. In 1934, he would design RCA's first plastic tabletop radio cabinet, the International, replacing what Vassos called the "wooden tombstone" type radios of the 1920s that preceded it, and he also would design the first living room console radios without furniture type "legs."

What Vassos called "wooden tombstones" were Philco's tabletop radios shaped in a Gothic arch, not unlike a tombstone. In 1932, there were a lot of them, including Models 36B, 71B, and 91B, all designed by **Clyde F. Schuler** (1892–1969), including a later Model 66MB.

Shuler was born in Pottstown, Pennsylvania, and educated at the Philadelphia Museum and School of Industrial Art. He earned a certificate in industrial drawing in 1911, a second for interior decoration in 1913, and in 1921, a diploma in interior decoration. He received his architectural registration for the state of Pennsylvania in January 1930 and opened an office with a partner, Schuler & Gurney, practicing domestic architecture and product work, such as for Philco. But by 1933 he had joined the faculty at the Philadelphia College of Art, where he would remain until 1952.

Major manufacturers mounted numerous pavilions or exhibits at the Chicago Exposition, many of which provided opportunities for designers. Alfonso Iannelli, the Chicago sculptor/industrial designer mentioned earlier, designed pavilions, rides, entrance gates, and ornaments for the fair, including the Havoline Thermometer building. Norman Bel Geddes had been appointed "designer of illumination" for all buildings and grounds in 1929, and he also designed the Aquarium restaurant, the Seats-Surrounded by Stage Theater, the Water Pageant Theater, the Island Dance restaurant, the Arial restaurant, the Divine Comedy Theater, the Open Air Cabaret, and the Temple of Music.[4]

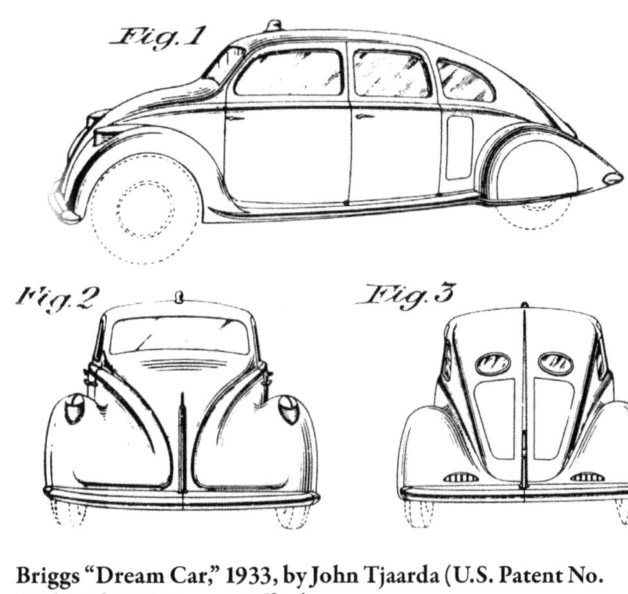

Briggs "Dream Car," 1933, by John Tjaarda (U.S. Patent No. D94,396, U.S. Patent Office).

Walter Dorwin Teague designed the Ford Rotunda, which featured the "Briggs Dream Car," a rear-engine prototype aimed at Ford's need for a "small Lincoln," designed by **John Tjaarda** (1897–1962), who had been hired as chief of body design in 1932 by the Briggs Manufacturing Company, Detroit's largest body supplier, which supplied body designs to Ford.

Tjaarda was born in Arnheim,

the Netherlands, of a titled family in the Sterkenburg area, as Joop "Jan" Tjaarda van Sterkenburg. He trained in aeronautical design at the Alexander Klemins School of Aeronautics in London and served as a Dutch Air Force pilot before immigrating to America in 1923 and changing his name to John Tjaarda. He initially worked as a design assistant with Locke & Company in Rochester, New York, a manufacturer of bodies for Stutz, Pierce Arrow, and Packard. In 1926, fascinated with aerodynamics, he developed streamlined automotive designs in clay models he called the Sterkenburg Series, after his family's home in the Netherlands. These would become the basis of his Dream Car for Briggs. He had worked for Harley Earl's Art and Colour Section at General Motors in 1927, for Duesenberg from 1928 to 1930, and had returned to Harley Earl from 1930 to 1932.

The Briggs car, although much sleeker, more streamlined, and more refined, bore a striking resemblance to the 1932 prototype of the Type 32 Kleinauto, designed by Dr. Ferdinand Porsche (1875–1951) in Germany, which would later in the decade become the famed Volkswagen (People's Car) subsidized by Hitler. Other auto manufacturers displayed concept cars at the Chicago Exposition, such as Cadillac's V-16 limousine and Pierce-Arrow's Silver Arrow, for which it used the byline, "Suddenly it's 1940!" But Packard won the best of show.

Another unusual car design appeared and was demonstrated at the exposition by **R. Buckminster Fuller** (1895–1983). It was a prototype of his teardrop-shaped Dymaxion Transport Unit, which looked a bit like an aircraft without wings. It had a rear engine that drove two front wheels, with a third wheel in the rear that was controlled by the driver with the steering wheel to change directions. Three such prototypes were built, but a fatal accident with one in 1935 would doom the project. Fuller was one of the most intellectual of inventor/designers.

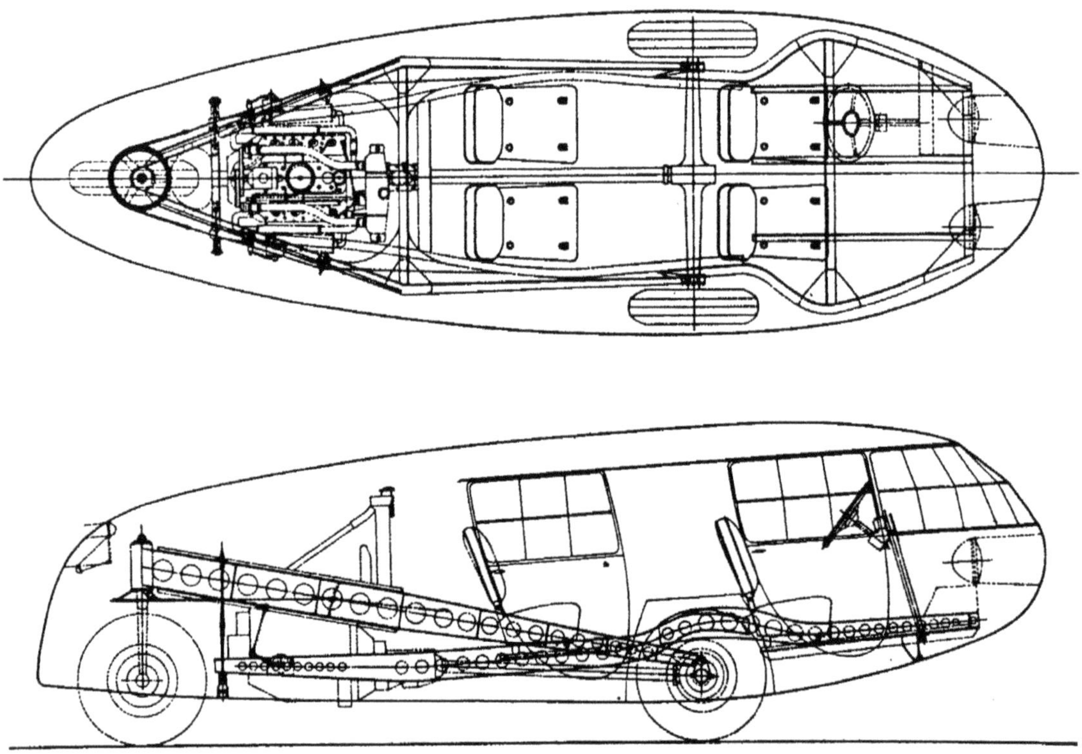

Dymaxion Transport Unit, 1933, by Fuller.

Born in Milton, Massachusetts, he attended Milton Academy and studied mathematics at Harvard from 1913 to 1915. He served in the U.S. Navy during World War I, and afterwards worked for the Armour Company of the meat-packing industry in Chicago. By 1927, at age 32, he was bankrupt, jobless, and depressed. But he resolved to think independently to find out what a single individual could contribute to a changing world to benefit all of humanity.

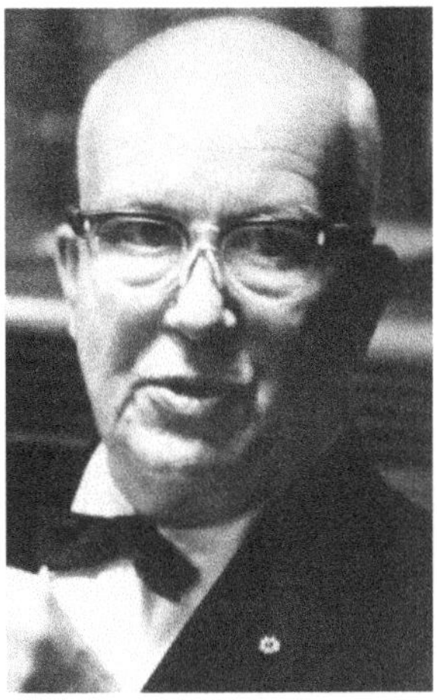

R. Buckminster Fuller, ca. 1960 (courtesy Industrial Designers Society of America).

By 1928 he was living in Greenwich Village in New York, decorating Romany Marie's popular café in exchange for meals, and displaying concept designs for his mass-produced house and bathroom, which he called Dymaxion House and Dymaxion Bathroom, respectively. He coined the term Dymaxion from the words "dynamic," "maximum," and "tension." At the café, he met Japanese-American artist and landscape architect Isamu Noguchi (1904–1988), and they teamed up to model the Dymaxion car based on the recent work of Aurel Persu, the Romanian aeronautical engineer mentioned earlier.

Major electrical equipment manufacturers were also represented at the exposition. Donald Dohner, then the Westinghouse director of art, painted huge murals for the company's pavilion. One of the exhibit designers for the General Electric (GE) pavilion was **Raymond E. Patten** (1897–1948), who had joined the Edison GE Company in Chicago in 1928, and who supervised appearance design. In 1931 he had designed an Art Deco, tilt-type, Hotpoint Gazelle toaster for the company, as well as waffle irons, clocks, and kitchen ranges. In 1933, GE established an internal design group at its headquarters in Bridgeport, Connecticut, and named Patten as its head. Patten was born in Malden, Massachusetts; attended Harvard; and studied engineering at MIT. His first position was with the Hume Body Corporation in Boston. He later designed custom auto bodies for the Dayton Wright Company and spent five years at the Packard Motor Car Company before joining Edison in Chicago.

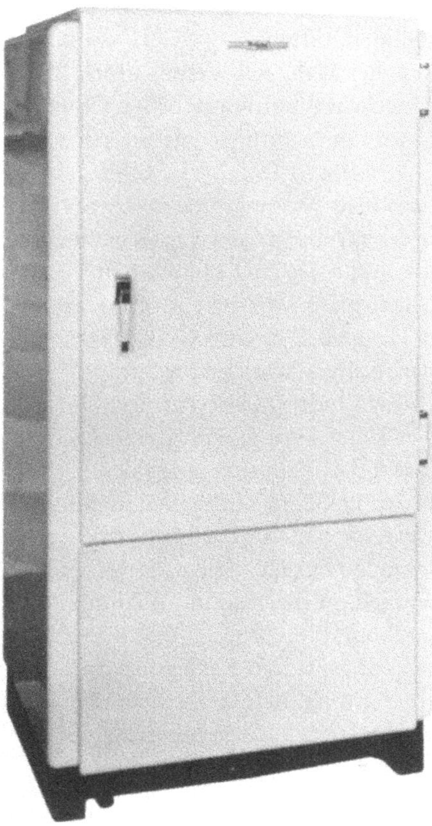

Under Patten's leadership in 1933, GE engaged Henry Dreyfuss to design a new flat-top refrigerator under the Hotpoint brand name to replace its clumsy 1927 monitor-top model, designed by engineers with an exposed cylindrical compressor on top, the name suggestive of the Civil War ironclad "cheesebox on a raft" warship, the *Monitor.* Dreyfuss enclosed the compressor and located it beneath the cooling compartment, rather than on top, creating the clean rectangular vertical box we are accustomed to seeing in

Hotpoint "flat-top" refrigerator, 1933 (courtesy Museum of Innovation and Science, Schenectady, New York).

refrigerators today. Dreyfuss also designed a modified version of the monitor top. Dreyfuss that same year had completely redesigned *McCall's* magazine and designed a Toperator washing machine for Sears, Roebuck & Company, with a replica of his signature on each of the 20,000 units sold within the first six months. Already, the names of prominent industrial designers were being used as sales features. It was a time when manufacturers were willing to do anything to increase sales.

Patten's earliest employees included Helen Hughes Delany, who restyled the GE electric stove line; George A. Beck, mentioned earlier, who had worked at Montgomery Ward and Sears, and joined GE in 1934; and **Gordon Florian** (1909–1984), who studied at Yale University of Fine Art in New Haven, Connecticut; at the Grand Central School of Art in New York; and had joined GE in 1931 as a market analyst. By 1934, the Hotpoint and GE brand names would be integrated, and in 1935, **Raymond C. Sandin** (1910–ca. 1986) would become a one-man industrial design department of the Hotpoint division of GE, and would remain as head of the department until his death. Sandin was born in Sweden, came to the United States in 1930, and studied architecture at the Illinois Institute of Technology before joining GE.

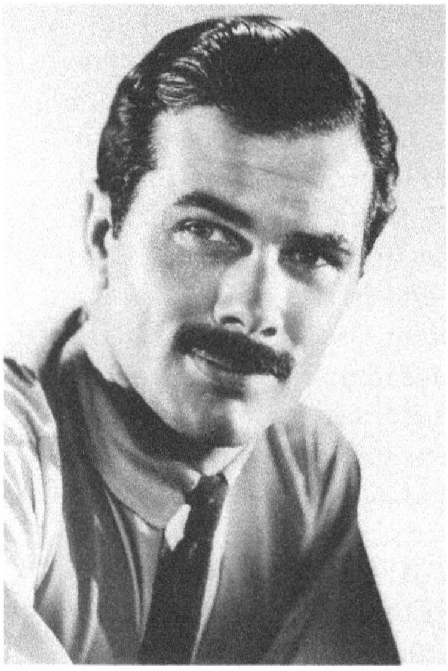

David Chapman, ca. 1940s–1950s (David Painter papers, Ryerson and Burnham Archives, Art Institute of Chicago. Courtesy Art Institute of Chicago).

Other exhibit designers on the design staff at the fair would become prominent in the industrial design field. One was **David Chapman** (1909–1978), a 1932 architectural graduate of the Armour Institute of Technology (which later became the Illinois Institute of Technology) in Chicago, who, at the same time he was working on fair exhibits, joined Montgomery Ward as an architect. Ann Swainson, head of Ward's Bureau of Design, appointed him head of product design. Chapman was born in Gilman, Illinois.

Another exhibit designer at the fair was **Jean Otis Reinecke** (1909–1987). Reinecke was born in Bourbon County, Kansas, and started his own sign-painting company at age 17 in 1926. He attended Kansas State Teachers College, then moved to St. Louis and attended Washington University's Art School. He began his career as a display artist at General Display Studios in St. Louis, and a designer of automated convention and World's Fair displays, where he soon became a partner and art director. He opened his own Chicago general display office in 1930 specifically to work on the fair. In 1934, he would establish a Chicago industrial design firm, Barnes & Reinecke, with partner James Barnes, with Reinecke providing the design talent and Barnes in charge of sales.

In the closing days of the Chicago Fair, on October 26, 1933, one of the final highlights was the arrival of the 776-foot German airship, the *Graf Zeppelin,* which circled Lake Michigan for two hours near the exposition site before Commander Hugo Eckener landed at the nearby Curtiss-Wright Airport in Glenview at 1:00 p.m. It remained on the ground for 25 minutes and then took off ahead of an approaching weather front, bound for Akron, Ohio. For some folks, it was not a welcome sight, since, with a huge swastika on its side, it was a grim reminder of the rise of Adolph Hitler to power in Germany earlier that year. In fact, German theoretical physicist Albert Einstein had just arrived in the United States nine days before, on October 17, a refugee from Nazi Germany. There would soon be many more. Nevertheless, the fair, which closed on November 12, was so successful that it was decided to reopen it again to run from May to October the following year, 1934. Over the previous 170 days, it had already attracted well over 22 million paid admissions.

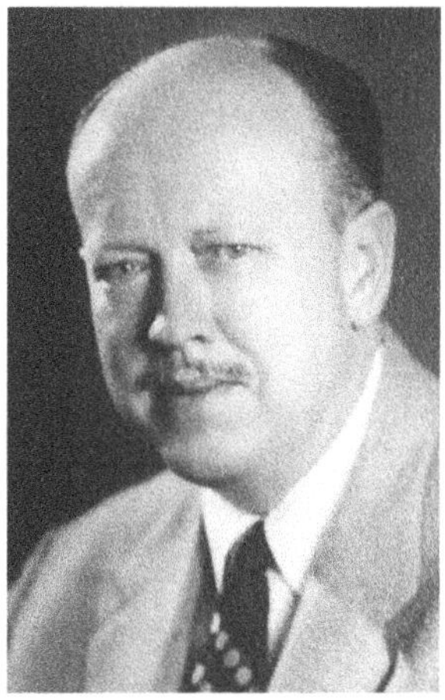

Robert D. Budlong, ca. 1950s (courtesy Merrily and Erik Holladay).

Before the end of 1933, **Robert Davol Budlong** (1902–1955) had joined the growing ranks of industrial designers and started his career with the Hammond Clock Company. He would begin a relationship with Zenith in 1934 as a radio designer and consultant that would last until his death. Budlong was born in Denver, Colorado, studied art at the Cummings School of Art in Des Moines, Iowa, and graduated from Grinnell College, Iowa, in 1921. He then studied art at the Chicago Academy of Fine Art and opened an advertising office in Chicago. Around the same time, there were several start-up firms in New York: Cushing & Nevell in 1933, and Jim Nash Associates in 1934, both of which would survive into the 1960s.

In 1933, Norman Bel Geddes exhibited his work in Detroit, Michigan; Milan, Italy; and the Louvre in Paris. He also designed a medallion to commemorate the silver anniversary of the General Motors Corporation that year.[5] Aside from his normal business routine, the rumblings of potential war in Europe inspired him to renew his interest in realistic games, and he developed what he called the War Game, a scale model simulation of a 20,000-square-kilometer battlefield on a 24 × 4 foot table, representing the coastline of two imaginary countries, complete with 9,000 cities and towns, mountains, valleys, tunnels, and bridges. It was much larger and more complex than his Nutshell Jockey Club of the 1920s.

Thousands of pins of various colors, sizes, and shapes stood in for 40-plus types of military units, including armies and navies of 28 belligerents. A complicated rulebook determined chesslike moves by a general staff regarding speed of movement and degree of destruction by attacks. The game ran once a week during winter months, with an evening of play commensurate with three and a half days of war. Players, consisting of brokers, architects, shipping magnates, and rare book dealers, took the game seriously, often sacrificing vacations, business, or family commitments. By 1935, Bel Geddes would put the game into storage because of time constraints, but his skills of miniaturization and three-dimensional landscaping would be valuable in a number of his future professional assignments.

But the real war against the Great Depression was still raging. Industry was still stalled, and most companies were still unaware that one of the reasons was the obsolete design of their products. By and large, they were also unaware of what to do about it. This was about to change dramatically.

CHAPTER 5

Industrial Design Unleashed

By 1933, eight years after the Paris Art Deco Exposition of 1925, which had sounded a wake-up call, warning that America was far behind Europe in modern design and therefore in international economic competition, there were perhaps three dozen industrial designers who were working in the so-called artless industries, a drop in the bucket compared to the 50,000 needed in the post–World War I consumer markets, as recommended in 1918 by Walter Sargent (1868–1927), then director of the Department of Fine and Industrial Art at the University of Chicago. Of course, World War I had dramatically suppressed the European advance in industrialization, and the Depression had stopped America's economy in its tracks.

At the bottom of the Depression in 1933, industry was stalled. Many manufacturers still resisted investment in new tooling for new designs because sales had declined dramatically, and they were in the red, struggling to stay in business. Many were bewildered by a situation foreign to their experience—a situation in which the functional but homely merits of their products were not enough to sell them. Those who sought qualified designers to design new products found that the few industrial designers that were available were too busy to take on new clients. But the vast majority of manufacturers, businessmen, and engineers were simply unaware of modern design or industrial design. They were not among the elite sophisticates who frequented art museums and international art expositions, or who read art magazines. They were practical people, concerned about the nuts and bolts of manufacturing, not art. Many of them were about to learn about industrial design and its leading practitioners for the first time, through an influential article that reached the entire business world.

The success of the Chicago Fair in 1933 was the first glimmer of hope for the future experienced by the public in three years. Now industry needed a morale boost, and it came in February 1934 when *Fortune*, the premier magazine of business, published a nine-page feature article about industrial design, which would have a profound effect on the new field, as well as on manufacturing and the economy itself. In fact, it was the first time the term "industrial design" was used in the popular press, rather than the traditional, more common term of "industrial art" for the practice of art in industry.

The article was titled "Both Fish and Fowl," intended to suggest that industrial designers were both artists *and* businessmen, a connection that most people, including businessmen, were unlikely to make. Artists, many people thought, lived alone in garrets, struggling to eke out a living, and selling their paintings on the street. They were rarely, if ever, thought of as businessmen. A cartoon accompanied the article on the title page, which showed a company president addressing his board of directors with the statement, "Gentlemen—I am convinced that our next new biscuit must be styled by Norman Bel Geddes."

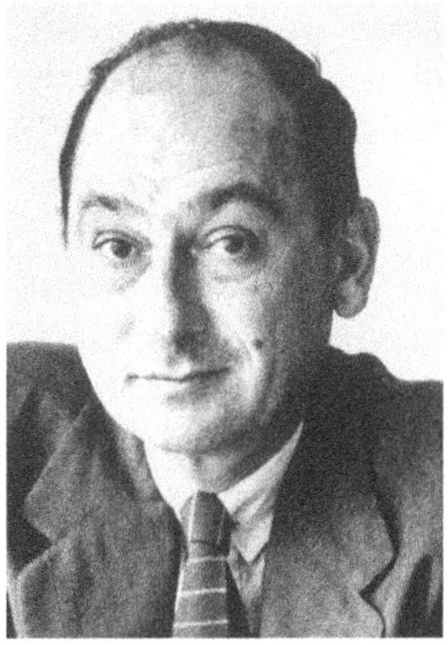

George Nelson, ca. 1950s (courtesy Industrial Designers Society of America).

The *Fortune* article was written anonymously, but is now known to have been authored by architect **George Nelson** (1908–1986), 26 at the time, and the winner of the 1932 Rome Prize in architecture. Nelson was then the assistant editor of *Architectural Forum* magazine, and would within a decade become a nationally known industrial designer. Nelson was born in Hartford, Connecticut, and graduated from Hartford Public High School in 1924, and from Yale in 1928 with a degree in architecture. In 1929 he was hired as a teaching assistant while obtaining his second bachelor's degree in fine arts at Yale in 1931. In Rome, he traveled throughout Europe and met many modernist pioneers, whom he interviewed for articles in *Pencil Points* magazine. Through these articles, he described the work of Walter Gropius, Mies van der Rohe, Le Corbusier, and Gio Ponti. In 1933 He returned to America to devote himself to writing. This *Fortune* article was probably his first for a major nonarchitectural magazine.

Nelson began the article with a quotation from Oliver M.W. Sprague (1873–1953), who had just in December 1933 resigned as special assistant to the secretary of the treasury in Washington in order to return to Harvard Business School and resume his teaching there after a leave of absence of four years with the government. Now he was free to express his opinions publicly, and rather than criticizing the administration, Sprague placed the blame of the deepening Depression squarely on manufacturers:

> Failure of industries to adapt policies designed to open up additional demands for industrial products is, in my judgment, the chief cause of the persistence of the Depression.[1]

In other words, manufacturers had failed to attract new consumers through the design of new and different products, precisely the service that industrial design offered. Nelson stated the essential case for industrial design:

> Whenever two products are equal in point of utility and price, the one that looks more attractive to the purchaser will be bought first. When the demand is sufficient to take only one of two products, the maker of the attractive product stays in business, the maker of the other goes bankrupt.[2]

After this do-or-die challenge to his business audience, Nelson briefly provided historical context by describing how the art industries traditionally and effectively used design, but that the engineer-run artless industries of the 20th century were unprepared and unstaffed to do so when style became important to their largely functional products. He then described in detail how industrial designers worked, and what benefits they could bring to business. The main thrust of the article focused on "a few leading industrial designers" and what they had done in the last few years by citing their previous experience, their clients, their fees, their typical achievements, and photo-portraits of ten of them. Those included were many who have already been described in this book. Quotes that follow are from the *Fortune* article.

Norman Bel Geddes, now a tousled, curly-haired 41-year-old, was described by Nelson as a "bomb-thrower" and the "P.T. Barnum of industrial design," but three pages are devoted to his background in stage design, his business practice, and his book, *Horizons,* including an entire sidebar page, "Story of a Stove," that was devoted to his stunningly modern and simple, but practical, stove concepts for the Standard Gas Equipment Corporation (SGE). The president of SGE, Frank Roberts, had walked into Bel Geddes's office and requested a new stove design within two weeks and offered to pay $1,500. Bel Geddes explained that he did not just draw pretty pictures, that study to produce a better stove would take a year, and required $50,000 as an advance of royalties. Shocked, Roberts conferred with his directors, but he decided to take the risk.

Bel Geddes put an engineer in the SGE plant, studied competitors' methods, and interviewed hundreds of housewives and store dealers. His report and recommendations numbered 300 pages. Photos in the article showed how Bel Geddes reduced the number of stove components currently stocked by SGE from hundreds to just 16 through standardization, which he demonstrated like a showman with colored block models. When the stove appeared on the market in January 1933, the first one all white, Robert's decision was vindicated: "Housewives wanted the Geddes stove. Utilities that had never heard about the line before welcomed it. Sales rose, doubled, held their gains."

In the same year as this article, Bel Geddes designed an Electrolux air-cooled gas refrigerator, two years before Loewy's Coldspot for Sears, as well as an Electrolux vacuum cleaner; a refrigerator for Servel, Inc.; a prefabricated service station for Socony-Vacuum Oil Company; metal and plastic beds for the Rome Manufacturing Company; cigarette cases, boxes, cocktail shakers, candlesticks, and ashtrays for the Revere Copper and Brass Company; and a Plymouth car design for the Chrysler Company. He also wrote an article entitled "Streamlining" for *Atlantic Monthly* magazine. All the while, he continued to design and produce theater sets and to revise theater architecture.[3]

In the article, Nelson praised designers, not merely for attractive and modern styles, but also for their contributions to manufacturing cost reduction and user convenience. Donald Dohner at Westinghouse found a way to save $40,000 on tooling costs by making a few changes to his model. He changed a metal vacuum cleaner to plastic, making it lighter, quieter, and more powerful, putting Westinghouse into the vacuum cleaner business. Finding that fat men operated most locomotives, he widened passages and spaced levers and wheels for fat hands. Dohner is shown as a serious-looking, slim, dark-haired 41-year-old with a "Hitler"-style mustache.

John Vassos, shown as a dark-haired, handsome 41-year-old, noted for his radios for RCA, his turnstiles for the Perey Manufacturing Company, and a Coca-Cola drink dispenser, also invented a hook for a wall telephone, which the company found patentable. Henry Dreyfuss, the youngest at age 30, a smiling young man with glasses, placed all the controls in one place, on the top, in his design of the Toperator washing machine for Sears.

The article featured detailed biographies and operating styles of key designers. Lurelle Guild, a balding 36-year-old, was described to be as "prolific in design as he was in writing. He redesigns about a thousand products a year, the majority of which are art-industry products, and makes an average of four finished drawings for each product." Guild claimed that he "has never designed a product that was a commercial failure, has never lost a client." Designs by Guild cited in the article included a refrigerator for the Norge Corporation, cooking utensils for the Wear-Ever Aluminum Company, and designs for Montgomery Ward for everything from stoves to roller skates. His coffeepot for Wear-Ever had a lower production cost but increased sales by 40 percent. His annual retainer fee was $25,000, his daily fees were $100 to $200, or alternatively, he offered flat fees of $300 to $25,000, plus royalties.

Walter Dorwin Teague, now 51 and shown with a mustache and wearing a dapper fedora, was described as "suave, able, and successful" and "is the generally acknowledged doyen of the profession" with an "unbroken record of commercial successes." His designs cited included cameras for Eastman Kodak, with whom he had a six-year retainer. In 1934 his famous Art Deco Kodak Baby Brownie camera was introduced, priced at only $1, and four million of them would be sold. He also had designed a furnace for National Radiator; a line of Taylor Instruments; a line of Pyrex ovenware; a mimeograph machine for the A.B. Dick Company; a dramatic Art Deco Bluebird radio with a blue, circular mirror face for Sparton; and an oil heater for American Gas Machine that increased sales by 400 percent. His retainer fees were $12,000 to $24,000, and his flat fees were $500 to $5,000.

Donald Dohner, whose department at Westinghouse reportedly cost $75,000 per year, was described as "among the dynamos of East Pittsburgh," and his design for a water cooler "captured 40 percent of the national market in a few months." His electric range "outsold previous models seven to one," a 700 percent increase, with lower production costs. His diesel-electric locomotive, intended merely for an exhibit, "was actually sold right off the exhibit floor." (Author's note: Dohner's diesel-electric design was destined to become Pennsylvania Railroad's famous GG-1 locomotive late in 1934. More on this later.)

Gustav Jensen, shown as a blondish, serious-looking 36-year-old, "generally regarded as the top man from the purely artistic point of view," was recognized for his design of his Monel metal sink for International Nickel Company, a telephone for AT&T, and a water heater for the L.O. Koven & Brothers Company. His retainer fees were cited as $500 to $20,000.

A brief biography of George Sakier, shown as a scowling, dark-haired 37-year-old, described him as "probably the purest type of industrial designer among those here described," and mentioned his work for American Radiator and Standard Sanitary Corporation, designing bathtubs, washbasins, vacuum equipment, and bathroom units stamped from metal, which "are the first prefabricated house units actually in production. The entire bathroom with the three major fixtures and the walls and floor can be assembled as a single unit, or the parts can be used separately. One hundred and thirty-three of them are now being put into a Washington, D.C. apartment house." His income was estimated at $15,000 to $25,000, plus retainers, from his design work. Also mentioned was his glassware work for Fostoria, with the result that "American open-stock glass for the first time took precedence over European, and finally he won the distinction of having his own designs for glass pirated in Europe."

Raymond Loewy, pictured as a dark-haired, mustached 41-year-old, was described as "the least publicized and one of the largest earners," with retainer fees from $10,000 to $60,000, plus royalties. His work cited included a globe-shaped radio for Colonial which registered a sales increase of 700 percent; his Hupp automotive designs; kitchen sink and bathroom units for Sears, Roebuck; and his duplicator (copy machine) for Sigmund Gestetner. Loewy was referred to as "he who discovered that all Americans like the color green and the smell of lilac."

Henry Dreyfuss was recognized for having already been written about in the *New Yorker, American,* and *Saturday Evening Post* magazines for his work in stage design, and in fact he was still designing one show on Broadway every season. His product designs cited included the Toperator wash machine for Sears, Roebuck; his Protectograph check protector for the Todd Company (introduced in 1935); and alarm clocks for Western Clock Company, but he was also designing interiors for Cities Service corporate aircraft and an insulated thermos jug made by the American Company while working on a "telephone of the future" for Bell Laboratories. His flat fees were $1,000 to $25,000, or an hourly rate of $50.

John Vassos was mentioned as a well-known writer of psychoanalysis, but also for his designs of a Coca-Cola dispenser; his turnstiles for Perey Manufacturing Company, which increased sales by 25 percent; and his radios for RCA. He generally charged $12,000 for retainer fees, or flat fees from $1,000 to $7,000. Harold Van Doren was described as "handsome," and his design cited was his scale for Toledo Scale, which permitted lower manufacturing costs and multiplied sales by 900 percent. His consultation fees were $100 per day, or flat fees from

$500 to $5,000. Joseph Platt's Vacuumatic fountain pen for Parker increased sales by 100 percent, with lower factory costs.

Jo Sinel's Accoustican hearing aid for Dictograph (1932) was cited for having resulted in a reduced price, with a net profit increase of 50 percent. Sinel had also just designed a typewriter for Remington, a Juicy Fruit vending machine, two lanterns for Dietz, and a Toastmaster Jr., none of which were mentioned in the article. In fact, in another 1934 issue, *Fortune* gave a full-page color spread to Sinel's packaging and trademark designs, for which he was also well known. Donald Deskey and Russel Wright were both mentioned as newcomers to the field. Deskey had just been retained by the Libby Glass Company in Toledo, Ohio, and was not only designing a champagne glass for them, but advising them on their entire line of products. Wright had recently designed elegant sterling silver flatware that the Museum of Modern Art would much later honor by commissioning the production of replicas.

Nelson concluded the article,

> Most manufacturers agree that the industrial designer has stimulated all those he has met in the company's organization, given them new ideas, new energy, a new attitude. Designers will not cure our industrial ills, for our ills lie deeper than design. But the success of industrial designers is symptomatic of a changing attitude on the part of the manufacturers, a growing recognition of the old American ideal of progress: cheaper and *better* goods for all.[4]

Nelson would establish a New York architectural design office in 1936 with William Hamby, which would continue until after the war, and Nelson would become a prolific and dynamic leader in postwar furniture design and establish his own office in New York in 1947.

The article had an immediate impact, not only on the small community of industrial designers, who benefited enormously from the national publicity and recognition, but on the large segment of the business community that had little or no previous knowledge of industrial design and how it could increase sales. Suddenly they made the connection between their primary problem of diminishing sales and the obvious solution of new designs to generate new demand. The traditional manufacturing image of industrial art as simply the addition of decorative surface embellishment, had been transformed into industrial design, the reshaping and modernizing of the entire product.

In a sense, this was the reestablishment of the artist as a multidisciplinary creative individual in the modern world, as emphasized in Nelson's title, "Both Fish and Fowl," not unlike the 19th-century craftsmen who were involved with the materials, costs, forms, engineering and production of artifacts, as well as the appearance. Industrial designers not only became the 20th-century craftsmen, controlling the form of mass-produced products, but could command fees beyond any craftsman's imagination. Predictably, the demand for industrial designers increased overnight.

The attraction for designers is not difficult to understand. In 1934, the average annual salary for corporate executives was $5,000, and the average laborer's pay was 50 cents per hour. One could buy a three-bedroom house from Montgomery Ward for $1,000, a men's dress shirt for $1, and an icebox refrigerator for $11.95. The average annual retainer fees of $50,000 and average hourly rates of $50 charged by industrial designers were incredible. In today's dollars, that would be $1 million annually and $1,000 per hour. But to manufacturers, higher fees for sales increases ranging from 25 to 900 percent, with an average of 350 percent, were worth it. Manufacturers could, and did, pay anything for that kind of boost to sales.

These fees were enough to attract some of the most talented people in other fields. Some automotive designers on Harley Earl's payroll at General Motors abandoned the field to enter the more lucrative world of independent industrial design consultants. Among these were

Carl W. Sundberg (1910–1982) and **Montgomery Ferar** (1910–1982). Sundberg was born in Calumet, Michigan, graduated from Northern High School, and attended Wicker Art School in Detroit. He began his career in 1927 designing custom car bodies under Ray Dietrich at Dietrich Body Company, and then worked in plastics design at Kurz-Kasch, Inc. He then joined Earl's Art and Colour Section at GM.

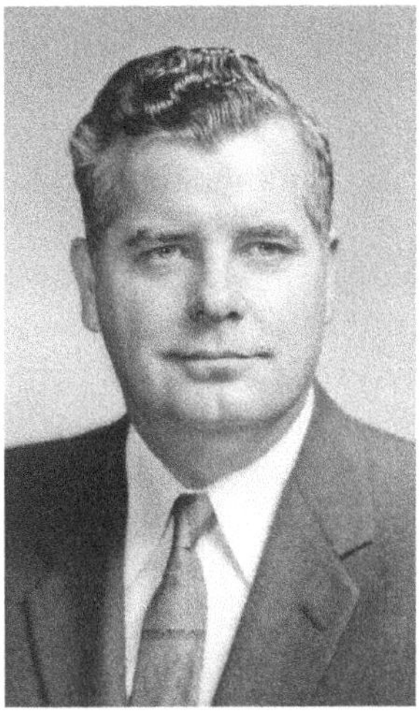

Carl W. Sundberg, ca. 1930s (courtesy Sundberg-Ferar, www.sundbergferar.com).

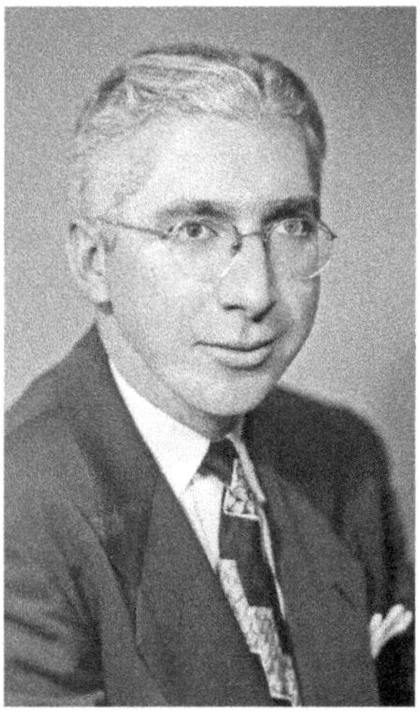

Montgomery Ferar, ca. 1930s (courtesy Sundberg-Ferar, www.sundbergferar.com).

Ferar was born in Boston and received a bachelor's degree in architecture at MIT in 1932, and took a year of study in Europe with a fellowship in 1933. He worked for several architectural firms in Boston before moving to Detroit to work for the Detroit Planning Commission, and he joined GM's Art and Colour Section, where he met Sundberg. In 1934 they both left GM to form the industrial design firm of Sundberg-Ferar, Inc., in Detroit, beginning a lifelong partnership.

Another automotive designer at GM, **John Richard "Jack" Morgan** (1903–1986), left GM to become chief product designer at Sears, Roebuck's first design department in Chicago in 1934, which was formed as part of Department 817, the Merchandise Testing and Development Laboratory. Morgan was born Juan Ricardo Morgan in Guatemala City, Guatemala, and moved with his family in 1913 to Canada. At age 11 he won first prize for a watercolor at the Canadian National Exhibition. He became a student of renowned landscape painter Thomas John "Tom" Thompson (1877–1917). As a painter, Morgan traveled abroad to Brussels, Berlin, Copenhagen, Oslo, Stockholm, Paris and London, and eventually became an accomplished artist with a studio in Wheeling, Illinois. He had two one-man shows in Chicago and exhibited in Palm Beach, Connecticut, and Pennsylvania.

But Morgan's love of automobiles inspired him to attend the Detroit Technical Institute, and in 1927 he joined the C.G. Spring and Bumper Company. In 1928, he was hired by Harley Earl's infant Art and Colour Section at GM, which by then numbered 50. Working in the Pontiac Studio under Frank Hershey (1907–1997), Morgan designed the first auto bumper to be integrated into the body. At Sears in 1934, Morgan designed the Sears Waterwitch outboard motor and would head Sears Department 817 until 1943, designing vacuum cleaners, kitchen cabinet consoles, and much of Sears' appliance merchandise, including the first "turtle neck" washing-machine tub. One of Morgan's key early designers was **Clare Hodgman** (1911–1992), born in Buchanon, Michigan, who had started his career in General Motors' Oldsmobile studio with Harley Earl in the early 1930s and had been previously working for Loewy.

Another designer, **Brooks Stevens** (1911–1995), like John Morgan, was also working on outboard motors, needed for a popular new outdoor recreational activity, power boating. His current client was Evinrude, and he had just opened his own office in Milwaukee in 1934. Twenty-three-year-old Clifford Brooks Stevens was born in Milwaukee and studied architecture at Cornell University from 1929 to 1933. He won a contest for designing a new trademark for his father's firm, Cutler-Hammer, and would now continue to design equipment and displays for the company as a consultant designer. His first product design for another firm was a glass peanut butter jar for Holsum Products.

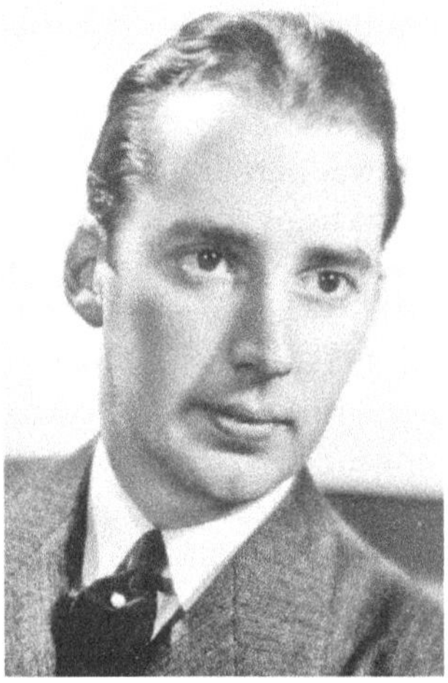

Brooks Stevens, ca. 1936 (Brooks Stevens Archive, Milwaukee Art Museum, Gift of the Brooks Stevens Family and the Milwaukee Institute of Art and Design).

The February 1934 *Fortune* article spurred industrial designers to toot their own horn. They organized an Art and Industry Exhibition at Rockefeller Center, sponsored by the National Alliance for Art and Industry, that featured 1,000 products by 100 designers, to illustrate "what designers are doing ... to help the engineer sell his mechanical devices." The exhibit included Raymond Loewy's iconic chrome-plated, teardrop, highly streamlined pencil sharpener (which was never produced); Henry Dreyfuss's check printer, a bathroom scale by Harold Van Doren; furniture by Gilbert Rohde; and a Wurlitzer radio by Russel Wright. By this time, the popularity of Art Deco and French Art Moderne was fading, and streamlining was in the ascendancy. Egmont Arens reported "an amazing response and welling up of national enthusiasm for 'streamlining,' which had captured the American imagination to mean modern, efficient, well-organized, sweet, clean, and beautiful."

In a more abstract manner, the Museum of Modern Art (MoMA) mounted a Machine Art Exhibition curated by architect Philip Johnson (1906–2005) from March 6 to April 30, 1934, celebrating the "abstract beauty" of shapes made by machines, and a poster on April 30, 1934, celebrating the abstract beauty of shapes made by machines. The exhibition catalog began with an ancient Greek quotation by Plato:

> By beauty of shapes I do not mean, as most people would suppose, the beauty of living figures or of pictures, but to make my point clear, I mean straight lines and circles, and shapes, plane or solid, made from them by lathe, ruler, and square. These are not, like other things, beautiful relatively, but always and absolutely.[5]

In the exhibit catalog, MoMA director Alfred H. Barr, Jr. (1902–1981), followed up on Plato's description of the beauty of geometry by stating, "Machines are, visually speaking, practical applications of geometry."[6] The catalog continued,

> The exhibition contains machines, machine parts, scientific instruments, and objects useful in ordinary life. There are no purely ornamental objects; the useful objects were, however, chosen for their aesthetic quality. Some will claim that usefulness is more important than beauty, or that usefulness makes an object beautiful. This exhibition has been assembled from the point of view that though usefulness is an essential, appearance has at least as great a value. The objects displayed were divided according to use in six categories:
>
> **Industrial units**: Machines and machine parts, insulators, cable sections, propeller blades [aircraft and ship], etc. [springs, ball bearings, insulators, car pistons, and a Sears Waterwitch outboard motor designed by John R. Morgan].
>
> **Household and office equipment**: Sink, furnace, bathroom cabinets [an Arco Panel lavatory in which a prefabricated mirror, cabinet, and lighting unit formed a floor-to-ceiling cabinet backing for a washbasin, by George Sakier], dishwasher, carpet sweeper, mirrors, fans, vacuums, and business machines [including a National cash register designed by Walter Dorwin Teague].
>
> **Kitchenware:** [Toaster, pots and pans, coffee urns, bowls, ladles, and the like].
>
> **House furnishings and accessories**: Objects used in daily life: tableware, [porcelain flatware,

tumblers], vases and bowls, smoking accessories, lighting fixtures, and furniture [including the Beta chair designed by Nathan George Horwitt, and billiard balls].

Scientific instruments: Precision, optical, drafting and surveying instruments [microscopes, slide rules].

Laboratory glass and porcelain: Beakers, hydrometer jars, Petri dishes, and boiling flasks.[7]

The listings of 400 items included the name of the item, the manufacturer, the designer, and the price. The manufacturers were treated as artists, not mere producers. Mounted on the walls throughout the exhibit, in clear black lettering, were the names of U.S. companies: Aluminum Company of America, U.S. Steel Corporation, American Radiator Company, Bingham Stamping and Tool, American Sheet and Tin Plate Company, etc. In this advertising-like environment, visitors recognized nearly all of the items and could handle and test them. The *New Yorker* wrote, "The place itself looks, more than anything else, like a very elaborate hardware store." It was a genuine shopping experience, as visitors purchased items such as chairs, billiard balls, and plumbing fixtures, demonstrating that in America, art and everyday objects fused into a single experience. Over 31,000 people visited the exhibit.[8]

MoMA's purpose, of course, was to establish machine-made products as a legitimate form of modern art, in the same way it had promoted that status for architecture in its 1932 exhibition, and to obsolete the Arts and Crafts movement, as Philip Johnson said, "to demonstrate ... that the decorative arts were no longer in existence, but that art could still be made without the handcraft approach—so it was an anti-handcraft show."[9] In fact, the Arts and Crafts movement had already expired. The famous Wiener Werkstätte (Viennese Workshops), founded in 1903, had just been closed, not by the Nazis, but because the high prices for its products had caused it to fail in its social cause of ensuring that the life of everyone was enhanced by everyday objects designed by artists. Now industrial designers were actually implementing that idealistic objective in America.

MoMA had another less obvious agenda, and that was to co-opt industrial design, this new art, under the leadership wing of architects, as had been the case in Europe. MoMA hated streamlining, which was harshly criticized by MoMA director Alfred Barr: "Their designs are aimed at sales appeal of the most superficial kind."[10] MoMA resented that advertising artists, set designers, writers, automotive and furniture designers were dominating industrial design and being heralded in the press and business world. MoMA wanted to take over the new profession for architects. This goal remained for the rest of the decade, when MoMA would launch a full-scale attack to dominate industrial design, which would succeed only in modifying the general style from streamlining to "organic" design. But the term "streamlining" still lives in our language, to describe the simplification and speeding up of things, as in phrases such as "we streamlined our office procedures."

The MoMA Machine Art Exhibition also established MoMA's collection protocol for the future, as MoMA acquired 100 of the manufactured objects in the exhibition into its permanent collection, a practice it would continue for decades, just as traditional museums acquired collections of paintings or artifacts. The exhibit boosted the sale of American-made goods, strengthened the growing alliance between artists and manufacturers, and broke down the wall between art and commercial products, just as Duchamp had done in 1917 with his urinal as art. When MoMA says it is art, it is art. Over the two months of the exhibit, more than 31,200 people attended, and the cultural elite and critics embraced it. Its effect on the popularity of modern design and on the acceptance of industrial designers rivaled the *Fortune* magazine article.

Meanwhile, in Detroit, George Walker had been working since 1929 for Dura, an automotive parts supplier, when in 1934 he opened his own industrial design office in Detroit and would soon design an upright vacuum cleaner for Apex, which would be marketed under the Montgomery Ward (MW) brand. But Walker would soon become involved with the automotive industry far beyond his imagination, as we will see later.

Car companies had been following the lead of General Motors and establishing internal design departments. Ford in 1933 began to update the appearance of its cars yearly, and to do this, Edsel Ford (1893–1943), Henry's son, established a styling department under the leadership of **Eugene T. "Bob" Gregorie** (1908–2004). Gregorie was born in New York City and raised in Cedarhurst, Long Island. In 1927, with no formal education, he started as a nautical draftsman at the Elco Boat Works in Bayonne, New Jersey, then moved to the more prestigious yacht designers Cox & Stevens in New York in 1928. In 1929, he followed his first love of automobiles by beginning at Brewster & Company, maker of custom bodies for Rolls Royce. Later that year he joined Harley Earl's Art and Colour Group at GM, but the stock market crash cost him his job. After a time unemployed and staying with his family in South Carolina, he was offered a job as body draftsman at the Lincoln Motor Company in Detroit in 1931. He jumped at the chance.

In 1928, following the lead of GM, Chrysler had established an internal Art and Color Department of five, headed by Herbert V. Henderson, but the department was subordinated to Chrysler's powerful engineering staff, and designers were relegated to cosmetic details such as grilles and hood ornaments. Chrysler's genuinely new innovative engineering design, under way since 1927, was by leading engineers Carl Breer, Fred Zeder, and Owen Skelton. Breer had designed the successful first Chrysler in 1924 and claimed it was his idea in 1927 to "follow nature" by designing a scientifically engineered, streamlined car. It was radically innovative in engineering and esthetics.

Wind tunnel experiments led to an ideal streamlined form. To achieve the tapered aerodynamic tail of a teardrop, the engineers moved the engine up over the front axle and placed the rear seat within the wheelbase rather than over the rear axle. This more evenly distributed the weight between the front and rear wheels, making a more comfortable ride and better handling. Rather than a single flat plane, the windshield was comprised of two panes meeting in a V and sloping to the rear. The body was the first unitary construction, with the frame and body shell integrated into a lightweight but rigid structure.

After the basic body form had been determined aerodynamically, stylists under Henderson were allowed to add cosmetic Art Deco details, such as parallel, decorative chrome elements on the grille, bumpers, and hood vents. The grille looked like a chrome waterfall splashing over the huge hood. Norman Bel Geddes, the country's foremost industrial designer, was engaged as a consultant in September 1933, and he reduced the number and widened the vertical chrome bars on the waterfall grille, raked the windshield back, and added side vents for cooling. The modernistic interior had rubber mats instead of carpets, and the seats were mounted in a frame of steel tubing. Introduced to the public in 1934, the car was hailed as the first functionally streamlined production car, the Airflow, used for both the Chrysler and DeSoto new models. Bel Geddes happily promoted the design to the public.

But to many, the bulbous sloping, front end that blended into the fenders looked clumsy, front heavy, and crudely shaped. It looked a bit like Dr. Porsch's 1932 prototype Kleinauto on steroids, devoid of grace and beauty. The engineers had achieved an innovative structural design, but potential customers stayed away in droves because the design, particularly the waterfall grille, was too radical for customers accustomed to seeing essentially vertical front noses on cars. Chrysler would sell 18,000 units the first year, but only sales of Chrysler's conventionally designed models saved the company from financial ruin. The disaster convinced Walter Chrysler that engineering innovation alone was not enough, and he hired **Raymond**

Dietrich (1894–1982) to become head of the Art and Color Department, and at the same time elevated the responsibility and power of the department.

Dietrich was born in the Bronx, and at age 12, his father sponsored his apprenticeship as an engraver with Manhattan's American Bank Note Company. He earned money as a pitcher for numerous semipro baseball teams to finance his education in the arts. In 1913 he joined coach builder Brewster & Company, which sponsored him in study at the Andrew F. Johnson Technical School in Manhattan to learn advanced drafting and body design. Graduating in 1917, he first went to Chevrolet but within a year rejoined Brewster. As a "master craftsman," he finished designs in a closed room to protect secret techniques.

In 1920, he and Thomas Hibbard (1898–1982) left Brewster to start their own firm in Manhattan, LeBaron Carrossiers. They initiated the process of designing a whole car in detail on paper to achieve a harmonious appearance, a technique that later would become standard automotive as well as industrial design practice. In 1924, the firm merged with the Bridgeport Body Company, forming LeBaron, Inc., but in 1925 Dietrich sold his stake and joined the Murray Corporation of America, the company that built most of Lincoln's and Ford's production bodies. Murray and Dietrich shared ownership in a subsidiary manufacturing plant called Dietrich, Inc. Dietrich assisted Amos Northup in his design of the 1931 Reo Royale Eight, the predecessor of the groundbreaking 1932 Graham Blue Streak, mentioned earlier, but Dietrich was forced out of Murray in 1931.

At Chrysler, Dietrich was given the job of redesigning the objectionable hood of the ill-fated Airflow, which he did by extending the hood and adding a more conventional-looking, vertical grille. He also styled a conventional companion for the Airflow, called the Airstream, which would be produced in 1935–1936 for Chrysler and DeSoto, and marketed beside the Airflow, meeting the public need while hoping to sell enough Airflows to offset its development.

The Airstream would outsell the Airflow by five to one the first year, and nine to one in 1936. The market failure of the Airflow would go down in automotive history as the result of styling too advanced for popular acceptance, and the failure of engineers to grasp the subtleties of appearance design. The entire industry concluded that even meticulous and innovative engineering could not sell an ugly car. The Airflow failure would become a powerful argument for stylists, who possessed the elusive skill of designing for public emotions and tastes, rather than just for practical and functional purposes like engineers.

Despite the disaster of the Airflow, and despite the scientific fact that streamlining of cars had little effect on operation except at high speeds, streamlining would remain a popular style for the remainder of the decade, because the style was imprinted on consumers' minds as an icon of modernity and progress, which they hoped would take them out of the continuing Depression.

Streamlining was front and center on May 26, 1934, as the Century of Progress Exposition reopened in Chicago for its second year. Headlining the opening was the national demonstration of several brand new, streamlined trains. On May 26, the diesel-powered Burlington *Zephyr*, known as the "Silver Streak," set a speed record for travel between Denver, Colorado, and Chicago, Illinois, by making a "dawn to dusk" dash over the 1,015-mile distance in 13 hours and 5 minutes (normal trains took 25 hours) at an average speed of 77 mph, and arrived at the Century of Progress Exposition on opening day to appear on the Wings of the Century stage for display. The *Zephyr* was built by the E.G. Budd Manufacturing Company, Philadelphia, for the Chicago, Burlington and Quincy Railroad (CB&Q), commonly known as the Burlington, to promote and advertise passenger rail service. Christening was on April 18.

The *Zephyr* consisted of three cars: the power car, baggage-coach combine, and the coach-

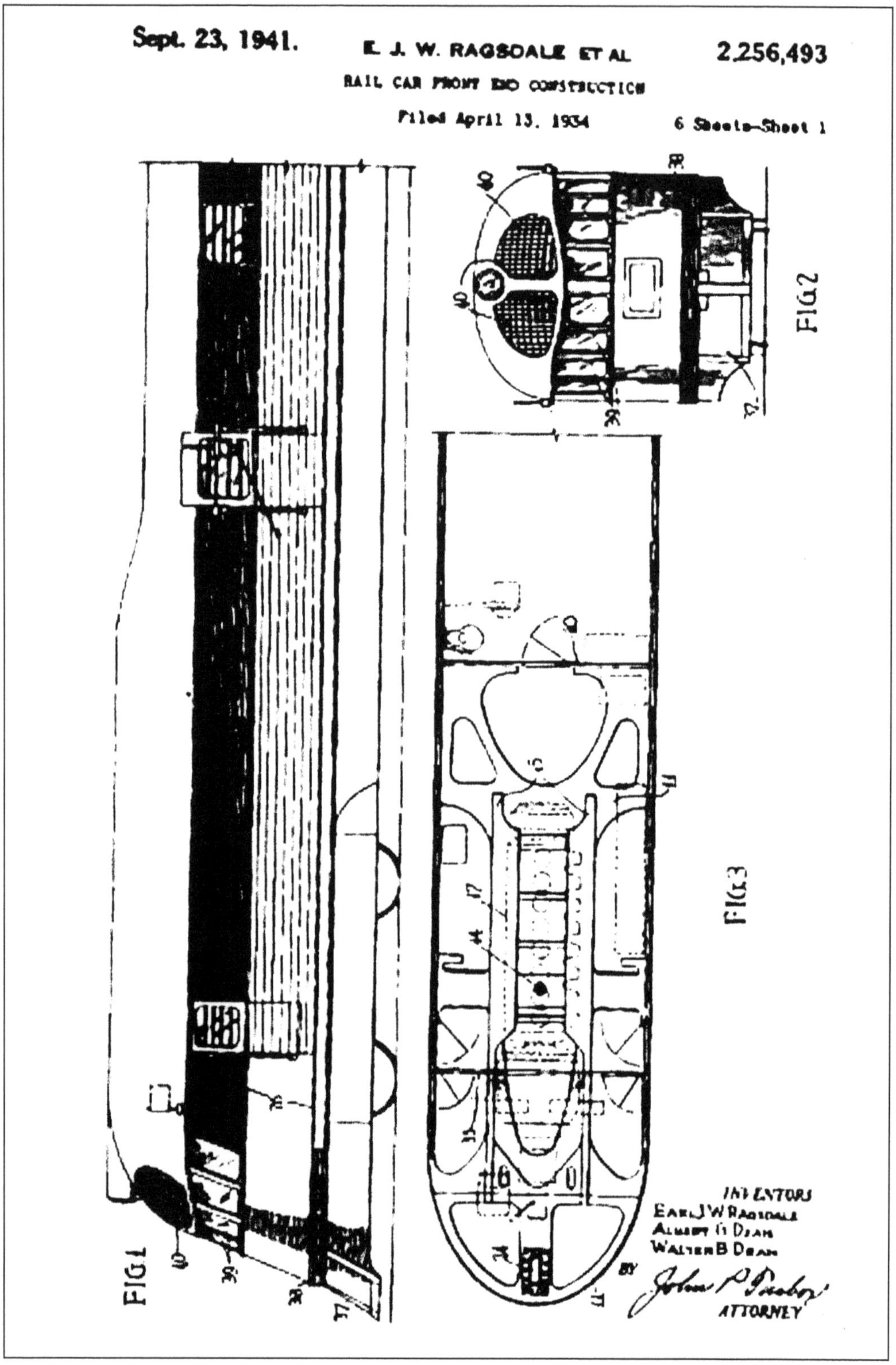

Burlington Zephyr, 1934 (Patent No. 2,256,493, U.S. Patent Office).

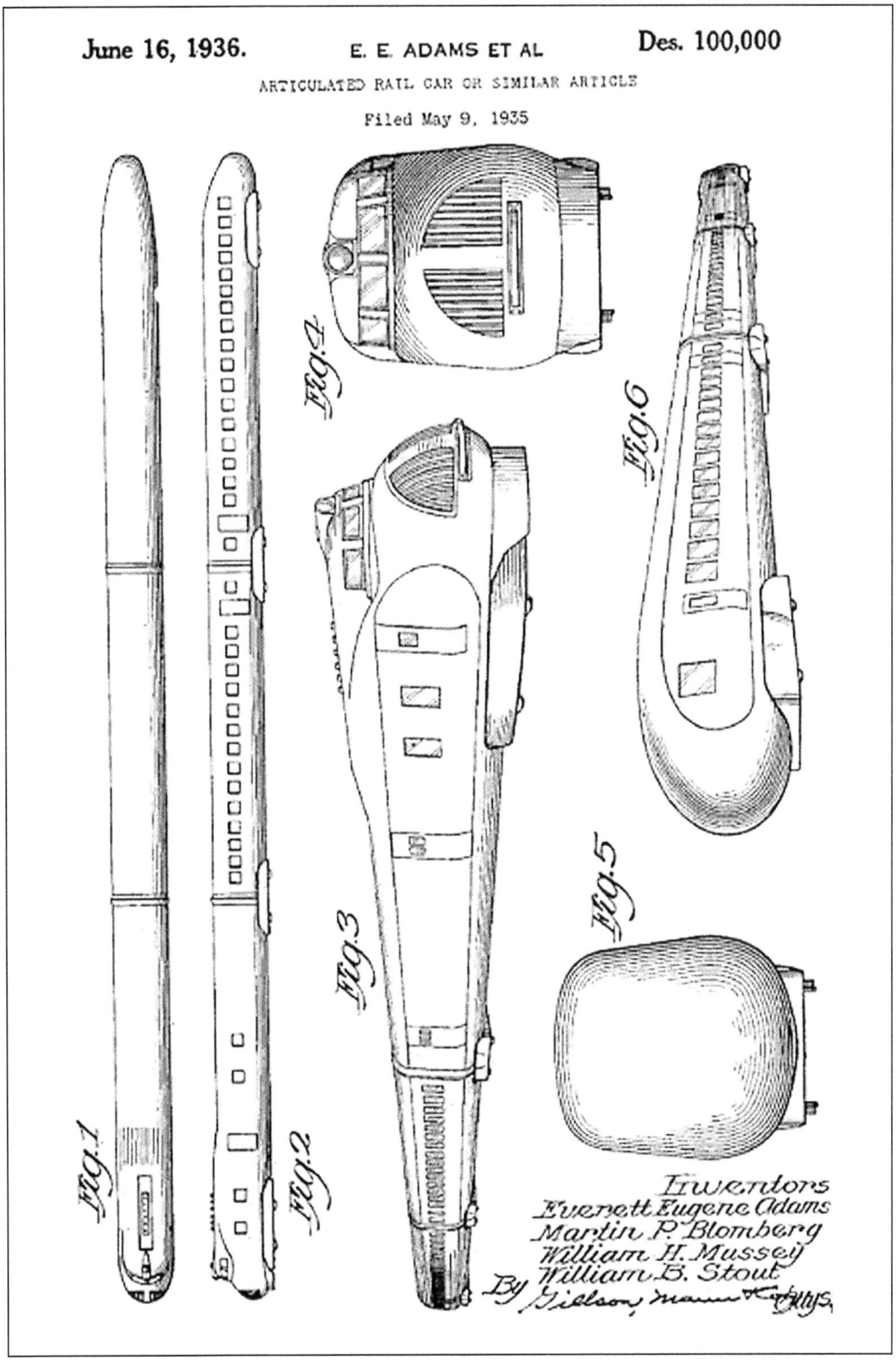

Union Pacific M-10000, 1934 (Patent No. D100,000, U.S. Patent Office).

observation car. It was made of bright stainless steel, requiring no painting or maintenance, and was fastened together using a new technique of electric spot-welding developed by Budd. The entire train weighed the same as a single conventional train car and reached 112 mph during its record run. A 600 hp, eight-cylinder, Winton engine powered a generator, and the electricity generated was fed to electric traction motors connected to the axles in the front wheel truck. There were four trucks, one between each car, and one at each end. The nose design was a slanted half cylinder, often referred to as a "bathtub shrouding" or "shovel nose," with forward-facing windows to provide full visibility for the operator.

The exterior streamline design was by aeronautical engineer Albert Gardner Dean, in collaboration with Philadelphia architects John Harbeson (1888–1986) and Paul Philippe Cret (1876–1945). After display at the exposition, the train was taken on a 31-state, 222-city publicity tour. More than two million people saw the train before it entered regular service in November between Kansas City, Missouri, and Lincoln, Nebraska. The very name, Zephyr, would soon be applied to products to capitalize on its popularity as a term of speed and modernity, such as Kem Weber's streamlined Zephyr digital clock design of 1934, and the Lincoln Zephyr of 1936.

The Burlington *Zephyr* was a hard act to follow, but there was another streamliner that tried to do so. The gasoline-powered M-10000 (its first design patent number was D100,000, followed by U.S. patents D100,001 and D100,002), produced by Union Pacific Railroad, visited the Chicago exposition after a tour of 68 cities that totaled 13,000 miles, everywhere attracting press attention and crowds that approximated one million. It was the first internal combustion, lightweight, streamlined, express passenger train in the United States, delivered to the railroad on February 12, 1934. Named on the design patent were Eugene Adams, Martin P. Bloomberg, William H. Mussey, and William B. Stout, all engineers at Pullman-Standard, the body developer.

There were three cars, the power and baggage car up front, and two passenger cars, totaling 204 feet in length and weighing 77 tons. All were built of Duralumin, a hardened aluminum alloy, and the bodies slanted inward at the bottom. The entire train was supported on just four wheeled trucks, located between cars, and at front and rear. The 600 hp engine, built by Winton, a subsidiary of General Motors, powered a generator of electricity that fed electric traction motors. The engine was housed in the front power car and featured a turret cab with windshields and a bulbous nose with two large air intakes for cooling. The body was painted in Armour Yellow with a Leaf Brown roof and undersides. Overall, the M-10000 design very closely resembled Norman Bel Geddes's Locomotive Number 1, his futuristic, never-produced design in his book, *Horizons,* in 1932, which also had three cars and a bulbous nose. M-10000, nicknamed the "Tin Worm," would enter regular service as the *City of Salina,* for the Kansas City-Salina route it served.

The Chicago Fair closed on Halloween, October 31, 1934, with a total two-year attendance record of over 48 million. Newspaper accounts described the last night as resembling New Year's Eve celebrations, with merrymaking running wild. For the first time in American history, an international fair had paid for itself, paying off the $10 million debt from a bond issued on October 28, 1929, the day before the stock market crashed.

But it was the modern trains exhibited there that would inspire the railroad industry to follow the automotive lead in modern design and draw industrial designers into the business.

CHAPTER 6

Recovery, Plastics, and Education

The two pioneering train designs connected to the Chicago exposition had a significant and immediate effect on the railroad industry. In November 1934, the Pennsylvania Railroad (PRR) hired Raymond Loewy to refine the Westinghouse design of the GG-1 design for mass production. Loewy had designed a trashcan for Pennsylvania Station in New York in 1932, and at the time had shown the PRR president a futuristic sketch of a streamlined locomotive.

The GG-1 was a class of electric locomotives powered by alternating current (AC) overhead wires. Twelve 385 hp traction motors drove the GG-1's 57-inch-diameter driving wheels on six axles. The body, over 76 feet long, was built in two halves joined with a ball-and-socket joint to allow negotiation of sharp curves, and each half had a raised control cab, each facing opposite directions, so it could be driven in forward or reverse.

In 1933, PRR had decided to replace its current P5a electric locomotives with ones that were more powerful and more streamlined. PRR asked General Electric and Westinghouse to design prototypes that were capable of 100 mph, a streamlined body shell, and a central control cab. Both companies submitted prototypes in August 1934, and both worked with PRR's Baldwin Locomotive Works as the design progressed. Features from both prototypes were refined and combined to produce the final design.

The Westinghouse prototypes were based on six plaster models made by Donald Dohner (mentioned earlier), Westinghouse's director of art in its engineering division, in 1933 when the project began. One of these models, shown in photographs, closely resembles the final design of what became the GG-1 prototype, which essentially was completed by August 1934, three months before Loewy was called in. By all traditions of attribution, Dohner should have been credited with the styling of the GG-1.

Loewy was called in for consultation on October 25, 1934, by PRR vice president and heir apparent Martin W. Clement to "enhance the GG-1's esthetics." A fee of $5,000 was agreed upon, but he had only about a week before the production order was to be placed, so his contributions on the existing prototype were limited to the color scheme, which became Brunswick Green, with five gold pinstripes running along the sides and curving down on the front and rear toward the ground. Loewy also claimed he had recommended a smooth welded body instead of the riveted one used in the prototype, but there is some evidence that the rivets were the result of rushing the completion of the prototype and that they were never intended for use on the production models. One hundred thirty-nine production models would be built between 1934 and 1943, and many remained in service until the 1970s.

In any event, Loewy was retained for $20,000 annually as the primary design consultant of PRR for the rest of the decade. Until his death, he publicly claimed the design of the GG-1

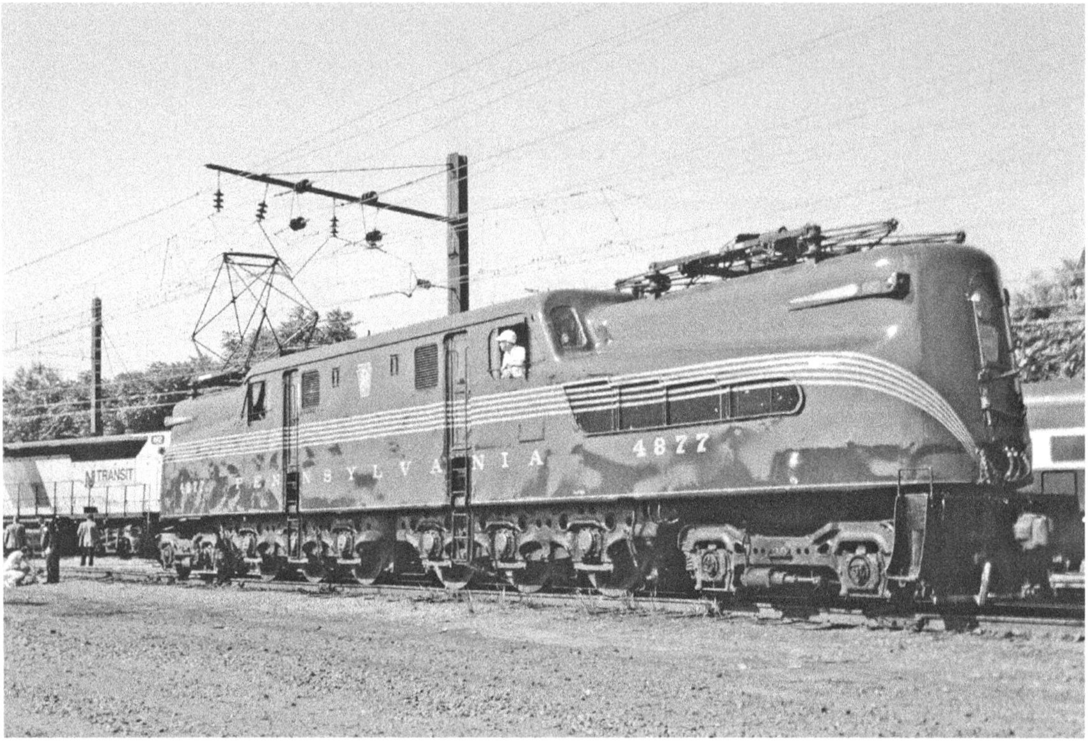

Pennsylvania Railroad GG-1 (*Classic Trains*, Summer 2009. Photograph by Karl Zimmermann).

as one of his most important achievements and was exclusively credited with its design in all literature pertaining to the GG-1. Loewy was its ultimate promoter, not Dohner, who was essentially a modest man, who would, within a month after he designed the models, leave Westinghouse to begin his career in industrial design education, for which he is better remembered.[1]

The main reason Donald Dohner never received credit was that after February 1934 he was no longer employed by Westinghouse. In fact he was fired as a result of the *Fortune* article on industrial design that month. The article, described in detail earlier, revealed the fact that his department at Westinghouse was costing the company $75,000 per year (nearly $1.3 million in 2013 dollars), a fact apparently not known by Westinghouse management, which was understandably shocked, and in the middle of a Depression took executive action to cut expenses.

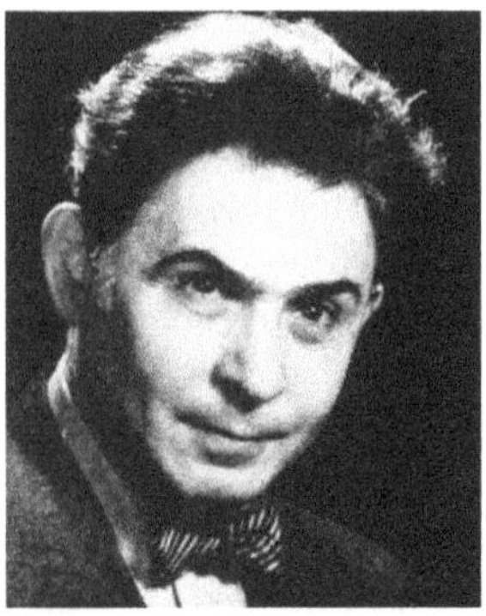

Alexander Kostellow, ca. 1950s (*Industrial Design* magazine, October 1954).

Ironically, this unfortunate turn of events resulted in a new career for Dohner, and a milestone for the industrial design profession. As you may recall, before joining Westinghouse in 1930, Dohner had taught design as an instructor at the Carnegie Institute of Technology (CIT) starting in 1929. Upon hearing of Dohner's loss of job, his old colleagues at CIT were overjoyed at the opportunity of welcoming back someone who was now a nationally known industrial designer, recently featured in *Fortune* magazine. Painting and Design Department painting instructor **Alexander Kostellow** (1896–1954) reportedly shouted to department head Wilfred Readio, "Donald's on the loose! Let's get him! It's [industrial design] the wave of the future! Painting is dead! Start an option [degree] in industrial design!"[2]

Alexander Jusserand Kostellow was born in Iran (then Persia). He studied in Paris and the University of Berlin and came to the United States in 1916, where he studied painting at the National Academy, the Art Students League of New York, and the Kansas City Art Institute, where he met and married Rowena Reed. Both came to Pittsburgh in 1929 to teach at CIT, he in painting, and she in sculpture.

Carnegie was not the first school with interest in industrial design. As mentioned earlier, there were a number of art schools already teaching it in various programs for engineers, and practicing designers were already lecturing at schools. In 1932, Pratt Institute in Brooklyn had engaged Donald Dohner as a "consultant and lecturer in industrial design." Students, who previously would have studied architecture, were now attracted to industrial design as an alternative, and indeed many did see it as "the wave of the future," as declared by Kostellow. The Chicago Art Institute offered industrial design courses in 1932, the same year the University of Cincinnati initiated an "Art and Industry" option in its School of Applied Arts, and in California, Kem Weber initiated a program in industrial design at the Art Center School in Los Angeles. He would teach there until 1943.[3] In 1933 the Detroit Society of Arts and Crafts had made national news by recognizing the automobile as an "art form," and at the Cleveland School (now Institute) of Art (CIA), **Viktor Schreckengost** (1906–2008) had initiated a pioneer program of industrial design in 1933, which would become well known for automotive design.

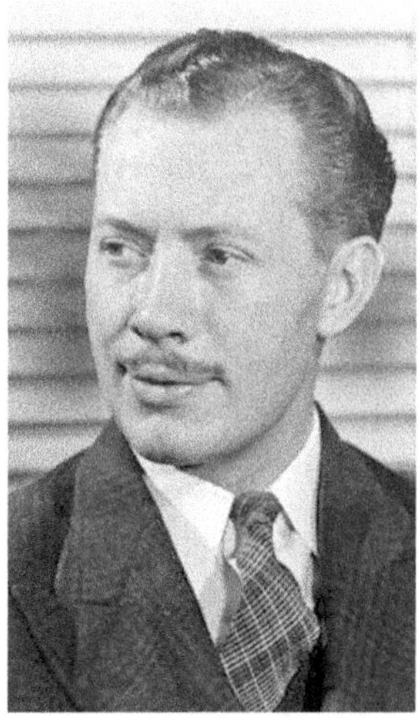

Viktor Schreckengost, ca. 1939 (courtesy David Nowacek and Gene Schreckengost).

Schreckengost was born in Sebring, Ohio, to a potter father and had studied design and ceramics at CIA, graduating in 1929. After studying at the Kunstegewerbeschule (Arts and Crafts School) in Austria, he joined the faculty at CIA and worked part time at the Cowan Pottery Studio. There he created his famous 28-inch-diameter Art Deco "Jazz" punch bowl that had been commissioned anonymously by Eleanor Roosevelt, who later ordered more for the White House and Hyde Park. Schreckengost soon became a consultant to the White Motor Company in Cleveland, and would later create an innovative truck design with the cab over the engine to provide more storage space in the back.

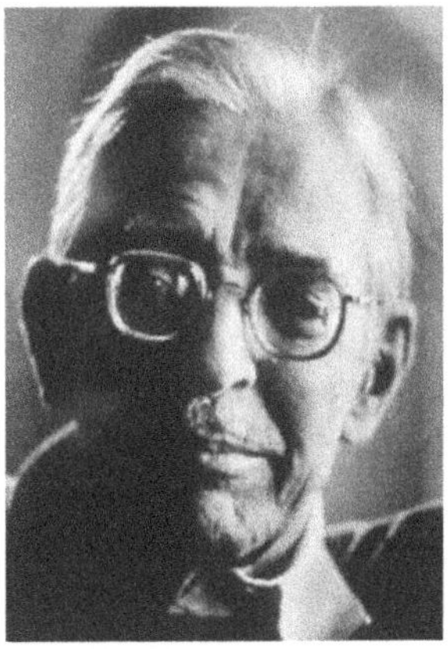

Robert Lepper, ca. 1940s.

At Carnegie in 1933, optional classes in industrial design, based on Dohner's 1930 syllabus, had been offered to students and were in progress as Dohner rejoined the faculty in 1934 and initiated a new degreed program in industrial design at CIT, the first school to do so. Assisting Dohner was Art and Design Department head, Wilfred Radio, and faculty members Kostellow, Russell Hyde, Fred Clayter, and **Robert Lepper** (1906–1991). Lepper had been hired to fill in for Dohner's position when he left for Westinghouse in 1930. Lepper was born in Aspinwall, a suburb of Pittsburgh, and graduated with a degree in painting from CIT in 1927. He spent a year or so in Europe soaking up modern art and joined the faculty at CIT in 1930, where he painted "machine art," such as his 1932 Crankshaft, a realistic image of a Lincoln 12-cylinder crankshaft with pistons floating in the air. Because of his interest in mechanical things, he was more or less drafted into the new industrial design degree program in 1934.

So in the fall semester of 1934, third-year students were offered an option for a degree in industrial design, along with other third-year options of illustration, interior decoration, or art education. The first class included three women and two males, all to become the first degreed graduates of industrial design in the world. Dohner took them to manufacturing plants, such as Westinghouse, Fostoria Glass, and ALCOA (Aluminum Company of America), to show them how things were made, and then instructed them to design new products based on what they had seen.

Elsewhere, modern design continued to be publicized. From November 5, 1934, to January 6, 1935, the Metropolitan Museum of Art mounted an exhibition, Contemporary American Industrial Art, by 20 architectural and industrial designers that featured entire-room arrangements by architects except for a full-scale replica of an industrial design office and studio designed by Raymond Loewy. Architects and museums related all industrial design to architecture because they believed that architects should dominate the new field, as was the case in Europe. Accordingly, many of the rooms were traditionally credited to name architects, who selected all the decorative furnishings to keep industrial design "in its proper place" under the supervision of architects.

However, this time, a number of industrial designers were invited to participate in the event, including Viktor Schreckengost (chinaware), Donald Deskey (an entire dining room), Gilbert Rohde (Herman Miller clock and furniture), Gustav Jensen (rugs, wall treatments), Peter Müller-Munk (silverware), Walter von Nessen (display case and metalware), Lurelle Guild (silverware), Russel Wright (metalware, floor lamps), Alfons Bach (furniture) and Walter Dorwin Teague (on the cooperating committee). Still, the products shown were all from traditional art industries, studiously ignoring the many larger artless industry products by industrial designers, such as the elegant gas range Teague had recently designed for the Floyd-Wells Company in Royersford, Pennsylvania, that rivaled Bel Geddes's SGE range in simplicity, whiteness, and modernity. But attempts by architects to supervise and control industrial design were too late. The horse of industrial design was out of the architectural barn. Industrial designers already owned the field of both mechanical and decorative, modern mass-produced consumer products.

Streamlined trains were still in the national news at the end of the year. On December 27, 1934, the New York Central Railroad (NYC) exhibited its new steam locomotive at Grand Central Station, with a streamlined shell painted in gunmetal gray and called the *Commodore Vanderbilt.* Named to honor the original *Commodore Vanderbilt* locomotive of 1870, it was the first to utilize streamlining by adding a streamlined shell over a steam locomotive, NYC's J-1E 4–6–4 "Hudson" No. 5344.

An undergraduate student at Cleveland's Case School of Applied Science, Norman F. Zapf (1911–1974), designed the shell for his 1934 senior thesis, "The Streamlining of a Locomotive." Zapf tested scale models in a wind tunnel and concluded that at typical passenger speeds of 75 mph, streamlining reduced drag by 30 percent and required 350 less horsepower than a nonstreamlined locomotive. Nearly all his suggestions were incorporated into the final *Commodore Vanderbilt.* The nose form was similar to the "bathtub" design of the Burlington Zephyr. Carl Kantola, an in-house New York Central employee, worked on adapting the design to the required mechanical modifications. A number of sources cite Henry Dreyfuss as being a design consultant for the *Commodore Vanderbilt,* but his contributions, if any, are unknown.

After being exhibited, the *Commodore Vanderbilt* departed on an exhibition tour of major cities on the NYC system, which was a smashing success, was highly photographed, and was surrounded by crowds wherever it went. After the tour, it was put in service hauling the *20th Century Limited* between New York and Chicago, the grandest, most elegant passenger train in America. For the rest of the 1930s, the *Commodore Vanderbilt* was the most popular Lionel O-gauge model train.

Although 1934 was one of the worst years of the ten-year Great Depression, it was the most exciting time for the new profession of industrial design, which became a household word as its leading practitioners became known nationally. Formal professional education in industrial design had begun. Manufacturers scrambled to catch up with the wave of modern design swirling around them by seeking new designs and new designers, hoping to extract themselves from the litany of declining sales reports. New design departments sprouted up in dozens of corporations. National spirits were rising due to the optimism generated by the Chicago Fair and the futuristic, streamlined trains crisscrossing the country. The future seemed to be the light at the end of the tunnel.

The fledgling practitioners of industrial design, now that they had established legitimacy in the world of design and in the public arena, were looking to legally protect their innovative, modern, American designs from the wholesale copying that had previously been common practice for European designs. That had been the primary concern of the American Union of Decorative Artists (AUDAC) when it was formed in 1927. Now, in 1933, to address AUDAC's accusations of design piracy, representatives and designers in the furniture industry joined together to found the National Furniture Designers' Council (NFDC) in order to draw up a code for enjoining piracy for the National Recovery Administration (NRA).

The purpose of the Roosevelt administration's establishment of the NRA in 1933 was to eliminate "cutthroat competition" by bringing industry, labor, and government together to create "fair practices" and to set prices. Industries were invited to submit codes of fair competition, and the furniture industry did so. But in 1935, the U.S. Supreme Court found that the NRA law was unconstitutional, ruling that it infringed on the separation of powers under the constitution. The NFDC disintegrated, but the seed for a furniture industry design organization had been planted, and in July 1935, Lawrence H. Whiting, director of the American Furniture Mart, called a special meeting of furniture designers, including those from the now defunct NFDC, and engaged Richard Bach, of the Metropolitan Museum, and Alfred Auerbach (1881–1944), design and merchandising consultant as well as editor of *Retailing* magazine, to serve as speakers for the meeting. The proceedings of the meeting are unknown, but the result, within a year, would be the forming of a furniture design organization, which will be described later.

That year, Al Lambach, mentioned earlier, returned to Chicago to work for Merton Gershun at the Chicago Furniture Mart, and also for the Tell City Furniture Company in Tell City, Indiana, where he designed radio and phonograph cabinets for Zenith, Clinton, Traveler, and Emerson. Because home radios of the day were traditionally enclosed in furniture for placement in living rooms, a number of furniture designers in the 1920s and early 1930s were actively designing in this field.

Furniture, as one of the traditional art industries, was still in the good graces of museums, such as Bach's Metropolitan Museum, and of course the architects as well, since many famous European architects, and equally famous American architects such as Frank Lloyd Wright, had been designing furniture for generations. A number of pioneer industrial designers had designed furniture too, including Norman Bel Geddes (Simmons Company, 1928); Paul Frankl (Skyscraper furniture, 1925); Gilbert Rohde (for Herman Miller Furniture Company, 1932–1935); John Vassos (RCA radio cabinets and his portable phonograph of 1935); Nathan Horwitt

(Beta Chair for the Howell Company, Geneva, Illinois, 1930); Kem Weber (armchair by Lloyd Manufacturing Company of Menominee, Michigan, 1934); Donald Deskey (since 1926); and Russel Wright (Haywood-Wakefield, 1934). But most of those above had designed in the modern styles, not the traditional historic styles of many designers in the furniture industry.

Throughout the 1930s, Wright and Walter Von Nessen had designed scores of modern metal accessories for the Chase Brass & Copper Company. In 1935, Wright had renamed his firm from Russel Wright Incorporated to Russel Wright Accessories, and designed a natural maple line of furniture called Modern Living, made by Conant Ball Company for Macy's, which was widely promoted. The popular line would later be called American Modern, produced in bleached wood "blond" maple, and would make Wright famous. He opened a showroom for his business in New York, with the help of businessman Irving Richards, under the name Russel Wright, Inc.

In 1934, one of the pioneer furniture designers, Paul Frankl, relocated to California and became an active interior designer, working in a style that would be called "California Modern" at his gallery on Rodeo Drive in Beverly Hills, where celebrities such as Fred Astaire, Katherine Hepburn, Walter Houston, and Alfred Hitchcock became his clients. Other designers were working for celebrities. Around this time, Margaret Bourke-White (1904–1971), the famous *Fortune* photographer, approached John Vassos to design her studio in the Chrysler Building in New York, as well as the interior of her apartment in Tudor City. He did, and they remained close personal friends until her death.

Furniture and furnishings were still the mainstream of the traditional art industries except that now modern design was replacing some of the traditional styles of historical periods. But the new mechanical, mass-produced products of the artless industries remained in artistic limbo as a category yet to be accepted by museums as legitimate art.

Museums, including the Museum of Modern Art, and architects still had serious reservations about other mass-produced, mechanical products in which industrial designers were exclusively excelling, such as stoves, refrigerators, sinks, toilets, radios, office machines, etc. Elitists balked at calling these items furniture let alone works of art. To the elite artistic world, these products were commercially embarrassing, far exceeding the profitability of furniture or the decorative arts, and far more financially rewarding to the designers than any architect could expect. Accordingly, many regarded this arena of design as pure commercialism, the enemy of fine art, and were reluctant to embrace such mechanical products on an equal footing with furniture.

Nevertheless, evidence of the influence of industrial design was everywhere in the consumer marketplace. Sears, Roebuck & Company had hired Raymond Loewy in 1932 to redesign its Coldspot refrigerator after Henry Dreyfuss had declined the project, due to his similar refrigerator project for General Electric. It was an unwritten code of ethics in early industrial design (and still is) that you didn't accept projects for products that were competitive with your own designs for another client. Upon accepting the job, Loewy described the current Coldspot as an "ill-proportioned vertical shoebox 'decorated' with a maze of moldings, panels, etc. perched on spindly legs high off the ground."[4]

Loewy's design was actually done by Clare E. Hodgman, of Loewy's staff, who after the project was completed, was hired by Sears. The design was developed carefully with clay appearance models and came out in 1935—a sleek, white, vertical box with three, vertical, parallel ribs running from top to bottom in the center of its front, accenting the rectangular chrome Coldspot name escutcheon at eye level. Loewy had been inspired by his design of the interior shelf pattern from the radiator of the Hupmobile, on which he was working at the time. Sears paid Loewy $2,500 for the design, but he had spent nearly three times that in expenses.

But sales of the Coldspot in 1935 would double to $1,388,000, and in five years would increase by fivefold. A highly appreciative Sears paid Loewy $25,000 for three future annual model changes for the Coldspot, each of which was more aggressively modern than its predecessor. Loewy's reputation was made. He celebrated by relocating his office to a penthouse studio at Fifth Avenue and 47th Street and opening his first office in London, England, the first American designer to go international. Recall that his first design in 1929 was for an English client, Gestetner. He now had 12 clients and was doing very well.

Sears hired **Robert E. Bourke** (1916–1996) in 1935. He was born in Chicago, went to Morgan Park High School in the southwest side, and as a kid was very interested in model airplanes. At age 11, he began to sketch model automobiles. He attended the Chicago Art Institute, cleaning floors and latrines to pay for tuition. He took courses in architecture and engineering at Armour Institute during the Depression and began working at Sears under Clarence Karstadt. His first assignment was to design a manure spreader by the Oliver Farm & Equipment Company. Jack Morgan headed the department. Just like his fellow employee, Clare Hodgman, Bourke would later work for Raymond Loewy.[5]

In May 1935, the French oceanliner *Normandie*, at the time the largest and fastest passenger ship afloat, completed her maiden voyage from Le Havre to New York in four days, three hours, and 14 minutes, winning the traditional Blue Riband (an unofficial accolade given to passenger liners for the highest average speed). The *Normandie*'s luxurious interiors were in the latest Art Deco and Streamline Moderne styles, and the *Normandie* was quickly heralded as a ship for the rich and famous.

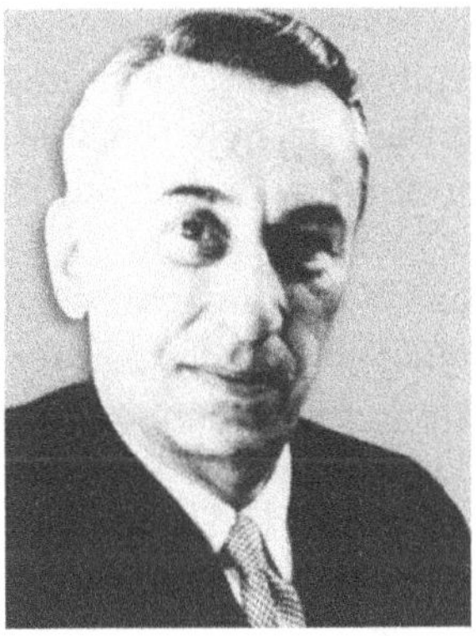

Peter Müller-Munk, ca. 1950s.

She was also the inspiration for an elegant, chrome-plated, brass pitcher designed by **Peter Müller-Munk** (1907–1967), the New York silversmith mentioned earlier. The Normandie Pitcher, shaped like the prow of its namesake, was made by the Revere Copper and Brass Company in Rome, New York, and would be produced until 1941.

Müller-Munk was born in Germany, studied as a silversmith at the University of Berlin, and immigrated to the United States in 1926. He worked at Tiffany's in New York as a metalworker from 1926 to 1928, exhibiting at the 1928 Macy's exposition. In 1929, he established his own silver studio and exhibited at the 1929 and 1930 Metropolitan Museum shows.

Now Müller-Munk was jumping into the nascent industrial design education field. At Carnegie Tech, in Pittsburgh, upon finishing the first year of the degree program he initiated, Donald Dohner was contacted by his former employer in the Pittsburgh public schools, James Boudreau, who had become dean of Pratt Institute's Art School in 1928. Boudreau wanted Dohner to organize and lead an industrial design program at Pratt, similar to the one he had just started at Carnegie. Having also worked with Boudreau as a lecturer and consultant at Pratt in 1932, Dohner accepted, and Müller-Munk was called in to replace him as head of the program at Carnegie.

In the fall, Müller-Munk and Alexander Kostellow would continue Dohner's curriculum and practice of taking students to tour local manufacturing facilities and having them design products based on the production methods they had seen. As a matter of fact, it is this unique accommodation of the client manufacturer's production and economic limitations that distinguished industrial designers from their predecessors, the pure artists and industrial artists. Pioneer industrial designers were developing the unique philosophy of industrial design as a multidisciplinary process that was not only art, but a balance of economics, engineering, materials selection, and human convenience considerations.

Their other distinguishing characteristic was their commitment to modern design, rather than the imitation of traditional period design styles, still clung to by many industrial artists, and the most modern style, now known as streamlining, was at the forefront of transportation design. On May 29, 1935, the first of a series of streamlined *Hiawatha* trains, the *Twin Cities Hiawatha*, began service from Chicago, through Milwaukee, to the twin cities of Minneapolis and St. Paul in morning and afternoon runs. Operated by the Milwaukee Railroad to compete with the Burlington *Twin Cities Zephyr* and North Western Railway's *Twin Cities 400,* the Class A 4–4–2 steam locomotive, built by the American Locomotive Company (ALCO), was designed by ALCO design consultant **Otto Kuhler** (1884–1977) with a nose form similar to the Burlington Zephyr, but without the front cab windows, the cab being in the rear as with all traditional steam locomotives. The nose was graced with horizontal stainless steel wings with a red engine number at its center. Kuhler also designed the famous Hiawatha "beaver tail" observation car at the rear, and all the interiors down to the napkins and draperies.

Kuhler was born near Essen, Germany. At age 19 he was commissioned to illustrate a catalog of steam locomotives. His auto body design for a Mercedes chassis won a gold medal in 1913. He served in the German army in World War I, after which he enrolled in the Academy of Art in Dusseldorf to learn etching. He was employed as a stylist with N.A.G. in Berlin and designed sets for the silent film *The Tunnel,* then became associate editor of *Der Motor* magazine. He styled car bodies for Snutzel Pére & Fils in Brussels and a 1914 roadster for Fiat in Italy. Kuhler immigrated to the United States in 1923 and worked as a commercial artist in Pittsburgh. After becoming a citizen in 1928, he opened a studio in Manhattan, promoting the streamlining of railroad equipment and working for the J.G. Brill Company styling trains and trolley cars, including a smooth front-end trolley called the Brill Bullet in 1931. In 1932, the American Locomotive Company (ALCO) commissioned him as a design consultant.

In April 1935, *Business Week* magazine announced that "Depression is a forgotten word in the automobile industry, which is forging ahead in production, retail sales, and expansion of productive capacity in a manner reminiscent of the 'twenties.'" At the Ford Motor Company in Detroit, the young styling department established in 1933 under Eugene T. "Bob" Gregorie had been struggling under the dominance of engineers, as the one at Chrysler had been. Ford stylists were subjugated to body engineers, who resented the stylist's role in design, and to Charlie Sorensen, who placed speed of production over beauty and style.

Both were under the powerful Henry Ford, who staunchly resisted modernization and change of any type. But like Earl, Gregorie had a corporate patron who understood the growing importance of automotive appearance: Henry's son, Edsel (1893–1943), who was a sophisticated patron of the arts, and who had an excellent eye for design. Edsel was determined to win control of styling for Gregorie's department and assisted Gregorie in battles with engineers and production supervisors. Unlike GM, which had a complex bureaucracy, Ford had only Henry to deal with, and only Edsel to counter him. So Gregorie needed only to convince Edsel of his designs.

Edsel had seen the styling disaster of the 1934 Chrysler Airflow by engineers, so when it was time to design the new, low-cost 1936 Lincoln, Edsel made sure Bob Gregorie was in full charge. The early concept for the car was the Briggs Dream Car, designed by John Tjaarda at Briggs for Ford at the 1933 Chicago Century of Progress Exposition, which was a rear-engine car not unlike a sleeker version of Ferdinand Porsche's Kleinauto then under development in Germany, with a front sloping hood that now reminded people of the discredited Chrysler

Airflow. Edsel was also concerned and thought the design was "too Teutonic," knowing the German origins of the Kleinauto.

Tjaarda was reluctant to change the design, so Edsel called Gregorie in to work with Tjaarda to resolve the dilemma. Gregorie decided to change the front end by relocating the motor to the front and enclosing it in a hood that was gracefully shaped like the inverted prow of a ship with a V-shaped grille. Gregorie's experience in boat design had contributed to an inspiring styling innovation. The fenders and teardrop rear end were retained, but with the new prowlike front end, the front fenders seemed to suggest water being sprayed up and to the sides from the prow. The final design would be introduced as the 1936 Lincoln Zephyr, capitalizing on the name of the popular streamlined Burlington *Zephyr* train. It would be hailed by many as the first commercially successful streamlined car, and its unique front-end design would be imitated by dozens of competitors over the next five years. It was indeed a watershed of automotive design and in fact remains a classic design today.

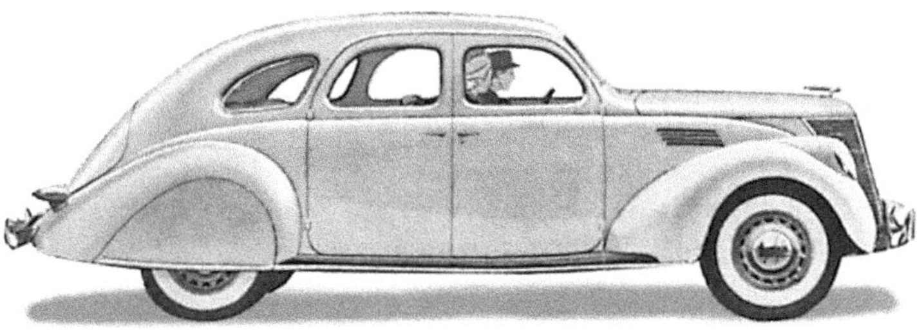

Lincoln Zephyr, 1936.

Another transportation marvel was introduced in 1935, the Scarab, engineered by aviation engineer William Bushnell Stout (1880–1956), with the exterior designed by John Tjaarda. Unlike any car of the time, the Scarab was what many call the first minivan, with a large interior compartment with a flexible seating system that could accommodate up to ten passengers. Stout envisioned it as a mobile office. To create the space, he did away with the normal chassis and drive shaft by using a rear engine and a unitized body structure, permitting a low flat-floor interior. The body design was without fenders, presenting a fully streamlined aluminum aircraft-type shell with built-in headlights. Although the Scarab was offered for sale at $5,000 (a Chrysler Airflow then sold for $1,345) and up to a hundred a year were to be manufactured, only nine were actually made.

The major star at the November 1935 New York Auto Show was the Cord 810 Winchester Sedan, designed by **Gordon M. Buehrig** (1904–1990) and made by the Auburn-Cord-Duesenberg Automobile Company in Connersville, Indiana. Cord was the brand name, after the owner of all three brands, Erret Lobban Cord, whose philosophy was to build innovative,

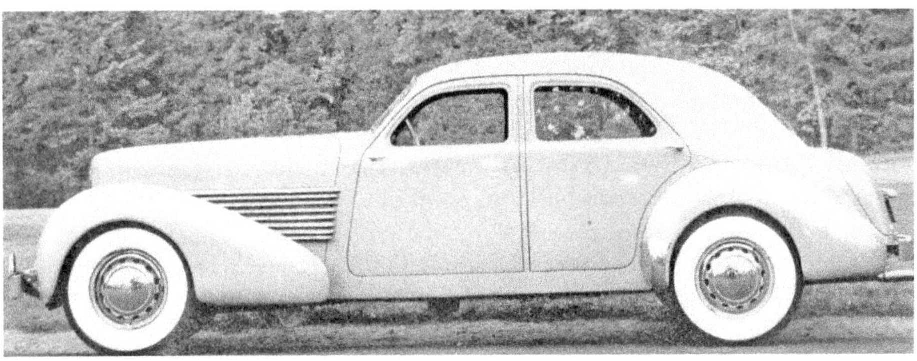

1936 Cord by Buehrig.

different cars. The Cord 810 was indeed innovative and different. It featured front-wheel drive, a rear-hinged hood, retractable hidden headlights, absence of a running board, no visible grille, and was offered with a supercharger. It was highly streamlined, with a rounded, boxlike hood featuring a grille of horizontal louvers that wrapped around the lower half of the box, which some dubbed the "coffin." The crowds at the New York show were so dense that visitors had to stand on the bumpers of nearby cars to get a look. Many orders were taken at the show, but the cars would not be ready for delivery until April 1936, due to problems of production. Over 1,000 were sold the first year. In 1936, Buehrig would also design the 1937 Cord 812, which in 1951 would be displayed by the Museum of Modern Art as a work of art. But by 1937, Errett Cord's empire would collapse and all production would cease, another casualty of the Depression. In 1936, Buehrig would go to the Edward G. Budd Company in Philadelphia as chief designer of prototypes.

Buehrig was born in Mason City, Illinois, and attended Bradley Polytechnic in Peoria before becoming an apprentice at the Gotfredson Body Company, Detroit, in 1924. He worked for Dietrich, Inc., in 1926, then at Packard, and in 1927 was one of the first employees hired by Harley Earl at the newly established Art and Colour Section at GM. In 1928 he worked for the Stutz Company in Indianapolis, and in 1929 became chief body designer for Auburn-Cord-Duesenberg, where he designed the Duesenberg Model J line of luxury cars. In 1933 he returned to GM, and in 1934 he returned to Auburn-Cord-Duesenberg.

Harold Van Doren and his partner, John Gordon Rideout, had developed a growing industrial design business in Toledo, Ohio. In 1933, they had designed a bright green, plastic, skyscraper-style table radio for Air King Products, a large Magnalite (cast aluminum with varying thicknesses, allowing for maximum heat conductivity) teakettle for Wagner Manufacturing Company (designed by Rideout), and plastic stemware for the National Silver Company. In 1934 they designed a Magnalite aluminum covered casserole for Wagner, and a trendsetting sheet-metal gasoline pump for the Wayne Pump Company. In 1935, they designed a lightweight, modern Sentinel Duplex retail scale for Toledo Scale, using the largest plastic molding of the time, and developed an entire line of streamlined children's bicycles, tricycles, scooters, and wagons for the American National Company, including the Skippy-Racer Scooter (1933), the Skippy Sno-Plane Sled and the Skippy Stream Line Racer Wagon (1934), and the Streamline Velocipede and the Pioneer Streamlined Tot Bike (1935). Now even kids could enjoy streamlined transportation!

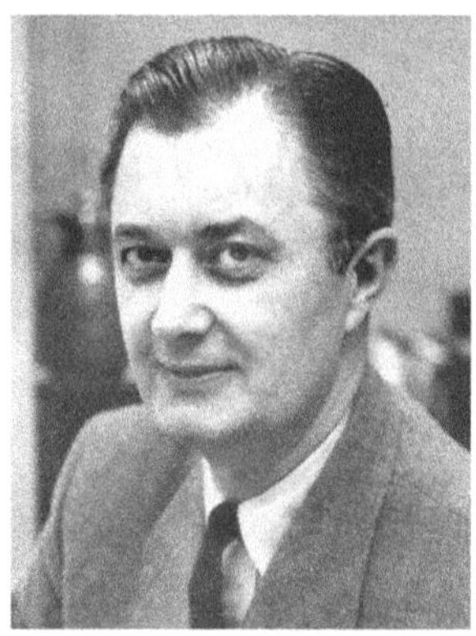

Arthur N. BecVar, ca. 1970s (courtesy Industrial Designers Society of America).

However, in 1935 Rideout left Van Doren to open his own office in Cleveland, Ohio, John Gordon Rideout and Staff, and hired **Arthur N. BecVar** (1911–2003), who had graduated from Notre Dame University in 1933 with a BFA, received an engineering certificate from Purdue University, and an MFA at the Institute of International Education, Charles University, Prague, Czechoslovakia, in 1933–1934. With Rideout gone, Van Doren renamed his Toledo firm Harold Van Doren & Associates. The reason for Rideout's move was not business, nor problems with Van Doren, but the more romantic result of his relationship with Alice Chalifoux (1908–2008), a concert harpist with the Cleveland Symphony Orchestra and head of the harp department at the Cleveland Institute of Music, whom he married in 1937.

The 1930s saw the development of all kinds of plastics, new synthetic materials with integral coloring that could reduce weight as well as replace many metal parts, thus avoiding the costs of metal machining, finishing, and painting. Before the 1930s, the only plastic used in products was Bakelite, a thermosetting phenol formaldehyde resin developed in 1907 by Belgian-born chemist Leo Baekeland in New York. But it was only available in black, dark brown, or dark maroon, not particularly attractive consumer colors.

The new plastics of the 1930s offered a full palette of colors. Typical was the urea formaldehyde called Plaskon used by Van Doren and Rideout in the white Toledo scales, mentioned above. Polystyrene and acrylic, the latter called by trade names Plexiglas or Lucite, had been developed in 1930. Polyethylene was developed in 1933, nylon in 1934 by DuPont, polyester in 1936, and melamine formaldehyde, called Melamine, and Dacron, both in 1938. Plastics transformed the look and feel of consumer products.

Not only that, but plastics revolutionized the manufacturing process. New injection molding machines melted a stream of small pellets of raw plastic into semiliquid form and forced it into complex steel molds. It quickly cooled back into solid form, the finished part was ejected, and the machine would begin the next cycle. Hundreds of parts could be produced within hours.

Industrial designer Belle Kogan, who at the time was designing clocks in plastics for the Warren Telecron Company, wrote articles in 1935 plastics magazines urging manufacturers to use plastics, which she described as

> a material with tremendous possibilities ... which no manufacturer, if he be alert and watchful of his competition, can afford to overlook. Radios, clocks, dishes, jewelry—all being developed today—have an enormous significance.

Kogan also pointedly advised manufacturers to study the "feminine viewpoint," citing the fact that

> thirty million women—all potential customers—constituting practically the entire buying structure of the nation—comprise a force, which cannot, or should not, be disregarded. The tastes of the American woman, her reaction to color and form, are of vital importance to the manufacturer.[6]

In another article, Kogan also advised designers to do their homework:

> It is the designer's job, therefore, to know materials, to know their qualities, their basic structures, their resistance to wear, and their cost—as well as to know markets and trends—in order to intelligently design for the manufacturer.[7]

Her articles were typical of others being written by industrial designers in trade magazines—stressing practical, technical, and cost-reduction results of design rather than simply artistic beauty. Kogan had an office in New York, producing houseware designs in silver, aluminum, ceramics, glass, wood and cloth for Quaker Silver Company and many other clients. In 1936 she would design a modern double serving dish in silver for Reed & Barton.

Regular commercial air transportation was just beginning, inspired by Charles Lindberg's solo flight across the Atlantic in 1927. The *China Clipper,* a Martin M-130 four-engine flying boat built for Pan American Airways, initiated mail service on November 22, 1935, from San Francisco to Manila. Its interiors, intended for long Pacific trips (59 hours, 48 minute flying time, with overnight stops at Honolulu, Midway, Wake Island and Guam), with sofa-type seating that converted into sleeping berths, lavatories, bar, kitchen, and pantry, were designed by Norman Bel Geddes in 1934. The twin-engine Douglas DC-3 airliner initiated commercial domestic flights shortly thereafter, on December 17, 1935, after refinement with development models DC-1 (1933) and DC-2 (1934). The DC-3 carried 21 passengers, and a transcontinental flight took about 15 hours, with only three refueling stops. People were astounded at the potential for future air travel.

Kem Weber was probably inspired to design his famous Airline chair in 1934 by earlier versions of the DC-3. Weber had labored for months to design a chair that could be sold disassembled and readily put together by the consumer, and the Airline chair was the result. Weighing 17 pounds and packed flat in a box, it was to sell for $24.95. The chair design failed to find a manufacturer, and so craftsmen, rather than machines, built 300 for the Walt Disney Studios in California. Weber is also noted for being the main architect of the Walt Disney Studios complex in Burbank, California. Mickey Mouse had been created in 1928.

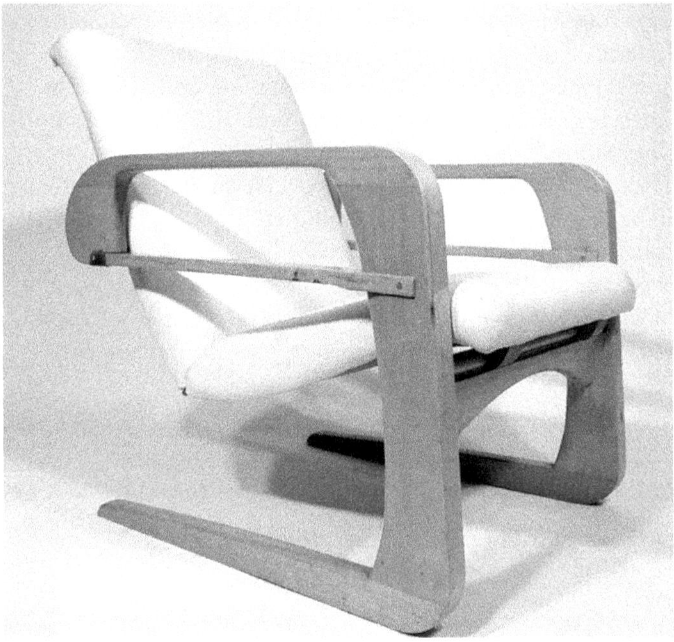

Airline chair, 1934, by Kem Weber.

At the same time that industrial design began to flourish, the Roosevelt administration's New Deal was pouring funds into new relief programs to help the unemployed, to provide social entitlement programs for the elderly, and to allow unions to organize. Industrial design, which was now driving industry, would become a potential political target of the newly empowered, organized American labor movement.

CHAPTER 7

Labor Design and Design Organization

With all the publicity and attention focused on the inevitable march of modern technology and design, as well as on the high unemployment levels, strikes and labor disputes of the time, it was perhaps inevitable that industrial design would become involved with government and the intensive labor activism of the Depression era. In April 1935, the Works Progress Administration (WPA) was created by an act of Congress upon the order of President Franklin Roosevelt as part of the Second New Deal. At the time, there were 20 million persons on relief. In order to employ the unemployed, the act provided $4.9 billion in funds for the construction of roads, bridges, and local public buildings, but it also employed musicians, artists, actors and writers in public works. A number of artist/designers, including some mentioned in this book, took advantage of WPA grants to paint murals in post offices, universities, and other public buildings.

In July, the Wagner National Labor Relations Act validated union authority. By August, the Social Security Act provided retirement benefits, the first federal assistance to the elderly. In September, the Hoover Dam (named after Herbert Hoover, president when it was begun in 1931) was dedicated, creating Lake Mead, the world's largest reservoir at the time, with President Roosevelt presiding. The Roosevelt administration quickly renamed it Boulder Dam, but the confusion between names continued. In 1947, the name would be changed back to Hoover Dam.

On December 2, 1935, in New York, a frigid and blustery day, several hundred men and women lined up outside the 35th Street offices of the WPA for a chance to enroll in the Design Laboratory, a pioneering school sponsored by the Federal Art Project (FAP), one of the WPA's cultural projects. It was the first institution to offer free, comprehensive education in modernist design. Its brochure claimed it had been "created to supply a hitherto unfulfilled and pressing need in America" for people with skills to devise, package, and promote consumer products that were styled to sell, thereby stimulating economic recovery. Clearly, this noble purpose was inspired by the incredible success of industrial designers in doing precisely that over the previous eight years.

By emphasizing "coordination in the study of esthetics, industrial products, machine fabrication, and merchandizing," the school would "train designers, not specialized craftsmen, by correlating thorough instruction in the general principles of design and the fine arts with shop practice." Two hundred twenty people enrolled that day, more that the school's administration anticipated, and there were still hundreds more on a waiting list.

The school was the result of political action by the Depression-era avant-garde, the unprecedented public arts bureaucracy of the New Deal, the militant industrial unionism of the Congress of Industrial Organizations (CIO), which had just been formed within the American Federation of Labor (AFL), the United Auto Workers (UAW), and the radical socialist politics of the Popular Front (a broad coalition of different political groupings, often made up of centrists and leftists).

The collaboration was an example of how modernism was viewed by some as a means to control cultural production by political action and thus prevent its dominance by business interests and capitalism. The idea of democratizing American culture was attractive to many elites of the time, who resented the new lucrative art form of industrial design being controlled by big business and, by the way, a handful of industrial design celebrities getting rich in the process.

The school organizers recruited well-known industrial designers such as Raymond Loewy, Walter Dorwin Teague, Donald Deskey, George Sakier, and Russel Wright to lend legitimacy to an advisory board by the use of their names, and lured furniture designer Gilbert Rohde to serve as director. It is likely that these men were attracted by the idealistic intention of training young designers, as stated in the brochure, and as desperately needed in the field, rather than by the hidden leftist ideology behind it.

This ideology was obvious in major study texts required by the curriculum, including *The Theory of the Leisure Class,* by Thorstein Veblen, published in 1899, which claimed that reform of industrial art had to be through the technology of machines, and highly influential sociologist/historian Lewis Mumford's book, *Technics and Civilization* (1934). Mumford considered streamlining as mere expressive and "outmoded decoration" and insisted that functionalist esthetics (such as the geometric abstract forms at the Bauhaus) was more appropriate for society, a view shared and promulgated by the Museum of Modern Art (MoMA), which venerated Bauhaus architects, and which would enthusiastically take up the cause of functional esthetics. The recent closure of the leftist Bauhaus by the Nazis as "a Marxist manifestation" generated much political sympathy throughout the liberal and progressive art community, which actively sought to bring exiled Bauhaus refugees to the United States to lead the modern movement they had initiated in Europe. Support of the Bauhaus was seen by some as a gesture of patriotic defiance against Nazism and war. Others saw communism as a similar deterrent to the evils of fascism, since Hitler's government was fervently anti-communist, and communists were fighting incessantly against Hitler and fascism.

Despite internal political and philosophical differences of faculty members, the school implemented a provisional three-year design curriculum in August 1936, which promised to "present the ... standards of taste and style ... in the so-called 'International Style'" (the term invented and promoted by MoMA starting in 1932 and based on the Bauhaus).

At the same time, an essay by the Federal Art Project (FAP) historian published in *The Magazine of Art* in October 1936 denigrated the "handful of super-men who are revolutionizing everything from hairpins to locomotives at a fabulous price," an unmistakable reference to the leading industrial designers.

Many students and faculty of the Design Laboratory joined militant organizations of cultural workers, including those where communists, socialists, or Trotskyites (those supporting the political and economic theories of communism advocated by Leon Trotsky and his followers, usually including the principle of worldwide revolution) predominated, and instigated strikes and demonstrations demanding autonomy of the school, calling for a "United Front against War and Fascism," all of which put them at odds with FAP administrators. In December, several students and faculty were arrested for public protests and demonstrations.[1]

Despite the labor unrest and strikes, 1936 was dawning with signs of economic recovery. The main economic indicators had regained the levels of the late 1920s, except for unemployment, still high at 11 percent, though this was much lower than the 25 percent of 1933. Industrial production had recovered significantly, and many businesses were getting back to normal, thus increasing design activity.

In 1936, Montgomery Ward volume rose from 25 percent to 82 percent over Sears.[2] By then, Anne Swainson was supervising 30 designers at Montgomery Ward's Bureau of Design, many of whom went on to become prominent industrial designers. Among the first was Dave Chapman, who had served as head of the Bureau's product design department. Chapman opened his own office in Chicago in 1936, which would become the largest industrial design firm in Chicago. Chapman had married his wife, Eileen, in about 1935. She was the daughter of the chief legal counsel for Montgomery Ward at the time.[3]

Famous celebrity industrial designers were being called upon to serve in positions of civic responsibility. In New York City, mayor Fiorello La Guardia had appointed Norman Bel Geddes to the Municipal Arts Committee of New York in 1935,[4] and Walter Dorwin Teague was appointed as one of the eight directors on the New York World Fair's Board of Design in 1936, as planning began for the 1939 event.

In 1935, Bel Geddes designed and produced *Dead End* in New York. He toured the city in police cars for atmosphere and recorded river and street sounds for four days for use in the play's background.[5] He had hired Robert Gruen (1913–1999) that year as a set designer, and Gruen would open his own glassware design office in New York in 1940. Also in 1935, Egmont Arens established his own design firm in New York, promoting what he called "industrial humaneering," emphasizing simplicity and function.

John M. Little of Minneapolis joined Van Doren & Associates in Toledo in 1936. Little knew Van Doren from Minneapolis, where Van Doren had been assistant director of the Minneapolis Institute of Arts (MIA) until 1930, and Little had opened an industrial design office there in 1931. In 1937 Donald Dailey (1914–1997) would join Van Doren's firm as design director. Dailey was also from Minneapolis and had attended art classes at the MIA. He had also studied marketing and engineering at the University of Toledo (1934–1936) and design at the Toledo Museum School of Design (1936–1937).

Henry Dreyfuss had been retained in 1934 by the Hoover Company to design vacuum cleaners. His annual retainer fee was $25,000 ($435,000 in 2013 dollars), and his first design, the Hoover Model 150 upright vacuum cleaner, debuted in 1936, beginning Dreyfuss's association with Hoover that would last for 20 years. That year, he also

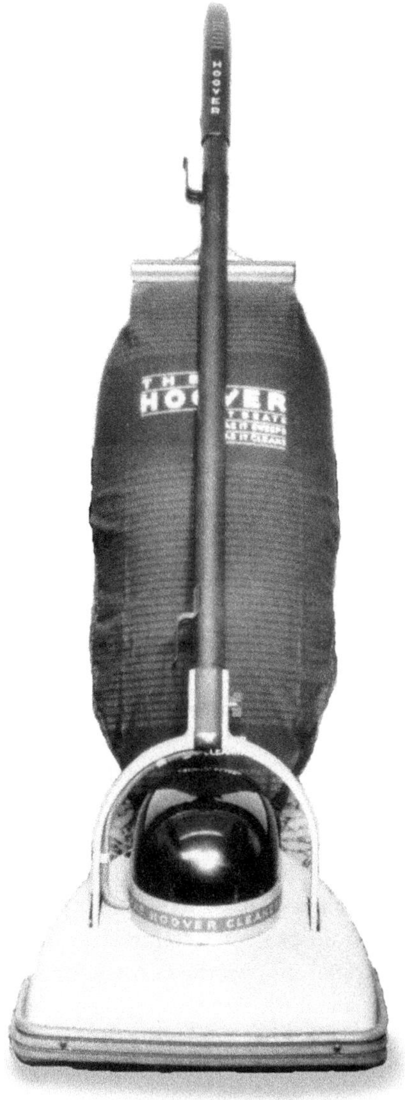

Hoover Model 150, 1936, by Henry Dreyfuss.

designed a typewriter for Royal, accessories for United Airlines DC-3s, and his first streamlined Mercury K-5 Pacific train design for the New York Central Railroad, marketed as the "Train of Tomorrow." The first Mercury in the series operated daily between Cleveland and Detroit starting in 1936.

For the locomotive, Dreyfuss covered the exterior pipes, whistles, and fittings with a smooth, "bathtub" cowl, with cutaway sides to display the driving wheels and rods, which were dramatized at night by concealed spotlights under the cowling on either side. Dreyfuss also designed the interiors on the train, including the coaches, kitchen, dining car, smoking lounge, parlor car, and observation car, all with arrangements designed to avoid the traditional "linear" look of railroad coaches and to have the atmosphere of a private club.

For example, the dining car was in three sections, two with conventional seating, but the middle section was made up of tables for two seated side-by-side, facing the center aisle. In the rounded-end observation car, Dreyfuss lowered the windowsills to maximize the view through four-foot-high windows. Instead of having seating arranged around the walls, facing in, he placed the seating in the center, facing out toward the windows. Later versions of the Mercury included the Chicago-Detroit Mercury and the Chicago Cincinnati Mercury. The first Cleveland-Detroit Mercury would last run in 1959.

Dreyfuss was not the only train designer. In February 1936, Raymond Loewy, with employee Carl Otto (b. ca. 1910–ca. 1986), hired in 1935, designed the Pennsylvania Railroad's K-4s streamlined shroud for steam locomotive No. 3768, hiding most of the mechanical details. Otto was born Carl Louis Otto in Michigan and was trained as an engineer at Michigan State University. He began his career as a draftsman with Duesenberg in Auburn, Indiana, and then went to New York to work for Norman Bel Geddes.

Loewy did not follow the "bathtub" streamlining shrouds of previous trains, but decided to retain the unique character of the traditional, cylindrical boilers of steam locomotives. He basically designed a bullet nose for a cylindrical boiler and placed it upon a streamlined platform covering the driver wheels that suggested the top half of a bullet. He would refine the design into the S-1 in 1939.

Loewy had also acquired the Studebaker account in 1935 and was retained for the design of exteriors of the 1937 models, at the same time as Helen Dryden, former art director of the Dura Company in Ohio, who had been retained at $100,000 per year for Studebaker car interior designs since 1934, worked on the interior. At the time, she was acclaimed as "the highest-paid woman artist." Loewy's exterior design, completed in 1936, featured a narrow, vertical grille with horizontal louvers, the top three of which extended rearward along the side of the hood. Headlights were teardrop shaped. Dryden was publicly credited in Studebaker ads as "the stylist" for the 1937 Studebaker, but she was dropped as a consultant that same year. Loewy would continue to design Studebaker cars until the company went out of business in 1961.

In 1935, Walter Dorwin Teague pioneered in the branding of gasoline service stations when he designed his first for Texaco, which included not just the logotype, but a functional, rain-protected architecture and layout that would become a standard benchmark for all service stations over the next two decades. In 1936, he also designed the Kodak Bantam Special camera in classic Art Deco style for his longtime client.

That same year, Brooks Stevens designed a custom Zephyr "Land Yacht" for William Woods Plankinton, consisting of a streamlined truck cab and 17-foot hitchable trailer with an electric icebox, full kitchen with electric stove, bathroom with shower, living room, hunting and fishing equipment, and sleeping quarters for eight. The entire length of the rig was 45 feet, displaying a most modern and exotic mobile home for its well-to-do owner.

Norman Bel Geddes in 1936 was busy designing exhibits for the 1939 World's Fair, a truck for the Autocar Company, and the first all-metal soda dispenser for the Walter Kidde Corporation, as well as exhibiting his designs at the Museum of Science and Industry in Chicago and designing stage sets for *The Eternal Road* in New York.[6]

In June 1936, Carnegie Tech graduated the first five students ever to receive a college degree in industrial design. When degrees were awarded alphabetically at graduation, the first to receive a degree was Maud (Bowers) Rice (1913–1998), a Philippine national. Donald Dohner was now at Pratt Institute in Brooklyn, starting a three-year program in industrial design, with the first graduates due in 1939. Peter Müller-Munk now headed the Art and Design Department at Carnegie. Alexander Kostellow, of the Carnegie industrial design faculty, and his wife, Rowena Reed, would depart to join Dohner at Pratt in 1938. Kostellow would soon become a prominent and influential faculty member in the Pratt program.[7]

In 1936, the Homer Laughlin China Company introduced a line of mass-produced ceramic tableware for the home called Fiestaware, designed by Frederick Hurten Rhead (1880–1942). The brightly colored, mix-and-match, modern line was highly successful in the market and is still a classic design valued highly by collectors. Rhead had emigrated from the United Kingdom in 1902 and worked for several ceramics factories, including Roseville, where he was director from 1904 to 1908; Jervis Pottery on Long Island; and from 1917 to 1927, he directed research at American Encaustic Tiling Company in Zanesville, Ohio. From 1927 until his death, he would remain as art director for Homer Laughlin.

In July 1936, the Spanish Civil War began, and in November, Franklin Roosevelt was elected by a landslide to his second term as president of the United States. Unemployment was still at 14 percent. Roosevelt would initiate payroll taxes to pay for Social Security, and with the threat of unrest in Europe, he would establish the Neutrality Act intended to avoid any involvement in taking sides in disputes between nations.

In Chicago, Lawrence H. Whiting, director of the American Furniture Mart, invited a group of furniture designers to a luncheon at the Furniture Club to hear speakers Donald Deskey, Gilbert Rohde, Kem Weber, and Leo Jiranek, all mentioned earlier. Possibly influenced by the Wagner Act of 1935, which established federal rights of workers to organize unions, to engage in collective bargaining, and to take part in strikes, attendees at the 1936 meeting at the Furniture Club agreed to form a new organization: the Designers' Institute of the American Furniture Mart (DIAFM).[8]

This appeared to be intended as a union of furniture designers, since a main requirement of membership restricted member design activities to furniture design, a restriction that was not appreciated by many industrial designers, who were designing everything from locomotives to wash machines. As we noted earlier in the formation of the Design Laboratory, this was a time of intense labor unrest. From September 1936 to May 1937, half a million workers were involved in sit-down strikes nationally. A strike against Republic Steel resulted in four deaths and 84 injuries.

By the spring of 1937, the WPA designated the Design Laboratory as one of the many programs to be discontinued, as the school struggled with layoffs, violence, and confusion. A radical, leftist, professional trade union, the Federation of Architects, Engineers, Chemists and Technicians (FAECT) came to the rescue as the new sponsor and to provide direction of the school. In the summer session, the curriculum was revamped along the lines of the Bauhaus Foundations Course and was turned into a four-year BA and BFA program comparable to standard universities. Enrollment increased to 400, and administrators hoped it would become a "labor university."

Other private educational institutions of design were being established or initiated at the same time. The University of Illinois at Champaign-Urbana established a BFA program of industrial design in 1937. Cooper Union in New York offered a four-year certificate in plastic design in 1937, with instructors Paul Feeley and Lila Ulrich, a Bauhaus graduate.

The California Graduate School of Design in Pasadena organized a two-year graduate program in industrial design with advice from Richard F. Bach of the Metropolitan Museum of Art, Royal Bailey Farnum of the Rhode Island School of Design, Peter Müller-Munk of the Carnegie Institute of Technology, and architect George Howe. It was the only graduate program of industrial design in the country. Graduate engineers were taught the principles of product engineering and industrial design.

Architect Hunt Lewis was named initial director of the school, and he invited as faculty members Walter P. Baermann (1903–1972), a 1932 émigré from Munich who had served as educational director of the Springfield (Massachusetts) Art Museum; Karl With (1891–1980), a German art historian and educator who had fled the Nazi regime in 1933; and Kem Weber, who also taught at the Art Center School of Design since 1932, and who had recently designed his famous 1934 Airline chair. Baermann was soon appointed director, and he remained in that position until 1941, when the school was merged into the graduate engineering curriculum of the California Institute of Technology (CalTech) and renamed the California School of Design.[9]

At the Cranbrook Academy of Art near Detroit, Charles Eames (1907–1978) joined the faculty in 1937 and headed the department of experimental design to explore new materials such as plastics and plywood for furniture. Eames was born in St. Louis and studied architecture at Washington University from 1924 to 1926. He worked with an architectural firm in St. Louis from 1925 to 1927, and then as a partner in the firm Gray & Eames from 1930 to 1937. At Cranbrook, Eero Saarinen (1910–1961), the son of Cranbrook's founder, Finnish architect Eliel Saarinen (1873–1950), had graduated from Yale in architecture in 1934 and joined his father in 1935. In 1937, Eero began working with Eames, along with Italian émigré Harry Bertoia (1915–1978).

With the expansion of industrial design education and the intense publicity generated by the new profession, men and women were entering the field in scores. Counting the automotive, railroad, airline, appliance, and furniture industries, the population of industrial designers was probably in the hundreds by this time. The purpose of this book is to focus on those pioneers who were active in the profession before 1935, so while important designers from this point on will be mentioned, their biographies and careers will be minimal, and the text will focus on those earliest dozens of machine designers that are of specific interest in this book.

German refugees from the Bauhaus were flocking to America. Bauhaus master Josef Albers (1888–1976) and his wife, Anni (1899–1994), had arrived in 1933 to become faculty members at Black Mountain College in North Carolina. Walter Gropius, former director of the Bauhaus, who had opened an office in London after his resignation from the Bauhaus in 1928, arrived in the United States in 1937 to become a professor in the graduate program at the Harvard School of Architecture, along with his Bauhaus student, Marcel Breuer (1902–1981).

Làszló Moholy-Nagy, also of the Bauhaus but in exile in London at the time, became, upon the recommendation of Gropius, director of the School of Design at the Association of Arts and Industries in Chicago, which he opened in 1937 and promptly renamed the New Bauhaus: American School of Design. Hin Bredendieck (1904–1995), a former Bauhaus student; Georgy Kepes (1906–2001), a Hungarian designer who had worked for Moholy-Nagy in his Berlin studio; and Alexander Archipenko (1887–1964), a Russian artist and sculptor who had worked in Berlin and immigrated to the United States in 1929, all joined Moholy-Nagy at the school. The school attracted only 20 students and would close in 1938 under financial difficulties, but it would be reopened in 1939 as the School of Design, under the patronage of Walter Paepcke, president of the Container Corporation of America.

Many industrial designers were enjoying a busy year in 1937. Detroit industrial design consultant George Walker began designing Nash automobiles, a client he would work for until 1945. Brooks Stevens designed a Model B tractor for Allis-Chalmers, which was lighter and cheaper than its key competitor, Ford's Farmall, and a streamlined sales van for the Western Printing and Lithograph Company based on his 1936 Zephyr land yacht design.

Henry Dreyfuss designed a whitewall Double Eagle automotive tire for Goodyear. Bell Labs introduced the "Phone of the Future" that Dreyfuss had been working on since 1930—a black phenolic plastic 300 tabletop phone that combined the transmitter and receiver into a single combined handset resting in a horizontal cradle. It replaced the "candlestick" model that had been introduced in 1916, and which, in 1919, had a rotary dial added. The classic 300 would be produced until 1950.

Otto Kuhler, working with the Baltimore & Ohio (B&O) Railroad as art director of the *B&O* magazine, designed its logo and its 1937 streamlined, bullet-nosed, steam locomotives, known as the "Kuhler type," for B&O's Royal Blue train on the Washington to New York run. The locomotive, a 4–6–2 "Pacific" Class, was painted light gray and royal blue. *Time* magazine reported it as "swashbuckling" and as "symbolizing the new era in railroading." The design was virtually copied by Raymond Loewy for his 1939 S-1 locomotive. Unfortunately, the Royal Blue line only ran as far as Jersey City, and passengers had to be transported to Manhattan by White Motor Company buses, which Kuhler also designed.

Despite all the streamlined shrouds designed for steam locomotives, the days of steam engines were numbered. Diesel power would replace it. In May 1937, as opening ceremonies for San Francisco's Golden Gate Bridge transpired, the flagship of the Acheson, Topeka, & Santa Fe Railway, the *Super Chief,* began regular service between Chicago and Los Angeles. Departing Los Angeles at 7:30 a.m., it reached Chicago in 36 hours and 49 minutes, setting a record by averaging 60 mph, often reaching 100 mph. Called the "Train of the Stars" because of the many celebrities from Hollywood it carried, it had been under development since 1935, when General Motors' Electro-Motive Corporation (EMC) delivered two diesel-electric units, not much more than boxes on wheels. Just before regular service was initiated, the heavyweight standard cars were replaced with lightweight stainless-steel cars built by the Budd Company in Philadelphia, including a dome-topped observation deck in one.

The streamlined engine was distinguished by its raised, forward-facing operating cab, with split windows similar to an automobile or airliner. This was not unusual, considering that the General Motors Art and Colour Section department under Harley Earl executed the styling. A dramatic "warbonnet" color scheme of a red nose and cab with a bright yellow cartouche with Santa Fe logo was styled by Leland Knickerbocker from the GM Art and Colour Section. Philadelphia architects Paul Cret and his partner, John Harbeson, designed the Budd Company car interiors. The *Super Chief* would operate until Amtrak would take over the nation's passenger service in May 1971.

Raymond Loewy in 1937 considered himself an expert in locomotive design, and at the height of his involvement with the Pennsylvania Railroad, he published a 7.5 × 10 inch book with 125 illustrations, *The New Vision Locomotive,* showing the aesthetics of locomotive evolution, and including photos of his K-4 design as it was evolving into the S-1. That year, Loewy also designed a less dramatic cream separator for International, and a Purma Special camera for Purma Cameras, Ltd., an English company.

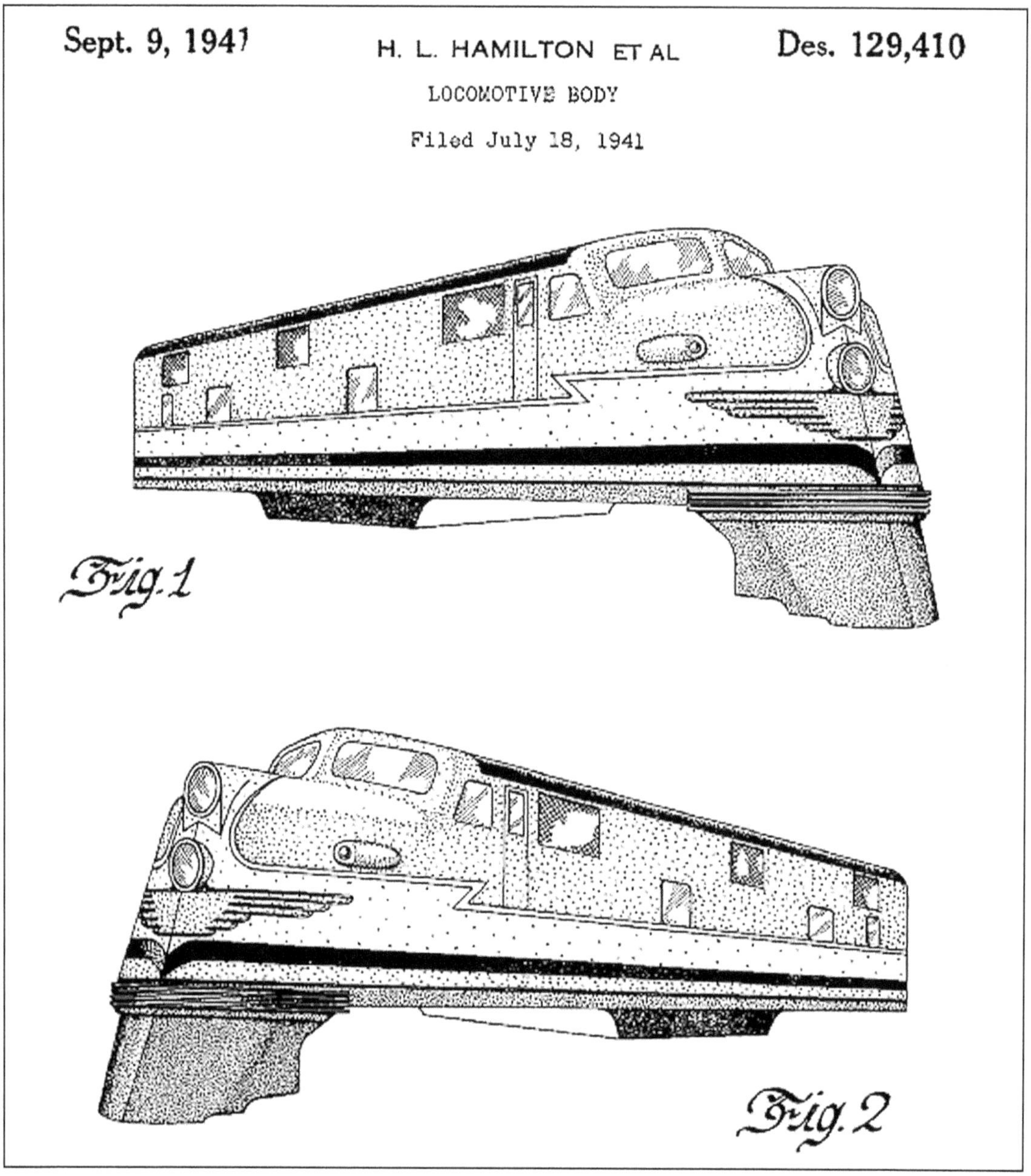

Santa Fe Super Chief, 1937 (Patent No. D129,410, U.S. Patent Office).

Peter Müller-Munk, who had headed the Carnegie Tech degree program in industrial design since 1935, now also began to practice professionally in the open market by designing an electric, two-speed kitchen Blendor (with an o) under the Waring name in 1937. The liquid-blending device, invented by Frederick J. Osius, was named after its financial backer, prominent bandleader Fred Waring (1900–1984), and was introduced at the National Restaurant Show in 1937 for bartender use, but it would soon become a popular home appliance. Müller-Munk's design, called the Miracle Mixer, was classic Art Deco, the base of the blender a cylindrical waterfall of chrome. By 1938, Müller-Munk would organize his independent design firm, Peter

Müller-Munk Associates (PMMA), in Pittsburgh, hiring many of his own industrial design students and graduates from the Carnegie Tech program.

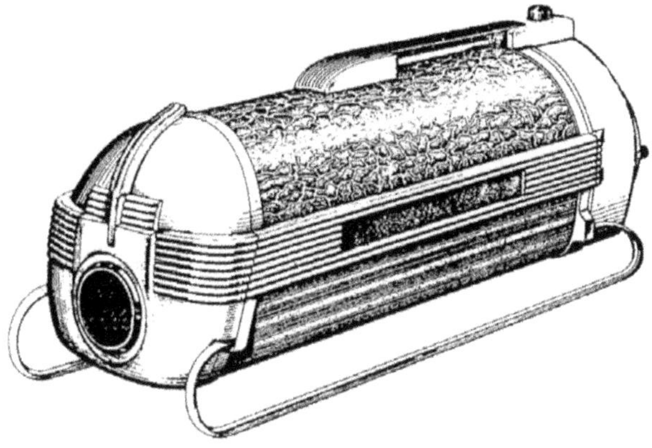

Electrolux Model 30, 1937, by Lurelle Guild (Patent No. D106,662, U.S. Patent Office).

Another classic appliance design was introduced in 1937, the KitchenAid Model K stand mixer, designed by Egmont Arens for the Hobart Manufacturing Company. Arens had designed previous mixers and coffee grinders for Hobart, but this design of the Model K, except for minor cosmetic changes, would remain relatively unchanged to the present day, a success story exceeding 75 years. Another classic designed that year was the Model XXX (30) tank vacuum cleaner designed in a modernistic style by Lurelle Guild for Electrolux. Millions would be sold between 1937 and 1954, the longest run of any vacuum cleaner design of the era, at a retail price of $69.95.

The long-term success of many product designs of the 1930s seems to conflict with the concept described in 1932 by Elmo Calkins of "planned obsolescence," which predicted that annual design changes spur consumer spending. But the longevity of certain product designs can be explained by several reasons. First, while many consumers are eager to obtain newer versions of a product, there are also some who are not and who will keep and use a favored product until it expires mechanically.

Second, many new, industrial-designed products, for the first time in the 1930s, acquired a shape, style, and configuration that seemed "right" for its assigned task, establishing what is called a "typeform," that is, a form that seemed to many the ideal for a particular type, or category, of product. Other manufacturing competitors, noting the popular success of a particular brand, would imitate the design, therefore increasing public familiarity and acceptance of the standard typeform until some substantially different, more popular, configuration replaced it as the new typeform for that particular product. The KitchenAid mixer survives because it is regarded as the ideal typeform for this particular category of products.

The power of historically successful designs is illustrated by the automotive industry, which has had dozens of different, successive typeforms over the years, but would remain nostalgic about car designs of the 1930s, which produced the first designs that were truly modern. Detroit in the late 1990s and early 2000s would develop a number of so-called retro designs, intending to recall that period, and which were intended to replicate the original typeforms. They included the (new) Volkswagen Beetle (1994), the Plymouth Prowler (1994), Chrysler's PT (Personal Transportation) Cruiser (1999), Buick's 2001 Blackhawk, Ford's 2002 Forty-Nine, and Chevrolet's HHR (Heritage High Roof) in 2006, all in 1930s styles.

But in the late 1930s, the problem was too many cars, and not enough roads. Since half of all households now owned an automobile, the U.S. road system was being challenged by an increasing number of cars on narrow, crowded roads. It was an obvious problem addressed by Norman Bel Geddes, on assignment for the J. Walter Thompson advertising agency in 1937. Based on his experience with his scale-model racetrack of the 1920s, and his war game of the early 1930s, he designed a model City of Tomorrow with expressways featured in a national advertising campaign for the Shell Oil Company. The model would be expanded to become the Futurama exhibit for General Motors at the 1939 New York World's Fair, which in turn would inspire the future federal interstate highway system. Jumping from highway systems to cocktail parties, Bel Geddes also designed a Skyscraper cocktail shaker and a Manhattan serving tray in 1937.

The automobile industry was transformed during the Depression. Car production fell from 5.6 million in 1929 to 1.4 million in 1932. Dozens of auto manufacturing brands in 1929 were reduced to 10 in 1935, which included the Big Three of Ford, Chrysler, and General Motors. GM, with five new models since 1930, led the pack and in 1936 was unionized. In 1937, GM's Art and Colour Section would change its name to the Styling Division. Harley Earl had a staff of 200 designers, cranking out annual model changes on the fleet of GM cars. Earl supervised five studios, one for each division (Cadillac, Buick, Oldsmobile, Pontiac, and Chevrolet), as well as 12 special studios.

Yet he wanted more designers, and so GM and its Fisher Body Division initiated a talent search by sponsoring and promoting a national, 1:12 scale model (one inch equals one foot) "Dream Car" design competition that awarded college scholarships to the best original student designs for a four-door sedan (later various different body styles were added). It was called the Fisher Body Craftsman's Guild, which would continue for 30 more years, encouraging a new generation of automotive designers, many of whom would end up in Detroit, and not a few in Earl's studios.

Actually, the contest was a major addition to an earlier Fisher Body Craftsman's Guild contest that was initiated in 1930, not to innovate designs, but to accurately replicate Fisher plans for a miniature scale model (1:15 to 1:18, where one inch equaled 1.25 to 1.6 feet) of a 19th-century horse-drawn Napoleonic coach, the trademark adopted by Fisher in 1923, and which appeared on all production Fisher bodies made for GM. There were two competitive categories: the Junior Division (ages 12 through 15) and the Senior Division (ages 16 through 19). Winners won prize money for a scholarship to a college of their choice: $5,000 for each of the top two winners in each category. That was more than double what a man typically earned in a year! The Napoleonic coach contest would continue along with the new model car competition until 1940, so applicants could choose either one to enter, but the Napoleonic contest ended at that time. Students would no longer be interested in historical craftsmanship, but in futuristic cars.

In 1935, **Amos Northup**, automotive designer for the Murray Corporation of America, the independent Detroit auto body manufacturer, had patented his famous "shark-nose" hood design, which generally resembled its namesake. The design would be copied by most manufacturers in the late 1930s, including the 1938 Graham, the 1937 Willys "windswept" design, the 1936 Lincoln Zephyr, and the 1940 Studebaker, Chrysler, Plymouth, and Ford. But in 1937, Amos Northup died at age 48 as a result of slipping on ice and fatally cracking his skull. His designs were a powerful influence as cars in the 1930s were transformed from boxes to the sleek forms that are familiar to us today.

Most GM production designs during the 1930s were not futuristic, dramatic or innovative, but followed the relatively conservative annual advance of styling. This changed in 1938, when Earl developed a more radical prototype Buick called only the Y-Job. Like the similarly sounding name iPhone by Apple in 2007 that predicted the future appearance of all handheld computer devices, the Y-Job predicted the shape and style of most postwar cars.

Why the name "Y-Job"? Most experimental cars were tagged with an "X," sometimes with a name or number. Just to be different, Earl decided to use the letter "Y," the next letter in the alphabet. The Y-Job was designed by Oldsmobile studio head George Snyder under Earl's direct supervision, with highly innovative styling features of hidden headlights, suitcase fenders, low horizontal grille, flush push-button door handles, electric windows, a power-operated electric convertible top, and a rounded fuselage body without traditional running boards.

Most unusual was the Y-Job's low, long silhouette with a 58-inch (less than five feet) height at the top of the split, angled windshield, A dozen closely-spaced, parallel and horizontal chrome strips (speed lines) adorned the lower half of both front and rear fenders, the chrome bumpers wrapped around the body corners, and a chrome "bombsight" hood ornament, which would become a Buick trademark, was the icing on the cake. The Y-Job was never shown to the public until after the war, but Earl used it as his personal transportation and it attracted a lot of public attention that way.

Some of its features would appear in the 1938 Cadillac 60 Special, designed by Bill Mitchell (1912–1988), who had been hired by Earl in 1935. But after the war, the 1950 Buick would look almost like a clone of the 1938 Y-Job. In 1938, Earl became a vice president of GM, and he would rule GM styling for the next 20 years. In 2001, Buick would recreate the Y-Job with modern advancements called the Buick Blackhawk, drawing extensively from the Y-Job and the 1939 Buick. It was entirely handmade, cost $475,000, and would be featured in the 2003 film *Bad Boys II.*

On June 15, 1938, Dreyfuss saw the introduction of his design of locomotive shrouds for the New York Central Railroad's 20th Century Limited, which was a refinement of his Mercury K-5 design of the previous year. The shrouds hid the assembly of boiler valves, pipes, and operating parts of ten new J-3, 4–6–4 Hudson-type steam locomotives. The shrouds, similar to Loewy's K-4 for the Pennsylvania Railroad, featured bullet-nosed boiler shapes, but with a center fin that resembled ancient warrior helmets. Dreyfuss also designed the cars using blue, gray, and silver, and like the K-5, they were designed with asymmetric furnishings to avoid the tunnel look of most train interiors. Probably the most famous American train, the 20th Century Limited was an express train that operated from 1902 to 1967 between Grand Central Station in New York and LaSalle Street Station in Chicago. The new train by Dreyfuss made the trip in 16 hours.

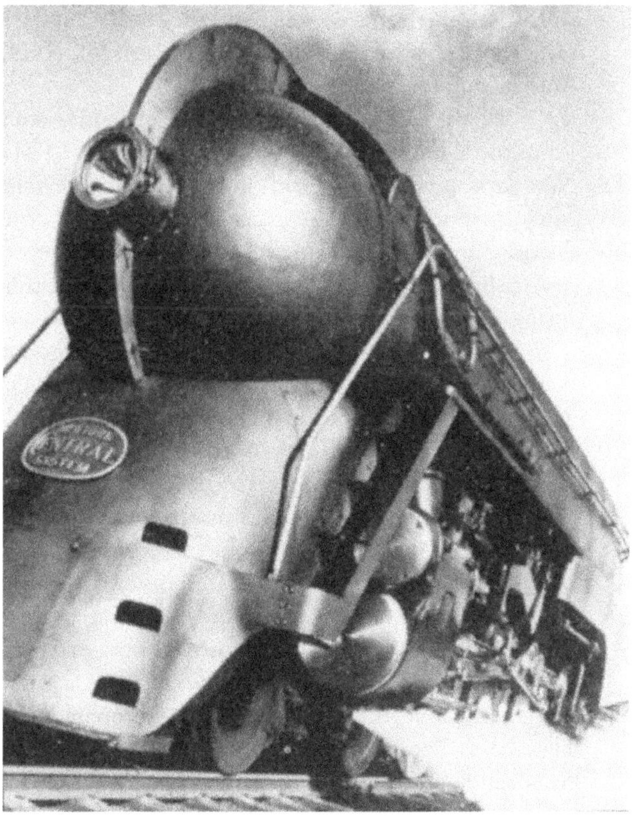

Pennsylvania Railroad's 20th Century Limited, by Henry Dreyfuss.

That same year, Dreyfuss also designed Model A and Model B tractors for John Deere, a new client he had acquired in 1937. Raymond Loewy was having a banner year in 1938, the year he became an American citizen. He opened a branch office in South Bend, Indiana, to service his major client, Studebaker. To head the group, he appointed **Virgil Exner (1909–1973)**, formerly with Harley Earl at General Motors, who had joined Loewy in 1937. Exner, born in Ann Arbor, Michigan, studied art at Notre Dame but had to drop out in 1928 due to lack of funds. He took a job as an advertising artist at an agency handling the Studebaker account. He was hired by Earl in 1933, where he became styling chief for Pontiac and designed Pontiac's famous "Silver Streak" hood ornament.

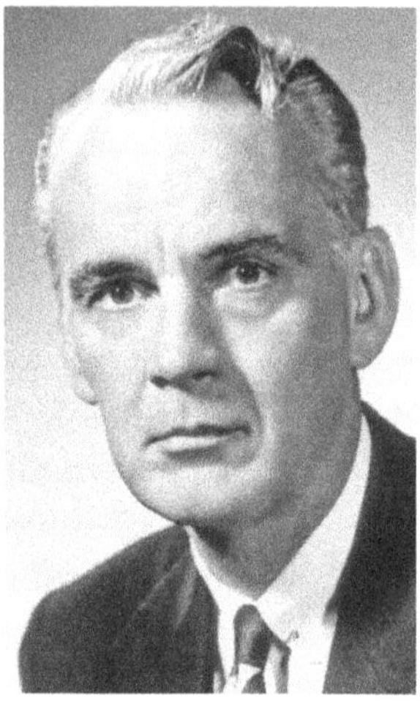

Virgil Exner, ca. 1953.

Loewy also hired Clare Hodgman, who had started his career in the early 1930s with General Motors in the Oldsmobile studio, and who had worked for Loewy on the Sears Coldspot refrigerator project in 1932–1935. The first task of the group was to begin work on the 1939 Studebaker Champion, a low-cost model. They also designed a new Studebaker "Big S" logo and emblem. Other Loewy designs that year included a Metro series of light- and medium-duty trucks and vans for the International Harvester Company.

Loewy had established a personal office in London in 1935, but now in 1938, he staffed it to function on a continuous basis, with Carl Otto (d. 1986) as head, along with Claire E. Hodgman. The new expanded office designed a logo for the British Petroleum (BP) Company, while the New York office designed the main hall for the ocean liner *Panama*. Hodgman worked on the Hillman Minx and the Rootes Sunbeam Talbot, and also designed conceptual racecars for the British *Autocar* magazine of March 1938. One of Otto's accounts was Standard Auto (Triumph).

In 1938, Brooks Stevens designed the first commercially successful steam iron for the Edmilton Company in Milwaukee, the Steam-O-Matic, which was introduced at Macy's in New York and Marshall Field's in Chicago. Viktor Schreckengost, head of the Cleveland School (now Institute) of Art, began creating stage and costume designs for Cleveland theater productions and continued to produce award-winning ceramic sculptures for Cleveland Museum of Art shows. Kem Weber was designing a line of tubular steel chairs for the Lloyd Manufacturing Company in Menominee, Michigan.

In Chicago, Ann Swainson, head of Montgomery Ward's Bureau of Design, designed a plastic radio cabinet, and her staff designed special china called Cascade, which was exhibited at the large-scale exhibition on modern design at the Metropolitan Museum of Art, and which became one of Montgomery Ward's most popular patterns. Swainson had traveled in 1937 to her native Sweden, visiting factories, department stores, and art schools, and was impressed with the Swedish government's part-time training of industrial designers in schools and factories.

Elsewhere in the United States, other corporations were hiring and engaging industrial designers as part of their merchandizing strategy, seeking increased sales and modern style.

In 1938, a major corporation was now expanding its industrial design activity into all visual aspects of corporate identity. Libbey-Owens-Ford (LOF) was formed in 1931 by the merger of Libbey-Owens Sheet Glass Company and another Toledo company, Edward Ford Plate Glass Company, which pioneered in automotive safety glass. In 1937, H. Creston Doner had been invited by LOF to leave his post with Goodrich in Akron, Ohio, to come to Toledo to direct the department of design there starting in 1938, a position he would hold until 1967. He supervised the architectural and interior design operations, as well as promotional campaigns for color and lighting effects; the design of LOF warehouses, showrooms, offices, and retail stores; and the design of a new corporate image for truck fleets, building signs, business forms, and advertising materials.

Meanwhile, in 1935, Libbey Glass Manufacturing Company in Toledo had been acquired by Owens-Illinois Glass Company and became a subsidiary. In 1938, Edwin W. Fuerst was appointed to head the design department at Libbey Glass, while his brother, Dwight Fuerst, headed the design area for bottling and packaging at Owens-Illinois. Edwin designed Greenbrier pattern stemware for Libbey.

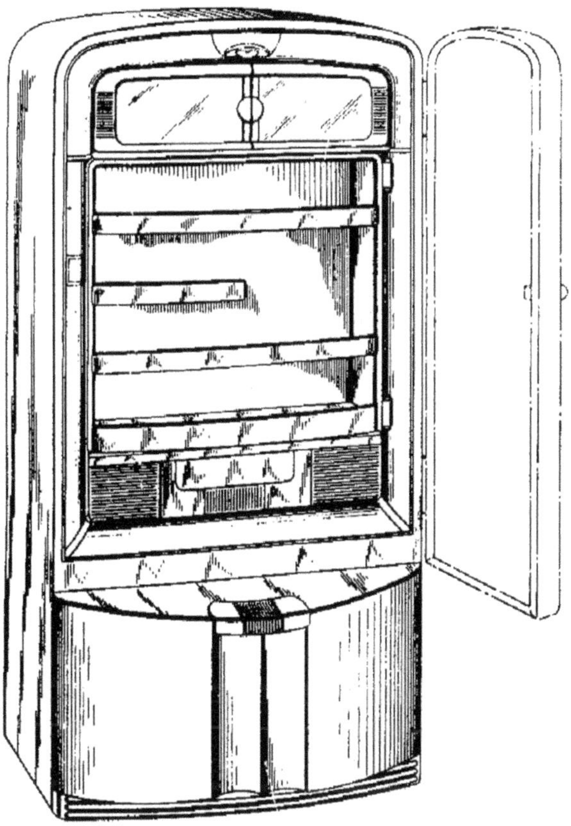

Philco refrigerator, 1938, by Harold Van Doren (Patent No. D131,103 U.S. Patent Office).

Elsewhere in Toledo, Harold Van Doren & Associates were also quite busy. Van Doren himself had worked as a color consultant to Libbey-Owens-Ford's Vitrolite division in 1937. But in 1938, Van Doren's office was now designing refrigerators for Philco, a Philadelphia company, and personal weighing scales and the Guardian Duplex Scale for the Toledo Scale Company.

Another industrial designer, Walter Dorwin Teague, also had clients in Toledo in 1938. He had been called in as a consultant to the Owens-Illinois Glass Company, which had acquired the Libbey Glass Company. Teague was to work with two divisions: bottling and packaging, and structural. He designed an Insulis glass construction block in a Circon pattern for the Owens-Illinois Glass Company's structural division, which then asked him to design the division's offices in the Chrysler Building in New York.

Additional design organizations began to be formed. In Chicago in July 1938, the Designer's Institute of the American Furniture Mart (DIAFM), founded in 1936 under the auspices of Lawrence H. Whiting, director of the American Furniture Mart, met for the first time since its founding, "at the instigation of Whiting, Alfred Auerbach, and Richard F. Bach of the Metropolitan Museum of Art."[10] The principal speaker this time was John Vassos, then head of design at the Radio Corporation of America (RCA) in New York.

In his address, Vassos urged that the organization divorce itself from the patronage and support of the Furniture Mart, because it restricted member work to furniture design. In fact, most members were furniture designers, and Vassos himself designed radios, which were often enclosed in wooden furniture. Vassos suggested the organization extend its practices beyond that of furniture design into consumer products, packaging, automotive design or any number of additional design opportunities then available.

The group, comprised of 45 cofounding members, agreed with him. Among them were certainly Alfons Bach, Leo Jiranek, Al Lambach, Ernie Swarts, Belle Kogan, and probably, although not confirmed, Lionel C. Algoren (1896–1984), Kem Weber, Donald Deskey, and Gilbert Rohde. They changed the name of the organization on the spot to the American Designers' Institute (ADI) to indicate that it was a national, not just a Chicago organization. Even though Vassos was not currently a member of the group before his presentation, the group elected him the first president of ADI, which would become incorporated in the state of Illinois in 1940. Belle Kogan was elected secretary and treasurer, and she set up meetings, often with speakers, keeping the minutes and other records. Her only help came from her own office staff. ADI also established a chapter in New York.[11]

ADI logo, 1938.

That same year, 1938, another industrial design organization was formed, also in Chicago, which would separate itself to escape from what it saw as furniture designer domination.

CHAPTER 8

War Drums and the New York Fair

According to sculptor/designer Alfonso Iannelli (mentioned earlier), a group of Chicago designers, including himself, Dave Chapman, "Chick" Waltman, H. Creston Doner, and Jean Reinecke, formed the Chicago Society of Industrial Designers (CSID), which limited its members only to the design of products for mass production, a much more attractive and permissive option to many industrial designers, who were in the process of trying to define their scope and marketability to industry in the broadest manner possible. Prior to and during this period, noted sculptor Iannelli had established Iannelli Studios in Chicago, and in addition to his sculpture projects, he designed consumer products, such as mixers, hair dryers, knife sharpeners, and hair clippers, for the John Oster Manufacturing Company, Sunbeam Electric Company, Wahl Eversharp Company, Mueller Products, and Fairbanks Morse Stoker.[1]

During 1938, the international situation was growing worse, almost by the day. In March, Germany annexed Austria to Hitler's Third Reich by means of the Anschluss (Connection), with little international resistance. In May, Hitler awarded "cultural prizes" to Fritz Todt for the Autobahn (Auto Way), the first "super highway," of which 3,000 kilometers had already been completed; to Ernest Heinkel for his He 111 bomber design; to Wilhelm Messerschmitt for his Me 109 fighter plane design; and to Dr. Ferdinand Porsche for his design of the new Type 38 Kraft durch Freude (Strength through Joy) or KdF Volkswagen (People's Car). Porsche had initiated the design in 1932 as the Kleinauto (Small Car), and Hitler subsidized its development when he came to power in 1933. Thirty sample cars were built in 1937 for promotional

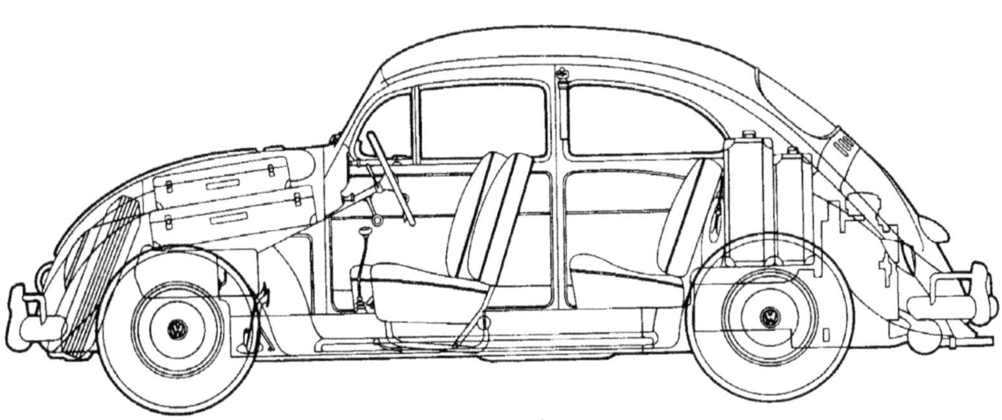

Volkswagen Type 38, 1938.

purposes, orders were taken, and Hitler introduced a savings scheme where the public could collect stamps that would eventually pay for the car.

So on May 26, 1938, witnessed by 70,000 people and 150 reporters, Hitler laid the cornerstone of the Volkswagen factory in a small village in Lower Saxony, named the car as described as above, and renamed the village KdF Stadt (KdF City). The first production was to start in September 1939, the same month that World War II would officially start. Then on July 29, 1938, Japan invaded Russia, and on September 29, neutral European nations agreed to German occupation of part of Czechoslovakia, called the Munich Agreement, which British Foreign Secretary Neville Chamberlain called "Peace in Our Time." Only a few Volkswagens were produced before World War II, and those were exclusively for Nazi dignitaries.

These dramatic events in Germany demonstrated to many that Hitler threatened world peace and that war was inevitable. To others, however, they demonstrated the incredible technological, organizational, and production capabilities of Germany. This initiated the reputation, still acclaimed today, of "German engineering." Perhaps Germans *were* "the Master Race," as Hitler claimed, and would lead the world out of the Depression with public works. Most people just wanted America to stay out of yet another European war, because the memories of World War I were still vivid. The U.S. economy had suffered a brief recession in early 1938, but with war already building in Europe, the economy began to recover, and by autumn, it was booming from orders from European countries for arms and war equipment.

Into this politically charged international atmosphere stepped the Museum of Modern Art (MoMA) with a highly publicized retrospective exhibition, Bauhaus 1919–1928, which ran from December 7, 1938, to January 30, 1939. As we know, numerous faculty and students of the Bauhaus had already immigrated to the United States after Hitler closed the school in 1933. The latest to arrive, in 1938, was Ludwig Mies van der Rohe, the last director of the Bauhaus, to teach design and to design new buildings in the new, MoMA-approved International Style for the Armour (later, Illinois) Institute of Technology in Chicago.

Herbert Bayer (1900–1985), a former Bauhaus print shop master who had resigned with Gropius and Moholy-Nagy in 1928, created Nazi publicity campaigns in Germany until he came to the United States in 1938 to work with Walter and Ise Gropius in designing the MoMA exhibition. He had managed to spirit many Bauhaus artworks out of Germany. Bayer would stay here after the exhibition as a consultant to major U.S. corporations, including the Great Ideas of Western Man advertising series for the Container Corporation of America starting in 1946.

The Bauhaus exhibit featured the sculptures, housewares, furniture, and architectural renderings of such Bauhaus luminaries as Gropius, Mies, and Marcel Breuer. The show and catalog, edited by Bayer with Ise and Walter Gropius, downplayed the school's Expressionist beginnings and excluded the political turmoil periods under Hannes Meyer and Mies. Mies reportedly was asked to take part but declined. The names of Bauhaus artists still living in Germany were suppressed for fear of reprisals against them, and Alfred Barr, MoMA director, stopped well short of condemning Hitler in his preface to the catalog; equally concerned that the show would be viewed as anti–German and that the Bauhaus would be seen as a "Jewish-Communist cabal," he proposed to include a statement specifying the small number of faculty who were Jewish. Gropius apparently dissuaded him from doing this, but had to be prevailed on in return to omit an assertion that the school under the directorships of himself and Mies "had always been non-political in character."

MoMA desired to demonstrate that the Bauhaus had already been reestablished in the United States. The "New Bauhaus: American School of Design," established in Chicago in 1937 by Bauhaus master Làszló Moholy-Nagy, would have been a perfect example to demonstrate this, and perhaps had even been planned by MoMA the year before. But the school had closed in the spring of 1938 and would not reopen until the fall of 1939, although this was not known at the time. It was no longer a good example for MoMA.

Students from the Design Laboratory School, based on Bauhaus principles, now urged MoMA to exhibit their work. By the summer of 1938, the left-leaning Federation of Architects, Engineers, Chemists, and Technicians (FAECT) had terminated its sponsorship of the school due to Depression cutbacks, and Gilbert Rohde had resigned as director. By changing its name to the Laboratory School of Industrial Design, it obtained a charter from the New York Board of Regents that permitted it finally to confer degrees, but most of its students and faculty continued their participation in Popular Front activities.

The school had struggled along on revenue from tuitions and by December was rewarded by its token inclusion of student work, tucked away in a corner of MoMA's Bauhaus exhibition, as MoMA's example of how the innovations of the Bauhaus were "gradually being introduced in the U.S."[2] MoMA's Bauhaus exhibit elicited many negative reviews from critics, who found it manipulative, chaotic and gadgety. Lewis Mumford's positive column in the *New Yorker* was a rare exception.[3]

The exhibition fit perfectly with MoMA's political agenda, initiated six years earlier at its International Exhibition of 1932: establishing the primacy of European architecture in the modern movement. But it also provided additional ammunition for MoMA's campaign to establish the Bauhaus as the original and "proper" source of the new profession of industrial design. If this was established, American industrial design could be described as a poor or misguided imitation of European innovation and could be "corrected" by MoMA to achieve maturity. This political objective stemmed from MoMA's basic assumptions (some would say biases) that guided its policies.

First, like virtually all museums in the United States for decades before, MoMA at that time believed that Europe was, and should be, the center and global leader of all artistic endeavors, particularly in modern art, architecture, and design. MoMA had touted the Bauhaus exhibition as modern design based on the "best of Europe," implying that America should imitate Europe. In fact, in the previous March, MoMA had exhibited the architecture and furniture designs of Finnish architect Alvar Aalto (1898–1976) to reinforce European superiority in these fields.

As a result of this Eurocentric template, European architects were regarded by MoMA as more significant than American architect Frank Lloyd Wright, even though Wright had inspired many of these same European architects nearly 30 years before and recently had designed a simple, modern Usonian ranch house community that would become a model for most U.S. postwar suburban developments. With similar logic, European Bauhaus designers were regarded by MoMA as more significant than American industrial designers, despite the fact that Americans were the first to actually design mass-produced products for mass consumption in a highly industrialized environment.

Second, MoMA, at that time, believed that architects had been, and should be, the leaders not only in architecture, but in all artifacts within a home, including furniture and accessories. Historically, this had been true for many famous architects, both in Europe and in the United States. This was certainly true at the Bauhaus as well, and since MoMA was under the control of architects, there was an assumption that because of this history, architects should rightfully be leading the new profession of industrial design. Many architects were particularly upset not only because industrial design in the United States over the previous ten years had been led, not by architects, but by advertising artists, stage designers, display designers, commercial artists, writers, and art critics, to name a few, but even worse, the leaders, such as Loewy,

Dreyfuss, and Teague, had become national celebrities and commanded incredible fees far higher than architects could hope for.

Lastly, at that time, MoMA, like many art museums, defined fine art as pure, aesthetic, individual expression, based on philosophy, theory, or rationality, and without any intent of commercial gain, mass production, or popular appeal. In fact, it was generally believed that art was good *only* if it defied popular tastes, traditions, and commercial interests. The Bauhaus fit this idealistic template combining artistic, counterculture, avant-garde, academic experimentation, historical context, and intellectual theory.

American designers did not fit this template. Their crassly commercial designs used popular "gimmicks" such as streamlining or skyscrapers, were acclaimed by the public through mass purchasing, and made lots of money for rich industrialists. What was there for a sophisticated, elite art museum to like? MoMA's curator of architecture and industrial art, John McAndrew, praised the Bauhaus industrial designs as 'timeless,' and "free of both modernistic and streamlined aberrations," but lambasted the designs of designers like Loewy, Dreyfuss or Bel Geddes as "not properly modern," for "imitating superficially the forms of modern art," and for "restyling almost everything so that what we owned might look old fashioned as soon as possible" (the accusation of "artificial obsolescence" surfacing).

These assumptions dominated MoMA policies for years, and put them at loggerheads with many practicing industrial designers and leaders of the profession, who believed that it was American industrial design practitioners of the Machine Age in America who defined the practical implementation, process definition, education, practice, and commercial success of industrial design in mass production and marketing. Industrial design in America was much more than aesthetic purity or 'fine art,' as MoMA narrowly defined Bauhaus contributions. It was the complex and iterative process of design: the advance planning of a product to include a reasonable balance of functionality, engineering and manufacturing feasibility, user emotional wants and ergonomic needs, manufacturing costs, appropriate retail prices, suitable materials, and *also* an artistic and attractive visual presentation to persuade consumers to buy. Fine artists did not do this.

MoMA's claim that the Bauhaus "invented" modern industrial design was a specious one. From Christopher Dresser (1834–1904) in the 1870s to the Bauhaus in the 1920s, there were dozens of designers and Arts and Crafts workshops that produced individual designs that could be called 'modern' in the sense of modern simplicity and style, but they were all produced by hand or in limited quantities. The Bauhaus itself, as mentioned earlier, was patterned after medieval craft guild workshops. Its so-called "invention" was based on an academic theory of Purism described by Reyner Banham, a respected international authority on architecture and design:

> [Corbusier stated in Esprit Nouveau in the early 1920s that] "Purism is evidence of the Law of Mechanical Selection" ... [which was] clearly intended to share the status of Darwin's Law of Natural Selection ... [This law of Purism] ... "establishes that objects tend toward a type that is determined by ... the ideal of maximum utility and the demands of economical production, which conforms to the laws of nature." Therefore they concluded that "objects of maximum utility and lowest price have simple geometric shapes" ... such as cylinders, spheres and cones. To most architects this would appear watertight, but to most production engineers it would appear too abstract to be useful, and demonstrably false in its outcome.[4]

In other words, Purism was seen as an absolute and irrefutable law of nature. Bauhaus designs were of bottles, jugs, pots and pans, glasses and pipes, handmade forms which approximate these simple geometric shapes. In fact, students were criticized when they strayed from such forms. There were no designs of complex mechanical devices having multiple parts or functions. The American machine designers of the 1930s were the first to create truly industrial designs, in collaboration with engineers, for industrial mass production and consumption, including electromechanical products at low prices for middle-class consumers and mechanical monsters such as locomotives. No other group of designers had ever accomplished this before. The terms "industrial" and "machine" were appropriate and unique adjectives for this group.

What did they think of Bauhaus design? Harold Van Doren, in his book *Industrial Design* in 1940, would write:

> The Bauhaus approach is a philosophy of life, as well as a method of design. It lacks, however, the realistic qualities that we Americans, rightly or wrongly, demand ... but it will be difficult, I believe, to acclimatize the esoteric ideas of the Bauhaus in the factual atmosphere of American industry.[5]

The American styles criticized by MoMA in the 1930s were popularly referred to as "streamlining," "Art Deco," "futuristic," "Art Moderne," "modernistic," "machine design," or "contemporary." Most used modern imagery that appealed to popular taste and public expectations of the times. What they were called was much less important than what they accomplished. They were new, exciting, and different from traditional designs, and together represented a uniquely American style that changed the look of the country's industrial products.

Because of this, for the first time, world industrial design leadership resided in the United States, not in Europe. This had been achieved, not by following any formal design standards, academic theories, or museum promotions, but by individual industrial designers getting down and dirty into industrial environments and determining what consumers actually wanted and needed. This was pure capitalism at work, restoring the American economy. Everyone was happy, except perhaps for MoMA and engineers.

Engineers were not all that happy with industrial designers intruding on their traditional turf in manufacturing firms. Because for decades they were the highest authority in most companies on how to design and produce practical mechanical products at the lowest possible costs, they were not pleased with the demands of industrial designers to change traditional production processes and materials or to add cosmetic costs for purely nonfunctional purposes. Many designers, including major ones like Dreyfuss and Loewy, had to fiercely battle in-house engineering cultures for their proposed design changes.

The day after the Bauhaus exhibition closed at the MoMA, on January 31, 1939, the Pennsylvania Railroad (PRR) completed the streamlined shell designed by Raymond Loewy for its S-1 class steam locomotive, the longest and heaviest rigid-frame reciprocating steam locomotive ever built, and the only one with a 6–4–4–6 wheel arrangement. The engine and tender was 140 feet long, almost half the length of a football field, and too big for many curves on PRR lines. It was intended to replace PRR's aging K-4s locomotives, designed in 1914.

Loewy had designed a streamlined shroud for the K-4s in 1936, and his design of the S-1 shroud was a refinement of that design, with its bullet-nosed cylindrical boiler with headlight at the tip, resting on a half-bullet "cowcatcher" base. The overall design was very similar to Otto Kuhler's 1937 streamlined, bullet-nosed, steam locomotive for the B&O Royal Blue line. The photos of Loewy standing by, or on, the S-1's base are among his most famous publicity poses. In a more human-scale result, Loewy's group in South Bend, Indiana, had just completed the design of the 1939 Studebaker Champion. The car was actually designed by Clare Hodgman, on Loewy's staff, who would also work on the 1940 line of Studebakers for Loewy.

As the Bauhaus exhibition ended in January 1939, America's designers were as busy as ever. Raymond Loewy designed the Farmall tractor for the International Harvester Company, one of his famous "before-and-after" transformations of an iron-wheeled contraption, looking a century old, into a clean and modern piece of farm machinery. This led to the design of a distinctive new sales center building for International Harvester, of which 1,125 would be built, and also to the design of the Tractractor caterpillar tractor.

By this time, the Barnes & Reinecke design office in Chicago had a staff that included David Painter (1913–2003), vice president in charge of design; James Teague; Fred Priess; and George Mendenhall. Reinecke had also served in 1937 as a part-time instructor at the New Bauhaus under Moholy-Nagy. Barnes & Reinecke's new client was the McGraw-Edison Electric Company, which had introduced the first "pop-up" toaster in 1926. Now Reinecke designed its Toastmaster model 1B9 for 1939 in a glittering, chrome-plated steel, streamlined box. It would increase sales dramatically, be widely imitated, and serve as the typeform for toasters well into the 1960s, by which time seven million Toastmasters would have been sold. In 1980, the company would change its name to Toastmaster. Barnes & Reinecke also patented an updated version of Hamilton Beach's classic soda fountain, Drinkmaster. There was an infinite variety of clients; for example, Robert Davol Budlong was designing accordions for the Scandalli Dallape Cellini Company.

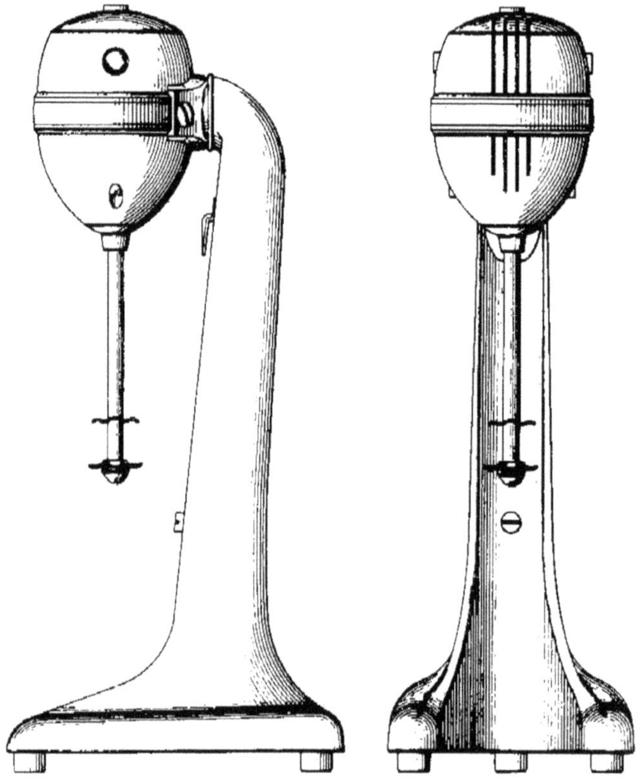

Hamilton Beach Drinkmaster, 1939, by Barnes & Reinecke (Patent No. D118,225, U.S. Patent Office).

Russel and Mary Wright were making headlines in 1939 with the introduction of their highly successful line of ceramic dinnerware called American Modern, made by Steubenville Pottery in East Liverpool, Ohio. Wright designed the line in 1937, but production had been delayed due to the 1938 recession, during which Wright attended New York University School of Architecture. The name American Modern

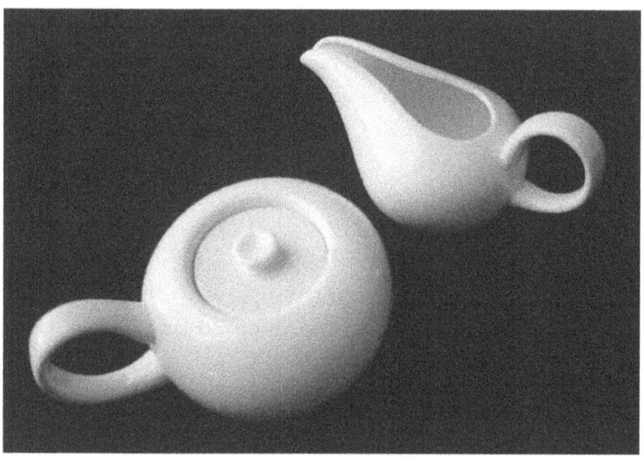

American Modern sugar bowl and creamer, 1939, by Russel and Mary Wright.

came from Wright's successful line of furniture of the same name still being produced, which was designed in 1935 under its original name of Modern Living. In 1938 Wright had established a "casual living" concept line of accessories. The new ceramic dinnerware forms were soft and organic, and glazes were in rich, earthy colors. American Modern dinnerware became the perfect gift for young brides. With Wright's name on the bottom of each piece, he quickly became a design celebrity, as 80 million pieces were sold. In 1941, American Modern would receive the American Designers' Institute award for the best ceramic design of the year.

Gilbert Rohde, who had resigned as director of the Design Laboratory in 1940, while continuing as director of design for the Herman Miller Furniture Company, now taught an industrial design program with Donald Deskey at the New York University School of Architecture, which they would continue until 1943. In 1941 Rohde designed a dramatic, transparent, circular coffee table for Herman Miller using plate glass for the top, resting on acrylic legs with brass tips. He would soon convince the company to discontinue production of all its period reproductions because of his belief that imitation of traditional design was insincere aesthetically. This was the first furniture company to do so.

In Toledo, Van Doren & Associates (VDA) designed a Speedway scale for Toledo Scale and a Maytag Model E Master Washer, the first to use white rather than the speckled gray and green finish common at the time. VDA had worked with engineer Thomas R. Smith to design the sleek, unified form that concealed the motor and hoses, increased capacity by 50 percent, and featured an insulated tub that kept water hot longer. After the war, demand would increase, and Model E's appearance went essentially unchanged for 44 years, becoming one of the most successful household appliances of the century.

Since 1936, in addition to their normal projects, many designers were deep in preparation for the New York World's Fair, scheduled to open on April 30, 1939. As the approaching fair dominated the news, *Vogue* magazine asked the leading industrial designers (Teague, Loewy, and Dreyfuss) to design fashions to be worn in the future. *Vogue* published the resulting designs on February 1 in an article entitled "Dress of the Future."

All the top designers would be represented at the fair. Planning had begun four years earlier. In 1935, at the height of the Depression, a group of New York City retired policemen decided to create an international exposition to lift the city and the country out of the Depression. They formed the New York World's Fair Corporation (NYWFC), elected former chief of police Grover Whalen as president, and invited a number of political and business leaders to plan, build, and organize the fair and its exhibits, with countries around the world taking part in the biggest international event since World War I. Walter Dorwin Teague was one of eight directors on the fair's Board of Design, and was not only designing four exhibits, one building, a pavilion, and a diorama for the fair, but had a full plate of consulting work.

One of Teague's projects in 1939 was designing glassware for Libbey Glass in Toledo, the design specifically to be used in the State Dining Room at the New York World's Fair. Teague designed a series of proposals, including a footed tumbler with a fluted, rectangular-shaped stem, inspired by the Ionic Greek columns on the front of the fair's Federal Building, and stemware with a rounded, fluted stem and cylindrical bowl. The first became the State Dining Room design, which had the etched crest of a spread-winged American eagle surrounded with thirteen stars. One hundred and twenty place settings were produced, with ten different styles of glasses and finger bowls.

This same pattern without the etching would become a stock Embassy pattern in the Modern American line of 1940, developed with Owens-Illinois' design department head since 1935, Edwin W. Fuerst. The second design, with the rounded fluted stem, would become the

Monticello pattern in the same line. That same year, Teague designed an internal grinder for the Heald Machine Tool Company, a tank vacuum cleaner for Montgomery Ward, a 620 Special for Kodak, an executive desk lamp for the Polaroid Company (designed by Frank Del Guidice), and a grand piano for Steinway & Sons.

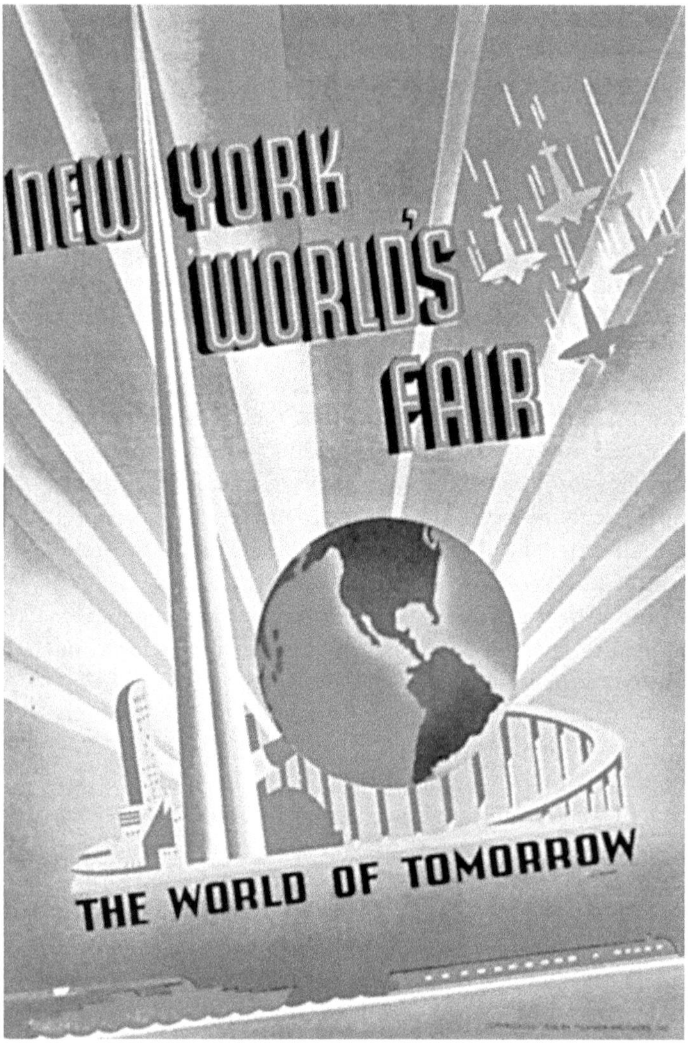

1939 World's Fair brochure (courtesy Victoria Matranga).

On April 30, 1939, the World's Fair in New York, optimistically titled The World of Tomorrow, held its grand opening attended by 206,000 people, with President Franklin Roosevelt giving the opening address. His speech was not only broadcast on radio networks but was also televised. NBC used the event to initiate regularly scheduled television broadcasts in New York City. An estimated 1,000 people viewed the Roosevelt telecast from about 200 television sets scattered throughout the New York area.

In the Radio Corporation of America (RCA) building, its theme and concept designed by John Vassos, RCA president David Sarnoff (1891–1971) in April also introduced the first consumer television sets, which were designed by Vassos and wrapped in a clear Plexiglas enclosure also designed by Vassos. Sarnoff's speech was titled "Birth of an Industry." He predicted that television would one day become an important entertainment and information medium. His crystal ball was 100 percent correct. Picture tubes in 1939 ranged in size from 5 inches to 12 inches, and the largest tube faced upward on top of a wooden cabinet with a raised, tilted mirror in back of the tube to view the picture horizontally.

The fair's official guidebook praised the captains of the Machine Age:

> The true poets of the twentieth century are the designers, architects, and the engineers who glimpse some inner vision, create some beautiful figment of the imagination, and then translate it into valid actuality for the world to enjoy.[6]

The theme of the fair was "Building the World of Tomorrow." "Builders of the World of Tomorrow" was emblazoned on all fair workers' uniforms. Technology was the magic ingredient that made it all possible, and fairgoers were confident that technology would take them into the future.

To people in 1939, technology meant ordinary stuff, like refrigerators, automobiles, trains, airplanes and radios; it meant building hydroelectric dams to produce electricity for rural areas; it meant Pyrex and Lucite, rayon and nylon; it meant new ways to build bridges and skyscrapers; it was about elevators and escalators; it was about speed, travel, labor saving and comfort.

Within the lifetime of people in 1939, technology like this had transformed everyday life for the better. Within a decade, technology had changed the shape of everything in the American home, including the home itself. These things are now so common and traditional that we tend to take them for granted as if they always existed, but they did not. They were created in the 1930s by technology, controlled by the "true poets" of the 20th century: the engineers, the scientists, the architects, and the designers. Compare 1939's technology to today's impersonal technology of space stations, supercolliders, and genomes, which are often quite remote to our personal, everyday lives.

The trademark symbols of the fair were a 610-foot-tall Trylon spire, which was connected by what was then the world's longest escalator to a 180-foot-diameter sphere called the Perisphere, both gleaming white and dominating the site at Flushing Meadows. The Perisphere housed a diorama called Democracity, which depicted a model future utopian city and suburbs of 2039 on the banks of a river. The city radiated outward in widening concentric circles around a single towering skyscraper at its center. A great hydroelectric plant on the riverbank powered the city, and the surrounding, rolling countryside was full of parks and towns.[7]

Designed by Henry Dreyfuss and Julian Everett, the diorama was viewed from above on a moving sidewalk, as if seen from an airplane at 7,000 feet, while a multi-image slide presentation was projected on the interior surface of the sphere, serving as a movie screen as large as an acre. In the diorama, miniature cars entered and left the model city via high-speed roads that connected to beautiful green towns clustered around factories where they worked. As night fell, the domed sky grew dark and stars came out, like a real night sky, and lights came on in the model city. Then a thousand-voice chorus was heard, and images of a dozen different groups of people (farmers, miners, engineers, etc.) were projected on the dome. The images repeatedly dissolved and were replaced by larger images, as if people were converging on Democracity. When they had arrived, shimmering light rose from the horizon to engulf the entire dome, then faded and the show was over. After the presentation, visitors exited the Perisphere on a 950-foot-long Helicline, an elevated spiral ramp that partially encircled the sphere three-quarters of the way around on the outside.[8]

The message for the future was clear. People would no longer have to live in a city. They would live in the countryside or garden apartments, around the city. Factory workers would live in green towns just like executives and white-collar workers. Democracity was a vision of modern suburbs and planned communities. Henry Dreyfuss Associates (HDA) also designed the fair's administration building and the AT&T exhibit, which featured "Pedro the Vodar," an early voice synthesizer robot.

The fair was divided into differently themed "zones," such as the Transportation Zone, the Communication and Business Systems Zone, the Food Zone, the Government Zone, the Community Interests Zone, the Production and Distribution Zone, etc. Different wall colors and colored lighting distinguished zones from each other. The Transportation Zone included the General Motors (GM) pavilion that featured a 36,000-square-foot Futurama exhibit designed by Norman Bel Geddes in the Highways and Horizons building. It was a diorama of a section of the United States, comprised of miniature highways, towns, waterways, 500,000

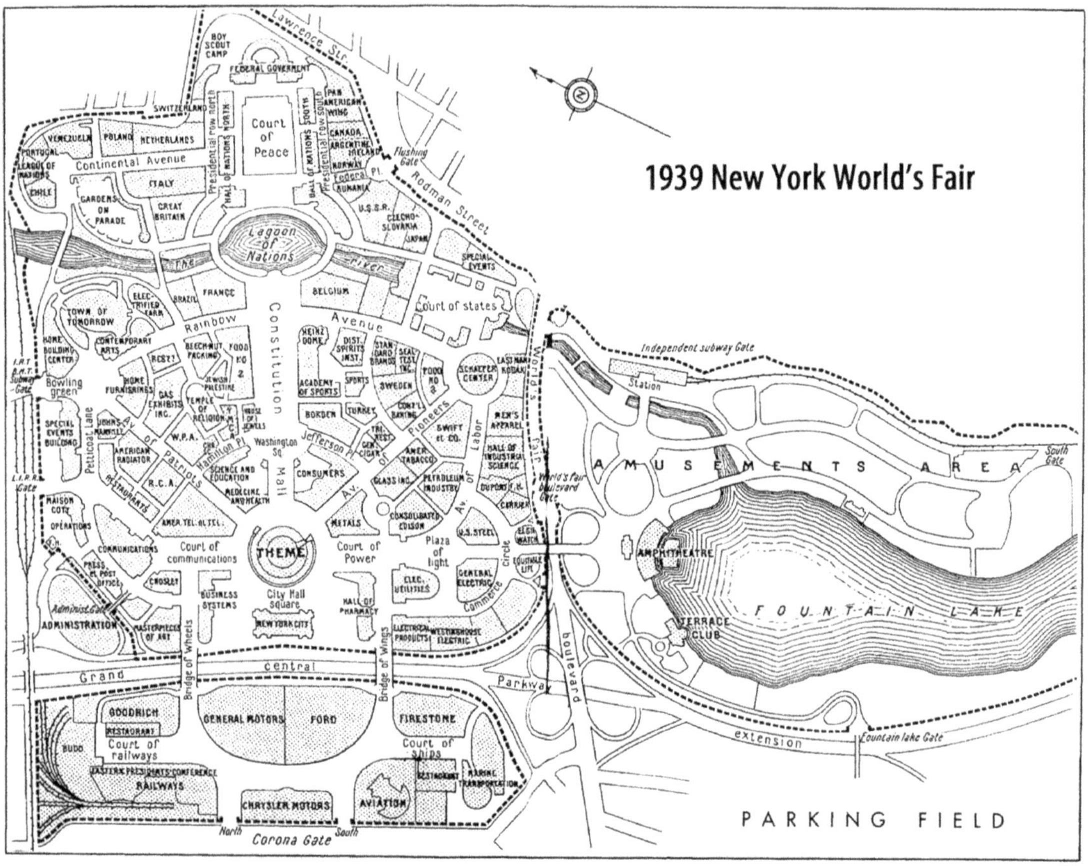

World's Fair map.

individually designed homes, 50,000 miniature vehicles, and a million miniature trees. Incredible details, such as "working" waterfalls, low clouds fashioned with chemical vapors clinging to mountainsides, clotheslines, and cow patties that many visitors never noticed, created a spectacular overall effect. Bel Geddes had perfected the diorama from his 1937 City of Tomorrow exhibit for Shell Oil, and from his earlier models of horse racing tracks and war games in the 1920s and early 1930s, described earlier.

Futurama was by far the most popular exhibit of the fair, with long, winding waiting lines in front of the enormous, high, narrow cleft flanked by a huge silver "G" and "M" on the modernistic, silver, GM pavilion shaped like a long, low triangle. Upon entering the bright red cleft, visitors descended a ramp to a darkened room with a large map of the United States on the far wall, and a narrator gave a talk about the country's overburdened road system, then announced: "General Motors invites you on a tour of future America. The moving chairs below the map will transport you into 1960."

Each chair had its own piped-in sound track with music and narration. The train of chairs was more than 1,500 feet long. The chairs moved smoothly as they floated on a rising and falling track that was 23 feet off the ground at its highest point, and each chair's sound system delivered the narration in perfect sync with the passing scenery. How was this done without computers or audiotapes? At the time, there were only two ways to record and play sound: vinyl phonograph records played on a turntable, or sound recorded on movie film.

At Futurama, this feat was accomplished by something called a Polyrhetor, developed by Westinghouse: a 20-ton, 12-foot-high cylinder holding 22 bands of movie film that recorded and played only sound, each individually triggered by the moving cars on a specific segment of chair track. At the end of the ride, the last you saw of the diorama was a typical city street intersection of the future, up close. Then the chairs curved around and you got off. The Futurama ride was over, and there you were, at that very intersection, in the heart of the GM complex. The intersection was life sized and in complete detail and exactly as it was in the diorama.

The visitor experience was an enthralling, simulated arrival into the future. The sense of arriving by airplane was especially powerful for the two-thirds or more of the public who had not yet traveled by air.[9] Bel Geddes also designed other fair attractions, including the exhibit and building of the National Dairy Company, the exhibit and building of the Wilson Packing Company, and the Crystal Lassies Theater (see below). During 1939, Bel Geddes also designed a full line of radios for the Majestic Radio and Television Company; radios for the Emerson Radio and Phonograph Corporation; and prepared, at President Franklin Roosevelt's request, a program for establishing a National Motorway Planning Authority.[10]

Next to the GM building was the Ford pavilion, designed by Walter Dorwin Teague. By this time, Teague's office had grown to 55 employees, including architects, engineers, artists, and industrial designers. The Ford pavilion featured a spiral roadway, like a coil of rope with wider coils at the bottom, with cars zooming up and down on part of the "Roadway of Tomorrow." Moving into the entrance hall, a large, round room, there was a vast mural in oranges, yellows, and blues jumping with activity: silver pistons that moved up and down, valves hopping in and out, a three-dimensional spark plug, factory production lines, crisscrossing conveyor belts, and smokestacks of the Ford River Rouge factory. Henry Ford's first engine and first car were displayed reverently. The first Ford Mercury was introduced, designed by chief designer Bob Gregorie and staff. During the fair, as fairgoers watched, actual full-size pastel renderings were drawn of Mercury, Ford, and Lincoln by Tucker Madawick (1917–2006), a newly hired Ford apprentice designer who had just graduated in Pratt Institute's first class of industrial designers.

In the next room was an immense wedding cake turntable displaying the "Ford Cycle of Production," a round white turntable, 30 feet high and 100 feet in diameter, with stepped sides, topped by a platform on which three blue Fords were mounted in a triangle, nose to tail, canted outward toward viewers. The slopes of the turntable were covered with small, colorful, moving models of stylized men performing the jobs that contributed to the finished product on top. They mined and harvested materials, refined and finished them, made parts and assembled cars. The turntable floated on pontoons on a concealed lake and was rotated with motors.

On the "Road of Tomorrow," also designed by Teague, visitors rode in late-model Fords on the tight loops of the coiled, spiral ramp under a succession of flying buttresses on a wide half circle on the outer edge of the hall that housed the turntable, tunneled straight through the entrance hall, and looped back down the spiral again. The spiral ramps were intended to show how traffic could be lifted to the express level and descend to street level in the heart of a city without wasting space. The private lounges Teague designed for the Ford building were often described as the most elegant interiors at the fair.[11]

Teague designed a number of additional exhibits for the fair, including the interior for the DuPont exhibit, which had a large sign: "From Miscellaneous Ores & Inorganic Raw Materials, Chemistry Makes ..." The multiple answers included the introduction of nylon, primarily to make women's stockings, which excited at least half of the population in the country. Teague also designed exhibits for the Kodak Company, as well as buildings for the U.S. government and U.S. Steel, the latter appearing to some as a big blue helmet or a steel hemisphere lifted

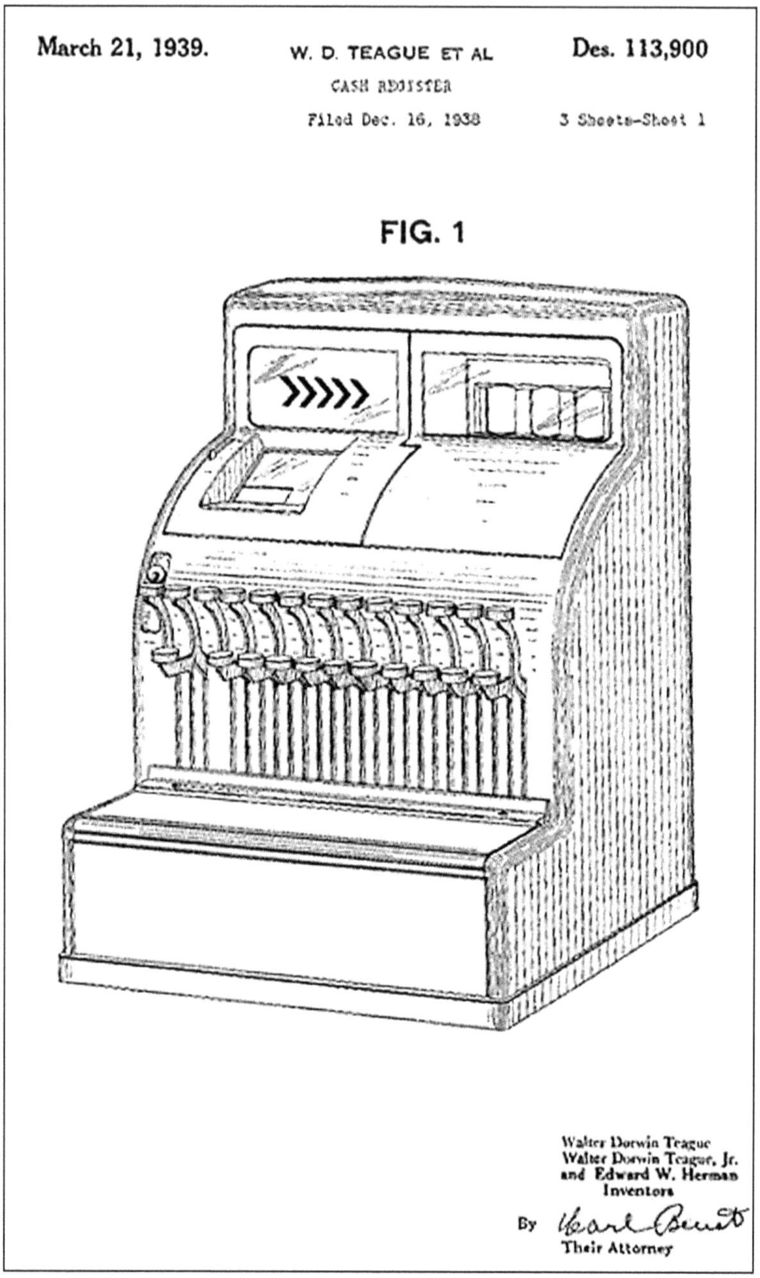

National Cash Register Model 100, 1939, by Walter Dorwin Teague Associates (Patent No. D113,900, U.S. Patent Office).

off the ground and supported by curving steel ribs of some sort running along the outside. Teague claimed that the steel was too interesting to hide beneath the stucco, so the intended concealment wall was eliminated. Teague designed National Cash Register Company's (NCR) new Model 100 and created a seven-story replica of the design on top of the NCR building, which was also displayed at the 1939 Golden Gate International Exhibition in San Francisco, which ran concurrently with the New York Fair.[12]

Teague's son, W. Dorwin, who worked for his father's firm since 1934, claimed he designed the Model 100 in 1935, but explained that the rule in all the early design offices was that every design was attributed to the head of the office, regardless of who actually created it. The Teague office was no exception. The explanation given for this practice by Teague was, "clients are paying for my services so my name must be on it, otherwise the client will think he is being cheated." Both father and son's names are on the patent.

Teague designed another masterpiece diorama, the largest in the world, which thrilled thousands and was located in the Con Edison building. Called the "City of Light," it was located along the concave wall of a gigantic room where people stood in a central viewing well to watch the show. Unlike other dioramas, it was viewed not in the round, but from the front, showing all of New York City, centering on the towers of Manhattan. Buildings to the rear seemed to be glued to the wall. As they stepped outward, in perhaps 15 ranks, they grew larger, and the ones in front were freestanding. The Empire State Building was 22 feet high. At the start of the show it was dawn, with buildings in pinks, roses, and grays against a pale blue background. The effect was breathtaking because of the flattened perspective, with the whole city compressed into perhaps a dozen feet.

There was much activity in the model city, with trains moving, cars crossing bridges, and elevators moving up and down in the towers. During the show, there was a thunderstorm with realistic light and sound effects, and as it darkened, all the countless tiny lights came on and the buildings turned deep blue. When the storm ended, a peaceful evening fell and still more lights came on. It was a magically beautiful sight.[13]

Gone with the Wind won the Academy Award for best picture that year, but fairgoers who saw *The Wizard of Oz* found Emerald City to be a familiar city of the future, just a bit more modern than the many model cities they saw at the fair.

Behind the GM and Ford buildings was the Chrysler Motors Exhibit building, designed by Raymond Loewy. Chrysler's building housed the Transportation Zone's focal exhibit, which featured Loewy's Rocketport exhibit. A futuristic rocket, shot out of a giant gun, would take passengers to London. Every twenty minutes, visitors were admitted to a semicircular area, in total darkness except for the spotlighted rocket on its launchpad. There were blinking lights, white and colored, along with a deep rhythmic sound, the whirling of motors, compressors, and high-frequency sound waves, conveying a sense of activity and suspense.

Then the sound of sirens: Liftoff! A blinding flash of hundreds of strobe lights and the roar of compressed air suddenly released conveyed the optical illusion of the rocket disappearing overhead into the blackness. Then total silence while a gradually diminishing point of light faded away into the stars.[14] Public statements by the fair's science director acknowledged that "rocket flight is at present impossible, but the problem is almost wholly limited to the invention of a proper fuel."[15] Within five years, rockets would in fact be raining on London, not from New York, but from Hitler's rocket bases in Germany. In the building, Loewy also presented models and drawings of "the Automobile of the Future," "the Taxicab of the Future" (an electric-powered, three-wheeled vehicle), and "the Motor Coach of the future." Loewy also designed the "House of Jewels" interior at another part of the fair.

The railroad exhibits in the Transportation Zone included Loewy's huge Pennsylvania Railroad S-1 locomotive and tender, absolutely stationary, but screaming at full 60 mph speed on special rollers beneath its wheels during the entire fair. At the fair's "Railroads on Parade" extravaganza, the S-1 met face to face with Henry Dreyfuss's 20th Century Limited, pitting against each other the two most dramatic, streamlined, steam locomotives in the world, and their two most famous industrial designers.

In the Production and Distribution Zone was a focal exhibit designed by Egmont Arens that had a large, concave movie screen showing simple, everyday activities, such as ordering food in a shop, and then all around it were scenes of the far-flung businesses upon which this simple act depended. More interesting was a display showing boxes devoted to each of the main categories of consumer spending; food, clothing, housing, health, etc. The government decided on four income levels: "Subsistence," a family income of $800 per year; "Maintenance, "$2,000; "the Good Life," $2,500; and "Luxury," $5,000. The average of these categories was $1,266, far below the "Good Life." In 2013 dollars, these arbitrary income levels would be $13,400, $33,500, $41,900, $83,900, and $21,200, respectively.[16] The claim in 1939 was that a third of the country makes less than "Subsistence," and nine-tenths, less than the "Good Life." In 2013, the median income, about $51,413, would be considered well above the "Good Life" using 1939 standards.

The focal exhibit for the Community Interests Zone, designed by Gilbert Rohde, made the point that people depend much more on each other than in the past. In a huge room with a semicircular wall, a series of scenes were arranged before it. One by one, each scene was lit up chronologically, and a narrator intoned, Scene 1, 1789: "No leisure, no time to play. Eight hours to sleep and sixteen to work." The last scene was 1939: "Eight hours' work, eight hours' rest, eight hours' leisure." The lights fade and a spotlight picks out the words, "For What?" An electric phonograph, according to Rohde, controlled the entire timing of the presentation.[17] Rohde also designed the exhibit for the Home Furnishings pavilion. That same year, the Herman Miller Company, for which Rohde headed design, pioneered in the use of showrooms for controlled furniture selling in New York and Chicago.

Russel Wright designed the focal exhibit for the Food Zone, which was by general agreement the most entertaining of the focal exhibits. The climax of the exhibit was a sixty-foot egg, inside of which was an avocado with five glowing jewels in its skin: a flock of lobsters flying inland from the sea, an aqueduct spilling roses into the desert, an eye blinking within a cave, and a can holding a clock racing backward. The jewels in the avocado symbolized carbohydrates, proteins, fats, vitamins and minerals. The lobsters signified that inlanders now eat fresh seafood. The blinking eye reminded all of "man's victory over night blindness through vitamin A foods." The aqueduct meant that irrigation made wastelands flourish, and the backward clock illustrated that "canning had perpetuated harvest time."[18]

The amusement area of the fair had a carnival-like feel that included pornographic exhibits, which were ignored by many attendees in favor of the showmanship in the main areas, but still were quite popular. One of these was the Crystal Lassies show, claimed at the ticket booth to be "The Lovliest Dancing Girls at the Fair." None other than Norman Bel Geddes designed the show. A nude manikin stood on a pedestal outside, and behind her the words, "Inside she's real. 15¢." A single dancer gyrated within an octagonal structure whose inner surfaces were completely mirrored. Two spectators could stand at each facet of the octagon and peep through one-way mirrors at the many dancers evidently cavorting within. The dancer's costume varied at different shows, but at times she was completely naked. Bel Geddes was probably demonstrating that he still had his original skill as a dramatic set designer.[19]

Germany and China were the only major countries not represented in the Government Zone at the fair. Germany was probably not invited, and China was being invaded by Japan. However, Japan was among the 60 countries that participated, all providing an array of cultural offerings to fairgoers, many of whom had never traveled overseas.

Due to its fall in 1938 to the Nazis, Czechoslovakia's pavilion was left unfinished. The international situation became much worse during the fair on September 1, when the Nazis invaded Poland, officially starting World War II. By September 3, Great Britain and France would declare war on Germany. The Soviet Union would sign a nonaggression agreement with

Germany in August and on September would invade Poland from the East. The Polish pavilion would be closed when the fair reopened in 1940. On September 27, 1939, general admission to the fair was reduced to 50 cents, initially just for October, but that price would be continued for all of 1940. On October 31, the fair closed for the season under the dark clouds of war. Twenty-six million attended that first year, but the fair was $23 million in the red.

The fair had closed, but its influence would last for ten years, when many of its predictions of the future would become the reality. Meanwhile, the Depression had become a thing of the past, and business, including the design business, continued to grow. Its leaders, the original pioneers, were becoming household words, and they were writing books about design.

CHAPTER 9

Organic Design and World War II

Lord & Taylor department store held a design competition in 1939 and presented $1,000 awards to Henry Dreyfuss, for "a washing machine elegant enough for a drawing room," and to Raymond Patten for an electric range "as sightly as a grand piano," according to the *New York Times*. Harley Earl started a styling school at General Motors in 1939, attracting designers interested in automotive work. Among the students accepted were Elwood P. Engel, Joseph Oros, George Walker, and Eugene Bordinat, all of whom became important car designers and executives after the war. Their pay while attending the school was $75 per month.

In Chicago, Sunbeam introduced its streamlined/modern-inspired C-20 coffeemaker and T9 electric toaster designed by Alfonso Iannelli as their "newest, most innovative products" in honor of the New York World's Fair. In 1940, Iannelli would end his collaboration with Sunbeam, for whom he had worked for ten years.

During the New York World's Fair, the Laboratory School of Industrial Design had published *A Design Students' Guide to the New York World's Fair,* supervised by a new faculty member, Paul Rand (1914–1996), who would become a famous graphic designer. The Guide studiously ignored the work of Bel Geddes, Dreyfuss, Loewy, Teague, and other mainstream industrial designers. Instead, students scoured the fair for the few functionalist architecture or designs scattered in obscure corners. MoMA curator McAndrew, who had recently joined the Laboratory's Board of Trustees, wrote an introduction to the guide praising the students for finding "good modern design among the pseudo-modern fantasies" and admonished visitors to be wary of the "soft corners and fungoid bulges on the buildings by some of our most celebrated industrial designers." It sounded like sour grapes from the jealous MoMA crowd.

Still struggling with funding in August 1939, the Laboratory School initiated a fund-raising drive with which to pay their new director, respected industrial designer George Sakier; improve salaries to staff; purchase new shop equipment; and relocate to larger quarters. A reception and exhibition was held on October 12, the same day Sakier assumed his post. The effort failed to generate sufficient funds, and in December, much of the school's equipment was auctioned off in a court-ordered foreclosure sale. About 50 students and faculty continued to meet voluntarily, but by May 1940, the demise of the school would be formally announced.[1]

That April, Germany had invaded Denmark and Norway, followed by invasions of the Netherlands, Belgium, Luxembourg, and France in May, the same month the New York World's Fair reopened, with many changes as a result of the war now raging in Europe. In June, the Soviet Union occupied the Baltic States. By July, the Battle of Britain had begun, as the Luftwaffe initiated the bombing of London. Ten exhibiting nations had declined invitations to remain, including the Soviet Union, which had just invaded Finland, but Finland, a Soviet satellite, stayed. "Zones" were dropped from the literature, and "For Peace and Freedom" became the new slogan.

The amusement area was emphasized as "the Great White Way," featuring a musical show called American Jubilee on a 270-foot-diameter, circular revolving platform. It was a sort of historical pageant. The Consumers' Building became the "World of Fashion." However, the backbone of the mall, Democracity and Futurama, remained as before. The fair closed forever on October 27, "amid war and crisis," according to *Life* magazine. Attendance was half a million on closing day, a combination of a carnival and a melancholy mood. Last-day crowds ripped out signs and picked flowers for souvenirs. Nineteen million more people attended than in 1939, but after two years, the fair was still in the red.[2]

About this time, Raymond Loewy received a cablegram from his English friends, the Ashley Havindens, who were enduring the terrible bombing in London. They asked if he could keep their two young children, Venise, 10, and Michael, 12, for the duration of the war. Loewy agreed, met the children at La Guardia Field, and sent them to school in Long Island. They returned home in 1944.[3]

Three industrial designers wrote books in 1940. Walter Dorwin Teague wrote *Design This Day: The Technique for Order in the Machine Age.* In it, he explained,

> We shall begin with the first source of all form, which is also an incessant variable. And this is the source of fitness, which is divided into three functions: Fitness of function, fitness of materials, and fitness of techniques. It is because of this principle of fitness that the ultimate form of an object—natural or man-made—is inherent in the object itself. It must be involved naturally to suit the function which the object is intended to perform, the material out of which it is made, and the methods used in its making.[4]

Teague recognized the interdisciplinary nature of problem solving, stating, "We see that we are not building little gadgets. We are building an environment, and we designers have to work with scientists, engineers, technologists, sociologists, and economists."[5] He was trying to correct the view that many held about industrial designers: that they were egotistical, independent, and obsessed with personal aggrandizement.

Norman Bel Geddes wrote his second book, *Magic Motorways,* using materials from his design of Futurama in the General Motors pavilion of the World's Fair, which, as you recall, was based on Geddes's 1937 City of Tomorrow exhibit for Shell Oil, as well as his futuristic concepts in his 1932 book, *Horizons.* These high-speed road concepts had been evolving since 1908, and the Automobile-Verkehrs und Übungsstrasse (Automobile Traffic and Training Road, or AVUS) implemented in Germany in 1921 was the first controlled-access road in Europe, which was also used as a racetrack. Nazis built the more extensive Autobahn starting in 1932. As auto traffic increased in prewar America, these concepts had become influential in the development of U.S. freeways and postwar interstate systems. In fact, both the Pasadena Freeway and the Pennsylvania Turnpike, the earliest such major systems constructed in the United States, were both opened in 1940.

The third book in 1940 was by Harold Van Doren, who wrote *Industrial Design: A Practical Guide.* It was essentially a how-to book on how to become an industrial designer, with detailed illustrations, specifics of presentation, model making, basic principles of design, educational programs, sketching and rendering techniques, mechanical drawing, design patents, and working with clients. In short, it was a textbook for students of industrial design. Van Doren was the first to formulate a good definition of industrial design:

> Industrial design is concerned with three-dimensional products or machines, made only by modern production methods. Its purpose is to enhance their desirability in the eyes of the purchaser

> through increased convenience and better adaptability of form to function; through a shrewd knowledge of consumer psychology; and through the aesthetic appeal of form, color, and texture.[6]

Van Doren also recognized that designers needed to suppress their egotism and celebrity status to blend in with industry:

> Once the industrial designer had made a dent on industry, it was perhaps natural for him to exaggerate his own importance to the scheme of things. Indeed, one might almost say that he would not be a good designer unless he had that sort of excited enthusiasm that makes salesmen sell and designers create. But in sober moments, he must have realized that, important as his contribution might seem to him, its relative importance might not be so great. As a rule, the artist is, and should be, only one of the gears in the train that includes management, sales promotion, advertising, engineering, research—all those departments making up the complex mechanism of modern commerce.[7]

These designers clearly recognized that industrial design had matured into a formal profession essential to the business world and wanted to document their personal leadership in the new field. They sensed that a new era was dawning and could see the expanding opportunities ahead.

Despite, or perhaps because of, the threat of war in 1940, the Depression of the 1930s was essentially over. People were spending again, a housing boom was in progress, 80 percent of American households were wired for electricity, 66 percent owned a car, 52 percent had electric washing machines and refrigerators, and 42 percent had vacuum cleaners. But now the problem became one of inventory shortages of household goods in the face of increasing consumer demand. As industry was preparing for increases in production to meet this new demand, the Defense Department was offering lucrative defense contracts for essential military supplies and equipment. Although still neutral, America was providing allies such as Britain, China, and France with large quantities of essential war materials, and many industries shifted to war production, leaving consumer needs unfilled. Many stores were limiting sales to their regular customers only.

Although the number of industries that converted to war materials increased during 1940, many consumer product manufacturers continued to produce at record rates, which meant that industrial designers were kept busy. Peter Müller-Munk Associates designed a Silex vacuum coffeemaker. Egmont Arens, along with Theodore C. Bookhart, designed a chrome, streamlined meat slicer for the Hobart Manufacturing Company. George Walker designed Eureka's vacuum Model D-171, which won a prize for beauty. Adolph "Al" Lambach was a staff designer for Admiral in Chicago, patenting a number of radio/television cabinets from 1940 until 1945. Gordon Florian became assistant director to Ray Patten at General Electric's Appearance Design department in Bridgeport.

Norman Bel Geddes designed the Patriot, a red, white, and blue plastic table radio for Emerson Radio and Phonograph Company. He also designed a refrigerator for the Kelvinator Corporation; interiors for the La Rue restaurant; redesigned the stage of Rockefeller Center Theater for ice skating; and designed costumes and scenery for *It Happens on Ice,* starring international skating star Sonja Henie.[8]

Brooks Stevens in 1940 had a staff of five, and 70 clients. His group designed an architectural Kuro jukebox for Chicago juke maker Gabel, and a red, streamlined research vehicle for the Johnson Wax Company in Racine as a tribute to the brightly colored office building that Frank Lloyd Wright had recently created for Johnson. In 1941, Stevens would design the Petipoint iron for Edmilton, with chromed side fins to facilitate cooling and to dramatize the streamlined appearance.

Henry Dreyfuss Associates (HDA) had designed the classic Big Ben alarm clock introduced in 1939 by Westclox, and in 1940, a dental instrument stand for the S.S. White Company. He also designed the Hoover Company's Model 60 upright vacuum cleaner.

Barnes & Reinecke, working with Minnesota Mining & Manufacturing Company (3M), designed the ubiquitous, clear plastic disposable Scotch tape dispenser we are still using today, as well as a cast-iron desktop tape dispenser. In 1940, Reinecke, then 39, was included in a feature article, "Today's Young Men," that interviewed 70 men who achieved notable success while still in their 20s and 30s.

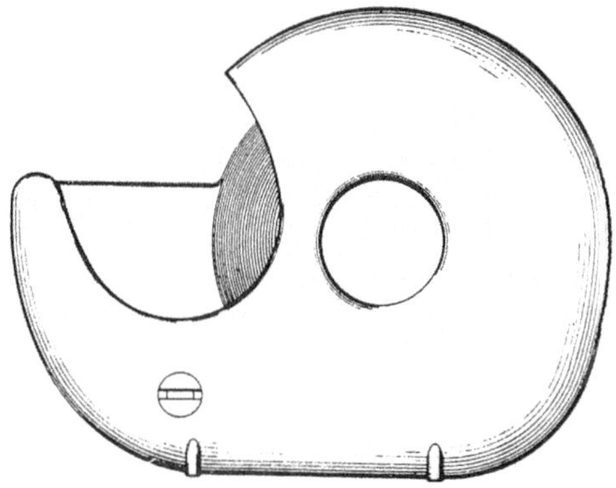

Scotch tape dispenser, 1940, by Barnes & Reinecke (Patent No. D118,629, U.S. Patent Office).

Robert Davol Budlong continued as design consultant to Zenith, designing many of Zenith's prewar radios, which would include the Universal Portable AC/DC radio with batteries in 1940, the Poket radio in 1941, and the Transoceanic Clipper radio in 1942. Zenith tried several times to offer him a full-time job as a Zenith employee, and as head of an internal industrial design group, but he wished to remain independent so that he could service other clients. However, he did relocate his office to the building on 333 Michigan Avenue that housed Zenith's showrooms.

At the Ford Motor Company, Bob Gregorie (and staff that included Tucker Madawick) designed a new Lincoln Zephyr Continental Cabriolet, with financial support by Edsel Ford. Edsel had conceived of it as a custom-built luxury car for himself, but its enthusiastic reviews convinced Ford to allow limited production of it in 1940. One of the most sophisticated designs of the time, it would become a classic and would win acclaim as a work of art at the Museum of Modern Art in 1951.

Donald Deskey designed a prefabricated weekend cabin called the Sportshack, with

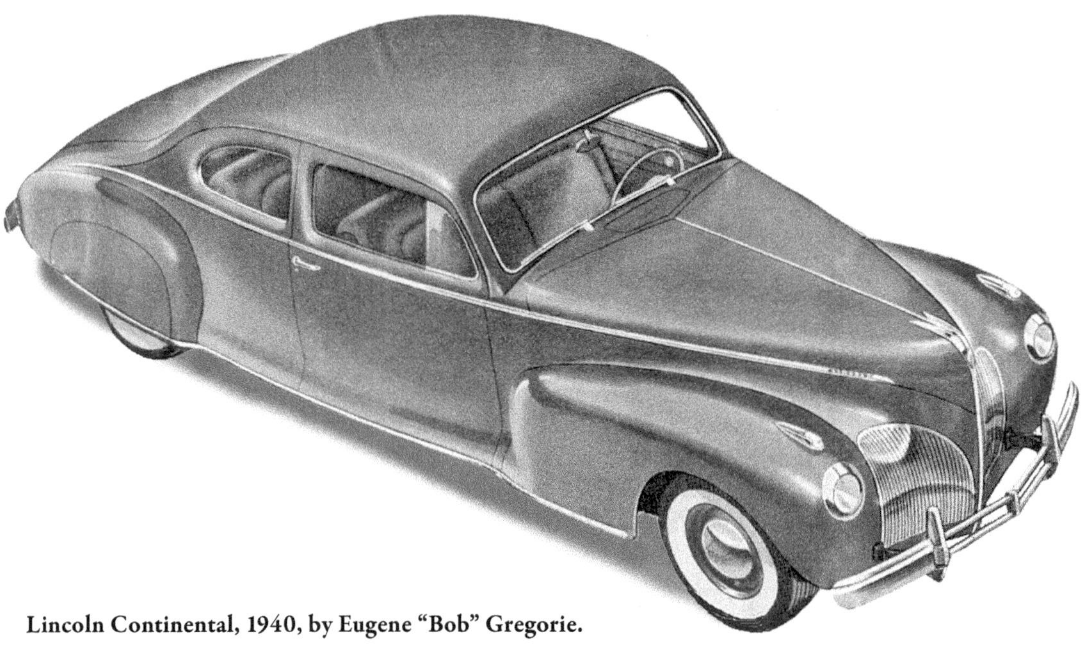

Lincoln Continental, 1940, by Eugene "Bob" Gregorie.

multifunctional furniture and walls of fir plywood striated mechanically to look like barn wood. It was displayed at the 1940 Metropolitan Museum of Art's annual series, Contemporary Industrial Art, with room displays created by architects and designers. Deskey would patent the specially treated plywood in 1942 and name it Weldtex. In about 1943, Deskey would found Donald Deskey Associates in New York, specializing in package design, which would operate until about 1975.

In March 1940, Loewy was visited in his office by George Washington Hill, president of the American Tobacco Company, who bet him he couldn't design a better package for Lucky Strike cigarettes, which was essentially a red target disk on one side with "Lucky Strike" on it, against a background of dark green. Loewy bet him he could in the amount of $50,000. In 1942, Loewy's new package design would be introduced, which repeated the red target on two sides, and replaced the green background with pure, shiny white. Loewy had won the bet. Lucky Strike explained the change with the public slogan, "Lucky Strike green goes to war." Raymond Loewy in 1940 also designed the Electrolux L300 refrigerator and was working on Greyhound bus designs.

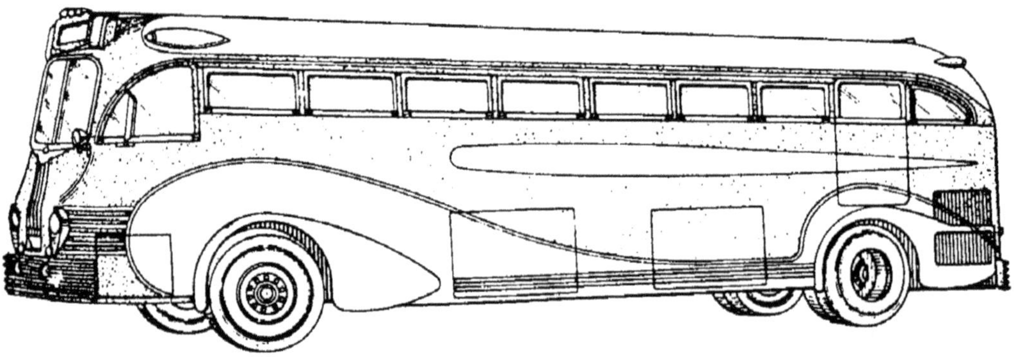

Greyhound bus, 1940, by Raymond Loewy (Patent No. D113,009, U.S. Patent Office).

Jo Sinel was still in New York freelancing for advertising agencies, designing calculators for Marchand and typewriters for Remington, as well as numerous design projects in packaging, interiors, and publications. He also taught industrial design courses at the Pratt Institute. But in 1940 he returned to California and would teach in a number of colleges, primarily at the California College of Arts and Crafts in San Francisco, where he had taught in the early 1920s—this time to teach a program along with package designer Walter Landor (1913–1995). Sinel, who reveled in variety and wanderlust, would also teach at other colleges in California, including Chouinard in Los Angeles, Claremont College in Claremont, and the Rudolph Schaffer School of Design in San Francisco.

Design schools were continuing to develop new programs in industrial design. Syracuse University sought the advice of Norman Bel Geddes, Lurelle Guild, and Henry Dreyfuss, and as a result revised and renamed an earlier "Design for Industry" program as "Industrial Design." Hin Bredendieck had left the School of Design in Chicago to initiate and direct an industrial design program at the Georgia School (later, Institute) of Technology. The Illinois Institute of Technology was formed from the Armour Institute in Chicago, which had been offering a design program under Mies van der Rohe since 1938. The handful of industrial design programs that had been established over the last six years had by 1940 graduated only several dozen designers, but these would provide the start of a second generation of postwar practitioners, the first generation with some formal university training in the new profession

Russel Wright and his wife Mary closed the home furnishings accessories business they had begun in 1935 and in 1939 formed a new company with Irving Richards called Raymor, which would continue to sell Wright products with exclusive distribution rights for five years. Their line of American Modern dinnerware had been included in the Museum of Modern Art's (MoMA) January 1940 annual exhibition, Useful Products under $10. Inspired by the success of this line and the deepening sense of national patriotism, Wright organized and initiated the American Way, a nationwide craft, design, and merchandising consortium of 100 designers to promote low-cost American style to industry, and American furnishings to consumers. The American Way was launched at a highly promoted event in December 1940 at Macy's in New York, featuring First Lady Eleanor Roosevelt. That November, her husband, Franklin, had been elected for a third term as president, violating the unwritten tradition of the two-term limit. His radio fireside chat in December urged the country to become the "Arsenal of Democracy" to support Britain, France, and China. Due to the growing public preoccupation with the war, the American Way would be abandoned within a year.

The Museum of Modern Art was still trying to establish its own credentials and authority in the now well-established field of industrial design. In 1940, it created a separate department of industrial design, headed by its first director, architect Eliot Noyes (1910–1977). Walter Gropius had recommended Noyes to MoMA, since he had studied architecture under Gropius at Harvard and worked for him after graduation. Noyes was born in Boston, and while a student, he worked as a curator at MoMA.

That same year, Edgar J. Kaufmann, Jr. (1910–1989), wrote to Alfred Barr, director of MoMA, proposing an Organic Design in Home Furnishing Competition. Kaufmann had been a resident apprentice in architecture at Frank Lloyd Wright's Taliesin East School and Studio from 1933 to 1934, and his father, Edgar J. Kaufmann, a wealthy businessman and owner of Kaufmann's Department Store in Pittsburgh, had commissioned Wright to design Fallingwater, Wright's famous 1936 modern house built over Bear Run Creek in Fayette County, Pennsylvania.

Kaufmann's proposal fit MoMA's strategy perfectly on a number of levels, and in fact was highly influential in determining the postwar direction of furniture design, and eventually the entire field of industrial design. First, MoMA saw furniture as a field in general that, except for Herman Miller, had yet to be dominated by modern design, particularly modern American design by pioneer industrial designers, as had the automotive and appliance fields. Most furniture makers were still imitating historical period styles, such as Colonial, Chippendale, Queen Anne, Arts and Crafts, etc. So if MoMA could establish and promote modern design standards for furniture, it could claim leadership in a category traditionally viewed as the historical province of architects.

Secondly, the use of the term "organic" was distinctly an architectural term used almost exclusively since 1908 by Frank Lloyd Wright, America's most famous architect, for his own work. It was derived from his mentor, Louis Sullivan (1856–1915), whose slogan "form follows function" became the mantra of modern architecture. Wright changed the phrase to "form and function are one," using nature as the best example of this integration. In his 1939 book, *An Organic Architecture: The Architecture of Democracy,* Wright wrote, "So here I stand before you preaching organic architecture to be the modern ideal ... determining form by way of the nature of materials."

But since 1932, when the Museum of Modern Art promoted the new International Style based on European architects, Wright was regarded by many U.S. architects as an old-school, irascible, irrelevant has-been.[9] So it was somewhat ironic that MoMA would use this term at all for this competition. However, the term served MoMA's purpose of relabeling 'modern design' with an obvious architectural term, rather than "streamlining," a common term that had been associated with American industrial designers during most of the 1930s, and which MoMA had repeatedly denigrated as "modernistic and streamlined aberrations."

In any event, the exhibition, Organic Design in Home Furnishings, ran from September to November 1941 at the Museum of Modern Art. Noyes organized and directed the competition and its terms, which included Noyes's definition of "organic design":

> A design may be called organic when there is an harmonious organization of the parts within the whole, according to structure, material, and purpose. Within this definition there can be no vain ornamentation or superfluity, but the part of beauty is none the less great—in ideal choice of material, in visual refinement, and in the rational elegance of things intended for use.[10]

One can almost read "streamlining" in place of "vain ornamentation or superfluity." According to Noyes, "The purpose of the contest was to discover good designers and engage them in the task of creating a better environment for today's living." Noyes defined design implicitly as teamwork, so all contestants except textile designers were to be teams of two or more designers. The competition itself was to be a business deal, a collaboration between museum, designers, and corporations. Each of the winning designers was to be awarded a production and distribution contract with a major American corporation. The exhibit and awards were to be sponsored by 12 leading department stores. In MoMA's political context of achieving architectural leadership through furniture, it would have certainly been helpful if architects, rather than industrial designers, would win such a competition. MoMA ensured that that would be the result.

As we know, a number of pioneer industrial designers over the last ten years had designed a number of acclaimed modern furniture: Norman Bel Geddes, Paul Frankl, Gilbert Rohde, John Vassos, Nathan Horwitt, Kem Weber, Donald Deskey, and Russel Wright, to name a few. But none of these were participants in the MoMA competition.

We do not know how competitors were chosen, but we do know there were 31, at least half (16) of whom were architects, based on a list in MoMA's 1941 exhibition publication, which listed six teams of architects competing in furniture design, the rest in fabric, weaves, prints, interiors and lamp design, as well as some Latin American designs. So it was no surprise that architects won the furniture design competition, as obviously preferred by MoMA.

The overwhelming winner was a team of architect/educators, Charles Eames and Eero Saarinen, both from the Cranbrook Academy of Art, for their designs of chairs and storage pieces. These architects and other winners were recognized at MoMA's 1941 exhibition, Organic Design in Home Furnishings, and their work was displayed as well in the accompanying publication. Other competitive furniture design teams of architects were Martin Craig and Ann Hatfield, Emrich Nicholson and Joseph Maier, Oscar Stonorov and Wilhelm von Moltke, Carl Anderson and Charles Bellah, and Harry Weese and Benjamin Baldwin.[11] After the war, Eames and Saarinen would receive contracts with Herman Miller Furniture Company and Knoll Associates, and a new era of furniture design, led by MoMA, would unfold.

In 1941, Harold Van Doren's East Coast clientele, in particular Westinghouse and Philco, required him to establish a second office in Philadelphia. Van Doren relocated there with Donald Dailey (1914–1997), leaving the Toledo office in the charge of Jack Little. Dailey soon designed a two-tone electric iron for Westinghouse. In Cleveland, Viktor Schreckengost had been designing a number of streamlined tricycles for the Murray Bicycle Company since 1938 and now designed for Murray a new concept appropriate for youngsters born in the age of the airplane: a Pedal Pursuit Plane, simulating a fighter plane. Raymond Loewy designed an electric razor for Schick and components for the 1941 Harley Davidson 74FL Knucklehead motorcycle.

Norman Bel Geddes designed a slicing machine for the Globe Slicing Machine Company, a cigarette vending machine for the U-Need-a-Pack Company, and a stapler for the Markwell Manufacturing Corporation. Based on his Futurama exhibit at the World's Fair, John Ringling North (1903–1985), of the Ringling Circus in Sarasota, Florida, had employed Bel Geddes in 1940 to design the circus layout of the midway, tents, posters, and costumes for the 1941 Ringling Brothers and Barnum & Bailey Circus. North's intention was to update and improve the "Greatest Show on Earth." Bel Geddes tinted the sawdust in rings of red, white, and blue; introduced modern lighting; added large-scale production numbers, and made the visual style of the circus more colorful and unified.

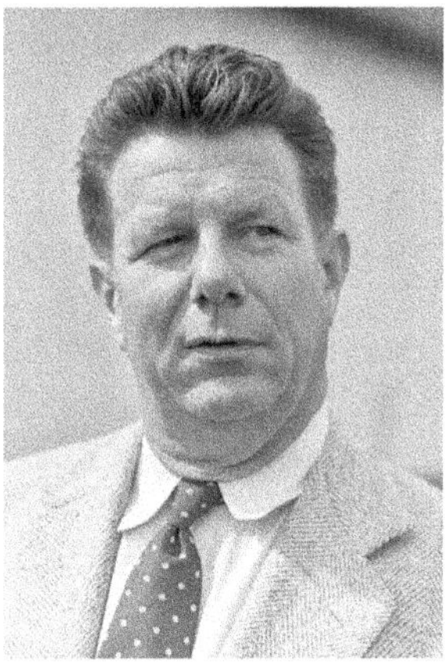

Norman Bel Geddes, ca. 1941 (State Archives of Florida. Photograph by Joseph Janney Steinmetz).

Joseph Platt and Marlin S. Baker designed the classic Parker 51 fountain pen, but due to wartime government restrictions on civilian pens, few were produced until after the war, when it would be produced from 1951 until 1972. It cost a quarter of a million dollars to develop, and Parker would spend $20 million on its sales and promotion.

Platt's main career was as an interior designer and art director. He served as style director for Marshall Field's department store in Chicago and decorated sets for Broadway plays and Hollywood films such as *Gone with the Wind* (1939), *Rebecca* (1940), and *Portrait of Jenny* (1948). Platt's wife, June, designed home furnishings under her husband's name, as well as continuing to write cookbooks, such as *June Platt's Plain and Fancy Cookbook* (1941).

Throughout 1941, the war in Europe continued to intensify. In April, the four Axis powers, Germany, Italy, Hungary and Bulgaria, invaded and dismembered Yugoslavia. Greece fell. In June, Germany invaded the Soviet Union, overran the Baltic States, and laid siege to Leningrad. In December, a Russian counteroffensive drove the Germans from Moscow suburbs into retreat. Finally, when Japan bombed Pearl Harbor on December 7, President Roosevelt declared war on Japan the next day. Within a week, Germany and its Axis partners declared war on the United States.

Now on a full war footing, Roosevelt's "Arsenal of Democracy" quickly moved into full speed. Most domestic manufacturers were overloaded with military contracts for aircraft, tanks, arms, munitions, and military equipment, essentially ceasing to produce consumer goods. The last passenger car, a Pontiac, was produced in February 1942. While General Motors was making tanks, the Styling Section under Harley Earl was designing camouflage for military vehicles and weapons during the war. Ford was making Jeeps; the Hoover Company was making bomb fuses and helmet liners; and General Electric was making gun turrets, howitzers, bazookas, and aircraft engines.

To avoid inflation, the Office of Price Administration (OPA) established price ceilings on manufactured goods still on the market, including refrigerators, irons, radios, ranges, typewriters, phonographs, and washing machines. Americans were told to "use it up, wear it out, make it do, or do without." Rationing was imposed on gasoline, shoes, butter, flour, fish, canned food, coffee, cheese, sugar, and meat. New home construction was stopped to make construction materials available for the military or government purposes. People started "victory gardens" in their backyard to provide food for the table and were urged to collect metal scraps, tin cans, paper, silk, nylon, and grease for the war effort. There were no products made of steel, chromium, tin, aluminum, copper, or nickel, or even some plastics such as acrylic, nylon, or Bakelite, all of which were required for the war effort.

Many young men were enlisting and being drafted into the armed forces and being replaced in the workforce, in many cases by women. Registration for the draft had begun in the summer of 1940 for unmarried males between the ages of 18 and 45, and male enrollment in colleges decreased dramatically as many potential students entered the services, as well as many recent graduates. Thousands of men and women relocated to new "defense towns" being built near major new manufacturing plants around the country to work on production lines of war goods. Everyone was affected, including those in the industrial design profession.

In 1942, Frederick Hurten Rhead, the designer of colorful Fiesta dinnerware in 1936, passed away at age 62 while still heading design at the Homer Laughlin Company in Newell, West Virginia. Fiesta had remained popular until the war, when the government took control of all available uranium for the atomic bomb, thereby taking away the vibrant red-orange uranium oxide used by Homer Laughlin in Fiesta. Without that dramatic color, sales of Fiesta declined gradually, and it would finally be discontinued in 1973. But in 1986, on the 50th anniversary of its introduction, the line would be revived at a time when vintage design would be embraced.

The American Designers' Institute (ADI) held a conference in New York, Design in the National Emergency, and promised that its members would be available to the government and manufacturers who needed assistance to meet difficulties. Leo Jiranek chaired ADI's committee, which submitted a report to the National War Production Board (NWPB) and the Office of Price Administration (OPA) suggesting how designers could best serve the country.

Many of the younger industrial designers, who would form the second generation of industrial designers after the war, chose to enlist or were drafted into the armed services and would serve bravely on the front lines, in the air, or at sea during the war. Others would use their artistic skills in the military to design camouflage, war-related models, instrument colors, or training devices.

Typical was Arthur BecVar, then 30, who left his design job at Revere Copper and Brass to work in the Navy's Ordinance Department. Charles (then 34) and his wife Ray Eames had relocated to Los Angeles in 1941 to work for MGM, but after the war began, they contacted the Naval Hospital in San Diego to propose a molded plywood splint to replace the padded metal splint then being used. Eames and Eero Saarinen designed it, the Navy placed an order for 5,000, and Eames formed the Plywood Wood Company to produce them in quantity. Evans Products would purchase the rights to produce them in 1943.

Although most pioneer machine designers were beyond draft age, nevertheless a few chose active military roles. Viktor Schreckengost, head of the design program at the Cleveland Institute of Art, and then 37, joined the Navy in 1943 as an officer, where he worked in radar detection systems, battle models, and watercolors of war devastation. He was made head of design in the Department of Training Devices, supervising other designers. Eliot Noyes, then 31, served as a major in the Air Force

John Vassos, then 57 and head of design at the Radio Corporation of America (RCA), enlisted in the Army Air Corps in 1942 as a lieutenant. In 1944 he was reassigned to Washington, D.C., to serve with the Office of Secret Service (OSS, the forerunner to the CIA). There, he developed camouflage techniques and produced a series of war posters encouraging vigilance on the home front. His most important job was to produce training manuals for the OSS, one of which was for the underground Greek forces fighting against German occupation.

He parachuted behind Nazi lines in Greece to train and act as the U.S. liaison with the Greek underground.

Other pioneer industrial designers enthusiastically engaged in services for the war effort. Henry Dreyfuss, Walter Dorwin Teague, and Raymond Loewy were called upon to design strategy rooms for the Joint Chiefs of Staff while working for the U.S. Army Ordinance Department and the U.S. Navy. One room featured a world map on a curved surface, 12 feet high by 25 feet wide, and the other was equipped for the projection of movies, slides, and diagrams. On a deadline of six weeks, the furnishings were built in the loft of an old brewery building in New York and then transferred to Washington for installation in the Public Health Service Building. The project served as a prototype for the planning facilities of many postwar transportation authorities and utility companies.

Dreyfuss designed and built four rotating world globes, 13 feet in diameter, one each for President Roosevelt, Winston Churchill, Joseph Stalin, and the Joint Chiefs of Staff. Due to a shortage of aluminum, the globes were constructed of laminated hoops of cherry wood. The Dreyfuss office was also assigned to re-study the cumbersome 105-millimeter antiaircraft gun. By reorganizing the components, the designers were able to reduce the critical field setup time from 15 to 3½ minutes.[12]

Loewy's office provided assistance to the Office of Strategic Services in the Department of Visual Presentation. Other work included new ideas for the Medical and Engineering Corps, for Ordinance, and for the Quartermaster General. For the Medical Air Corps, they modified a military glider into a field hospital that could be towed into a combat area, landed, and set up for operation. They also designed tents, swamp shoes, and a first-aid kit for airborne troops.[13]

Norman Bel Geddes's office developed photo recognition training programs for the military. Using experience gained in constructing his elaborate, recreational war game of the early 1930s, he prepared war-related scale models of military operations to prepare for invasions, such as a model-map of Tunis for the Army Signal Corps. He built models simulating accurate shorelines for conducting invasion instructions for combat officers; designed an aircraft identification system for the U.S. Army; advised a New York blackout technique in greater New York for the U.S. Army; designed self-camouflage techniques for U.S. Army engineers; designed electric, radio-controlled model aircraft for Strategic Command, Army Air Force, supervising their construction at Orlando, Florida; and designed a simulated tank battle in the USSR.[14]

Teague's organization worked for the Naval Bureau of Ordinance and completed more than 150 projects during the war. In 1942, Teague's son, W. Dorwin Teague, decided he should be working on something more useful for the war effort than on World's Fairs and vacuum cleaners. He joined the Research Engineering Department of the Eclipse-Pioneer Division of Bendix Aviation in Teterboro, New Jersey, as a junior engineer. His first assignment was troubleshooting the Boeing B-17 bomber's cabin supercharger. Dorwin would remain at Bendix working as an engineer until he rejoined his father's firm in 1953.

Egmont Arens advised on military instrumentation colors for easy readability; instructed how colors could be used to make bomb shelters, command centers, and factories more bearable and efficient; and suggested color-coding of the wiring, hydraulic lines, and the control systems of aircraft. Buckminster Fuller developed the Dymaxion Deployment Unit (DDU), a 20-foot circular hut constructed by the Butler Company of Kansas City using corrugated metal, which was insulated and had a porthole, a door, and a domelike ceiling with a hole for ventilation. Two hundred were built in 1942 for the Army Signal Corps to serve as radar huts. Kem Weber tried to interest the government in a panelized construction system he had invented for defense-

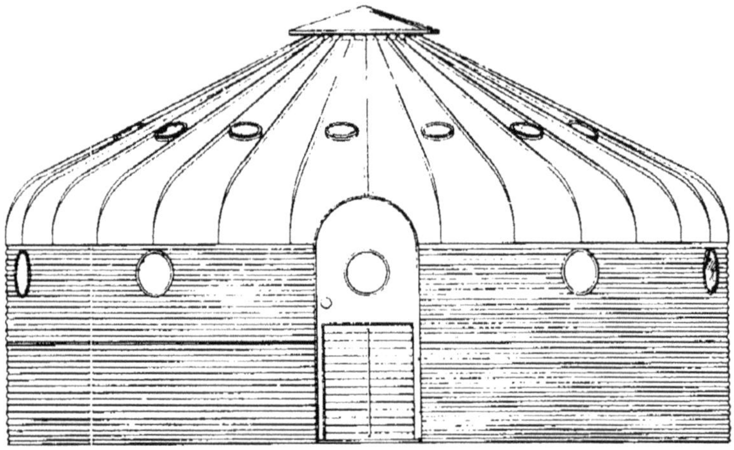

Dymaxion Deployment Unit, 1942, by Buckminster Fuller (Patent No. D133,411, U.S. Patent Office).

related housing projects but could not secure a contract.

Brooks Stevens designed military products including a 1941 snowmobile for use as "motorized cavalry," a paratroopers field kit, an individual water filter, field stoves, and dehydration units. In 1942 he came up with an ingenious plan for rebuilding existing Ford and Mercury passenger cars to increase capacity from six to twelve, for use as ambulances and personnel carriers for the Red Cross, defense plants, and army camps. In 1944, he collaborated with an inventor on a device for legless veterans, replacing foot pedals with a motorized steering column. Pulled back, it accelerated; pushed forward, the car braked.

In 1941, a patent disagreement between John Tjaarda and the officers of the Briggs Manufacturing Company, his employer since 1932, resulted in his employment being terminated, and many years of legal battles would follow. In 1942, Tjaarda organized his own design firm, John Tjaarda and Associates. He attempted to design, finance, and manufacture his own automobile, the Cortez, but due to lack of financing, it was unsuccessful.[15]

Locomotives were essential to national wartime transportation for troops and materials. Railroad operations had increased dramatically, so their development was not curtailed. In 1942, Pennsylvania Railroad introduced two prototypes of its T-1 class, duplex drive, 4–4–4–4 loco-

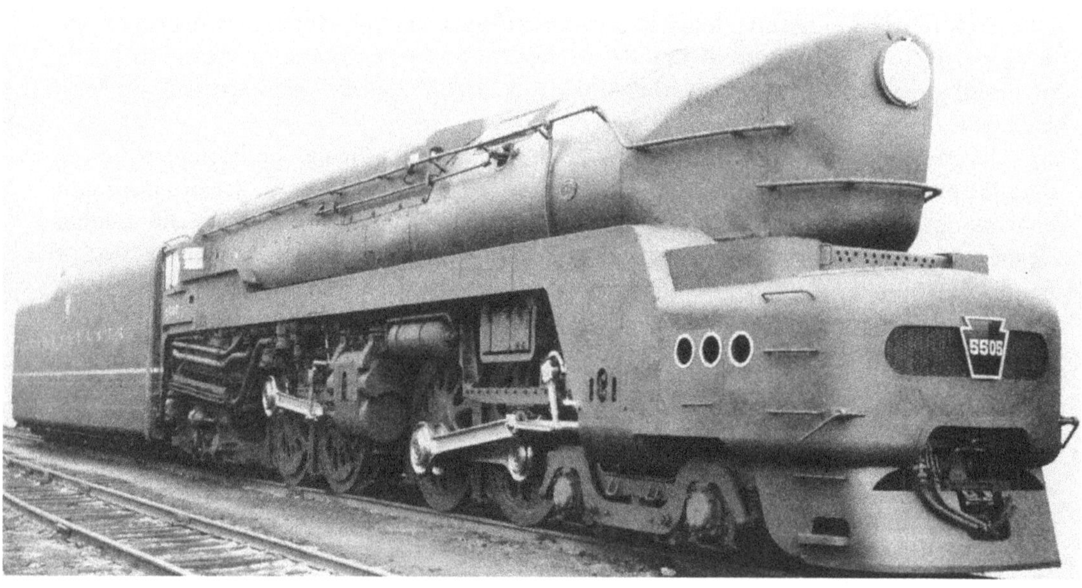

Pennsylvania Railroad T-1, locomotive, 1942, by Raymond Loewy.

motive, its last steam locomotive, and its most powerful. Weighing about a million pounds, it could haul 880 tons at 100 mph. Raymond Loewy designed it with a streamlined, vertical, prow-shaped nose. That same year, Loewy released several renderings of "A Conception of an Automobile of the Future" to the press, claiming to represent a "step in the direction of clean, uncluttered body styling." Both sketches showed a bullet-nose, aircraft-type spinner, a feature that was conceived by Bob Bourke that would later appear in the designs of both the 1950 Studebaker and the 1949 Ford.

In 1943 the interior design profession took notice of industrial design. Its magazine, *Interiors,* published since 1940 by Charles E. Whitney, initiated a section called "Industrial Design." It featured mass-manufactured products and was conceived and written by Donald Dohner, who had just resigned as head of Pratt Institute's industrial design program to establish a product and packaging firm in partnership with a fellow educator at Pratt, J. Gordon Lippincott (1909–1998), and with the Douglas T. Sterling Company in Stamford, Connecticut.

The new firm, Dohner & Lippincott, hired its first employees: two 1943 graduates of Pratt Institute, Budd Steinhilber (b. 1924) and Read Viemeister (1923–1993), who illustrated many of Dohner's design concepts for his articles. Tragically, **Donald Dohner**, at age 51, died accidentally of carbon monoxide poisoning on Christmas Eve, 1943, and his *Interiors* articles were taken over by his partner, J. Gordon Lippincott. The new head of the industrial design program at Pratt would be Alexander Kostellow, with the title "supervisor of industrial design and supervisor of design and structure." Kostellow would become an influential national voice in industrial design education and enhance the reputation of Pratt as a leading design institution. Nevertheless, it was Donald Dohner, who initiated and headed the first degreed industrial design program at Carnegie Institute of Technology in 1934, and who at that time taught Kostellow, then a painter, about industrial design. It would be Dohner who would be honored in a posthumous Special Recognition Award for Dohner by the Industrial Designers Society of America (IDSA) in 1998, and who would receive a formal citation recognizing Dohner as "The Father of American Industrial Design Education."

Walter Von Nessen also passed away in 1943, at age 54. After his death, his wife closed their studio, but she revived it after the war to continue production of his lamp designs, as well as innovative lamps of her own. Designer Stanley Wolf, who would purchase the business in 1954, joined her and would continue the Von Nessen tradition with his own designs, as well as designs commissioned from Elizabeth Kauffer and George Nelson in the 1960s. Today, Nessen Lighting in Mamaroneck, New York, is still producing Von Nessen's famous swing lamps of 1927, as well as a complete line of modern lighting fixtures.

By this time, the increasing advance of military operations overseas had made it clear that the war would eventually be won by the Allies, and that the invasion of Europe would soon begin. Many companies began to plan for postwar business opportunities with new products and materials. Material suppliers were at the forefront of these efforts. Durez Plastics and Chemicals, in order to stimulate consumer interest in plastics after the war, initiated a series of magazine advertisements and a promotional brochure featuring proposals of prominent industrial designers for innovative postwar products using Durez plastics, which stated,

> Industry may not solve the entire problem of unemployment in the immediate period following our inevitable victory, but by being ready it can contribute to the major part of the problem. Many pre-war products are out of date. Many production methods are obsolete, Tremendous advances have been made ... research has and will continue to unfold great opportunities.[16]

Participants included Jean Reinecke, Raymond Loewy, Egmont Arens, Peter Müller-Munk, Brooks Stevens, Dave Chapman, Carl Sundberg, Henry Dreyfuss and Donald Deskey.

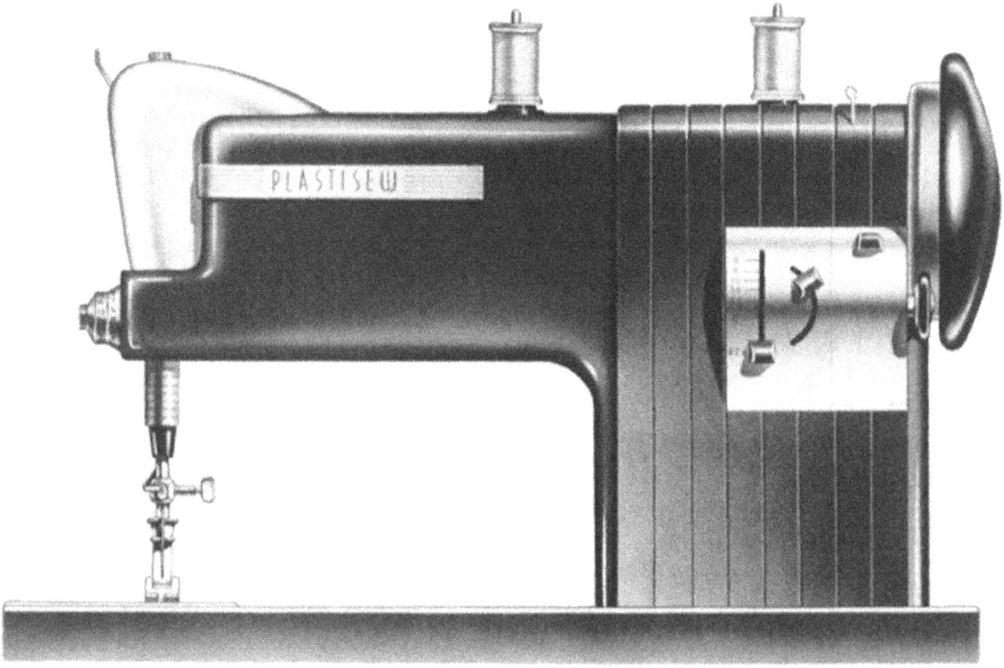

Peter Müller-Munk sewing machine concept, 1942, for Durez Plastics (courtesy Victoria Matranga).

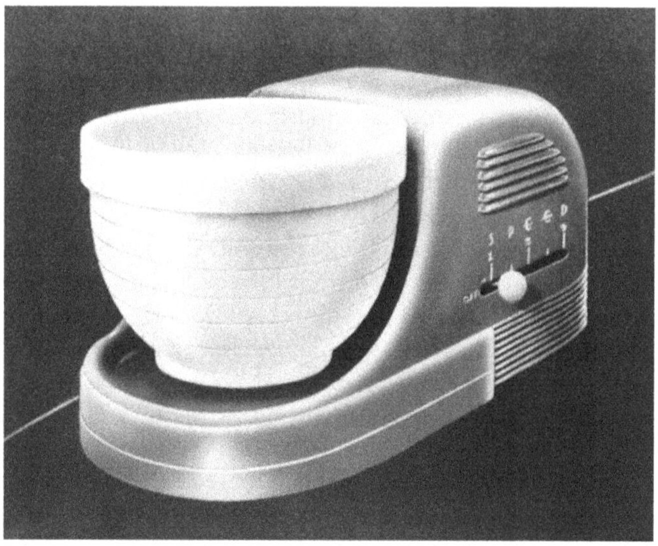

Brooks Stevens Supermix concept, 1942, for Durez Plastics (courtesy Victoria Matranga).

Most were highly futuristic in concept. Arens proposed a cylindrical refrigerator with sterilization capabilities, an ice cube maker, and a cold water faucet. Loewy proposed chairs made from molded plywood. Dreyfuss proposed a modern aircraft passenger interior. Sundberg proposed a sleek, modern bathroom. Müller-Munk proposed a lightweight sewing machine with lustrous, integral colors. Deskey predicted push-button and cordless telephones, intercom systems, and telephone directories replaced by microfilm. Stevens proposed a Supermix tabletop mixer, a plastic car, and a vaporizer, which was built after the war by the Spartan Company in Minneapolis to alleviate bronchial ailments. Chapman proposed a typewriter operating with electricity.

Many companies and design consultants needed no encouragement to initiate postwar planning and promotion of new products. They had been waiting for this opportunity for years.

Chapter 10

"The Club" and Postwar

Other companies not actively engaged in defense production were offering lines of new products for business expansion after the war. The Herman Miller Furniture Company, under the direction of Gilbert Rohde, introduced its Executive Office Group, the first integrated line of modular furniture that could be assembled in various configurations, the forerunner of postwar open office systems. Dave Chapman, Inc., advised a number of client companies in adapting to peacetime production, including Parker Pen Company, International Harvester Company, Brunswick-Balke-Collender, and Scovill Manufacturing Company. Norman Bel Geddes designed a postwar line of radio cabinets for RCA, Coca-Cola vending machines for Mills Industries, and domestic ranges for Floyd Wells Company. He also conducted development work on the Nash automobile.[1]

During the war, the Crane Company, a manufacturer of domestic bathroom fixtures, made steel valves for the U.S. Navy (each ship required 1,500 to 15,000 valves per ship). Because Crane had discontinued making domestic products for the duration of the war, it did not need the services of Henry Dreyfuss, its design consultant, but continued to pay his retainer so he would not work for a competitor. Dreyfuss set up a satellite office in the Crane building in Chicago to study the company's entire line. His staff chose products that should be carried over and designed new products to be introduced after the war. So when the war was over Crane would have a new line of products ready to go immediately.[2]

Libbey-Owens-Ford (LOF) had been challenged by *National Geographic* magazine to create a kitchen of the future. H. Creston Doner, head of LOF's department of design, accepted the challenge, and the result was the Kitchen of Tomorrow, built in the home of an LOF executive, which was featured prominently by major magazines and accompanied by a promotional campaign that included a short feature film by Paramount to be shown in movie theaters. Three full-sized models toured nationally for two years, starting in 1944.

The kitchen promoted the use of glass extensively, but also proposed appliances and furniture for a "peacetime revolution of convenience and comfort." The kitchen and dining area were joined in an open space arrangement, made possible by a dining table that flipped up to attach to a wall, a pass-through counter area, and a refrigerator accessible from both kitchen and dining area. The glass-front refrigerator and cabinets brought dinnerware and food into view. The traditional stovetop was replaced with a unified wall unit that included a glass-walled oven/rotisserie grill, built-in toaster, mixer/juicer/meat grinder, waffle maker, and digital clock/thermometer, as well as recessed cooking/baking vessels. Foot pedals controlled the faucet in the sink.[3] In 1944, H. Creston Doner wrote *Wartime Accomplishments of Libbey-Owens-Ford.*

As mentioned earlier, college campuses during the war were at a low level. By 1943, there were very few male graduates or applicants, since the draft swept up many high school graduates. Some schools contributed to the war effort, such as the faculty at Pratt, by designing camouflage for military installations. Others were planning for an expected postwar resurgence and determining how to educate students of the new profession, industrial design.

The American Designers' Institute (ADI) was under the leadership of its president, Alfons Bach, and its education committee, headed by Alexander Kostellow, and with members László Moholy-Nagy and Donald Dohner had surveyed its membership and solicited the advice of John Vassos, Ben Nash, and George Kosmak (b. 1905) to develop a prototype educational program for industrial design. In 1944, ADI announced its Educational Program for Industrial Design, a four-year program combining technical, business, and academic courses. It in fact would become a template for the many new programs of industrial design that would be initiated after the war.[4]

In 1944, many designers were repositioning themselves for postwar business opportunities. Harold Van Doren expanded his firm, already with offices in Toledo and Philadelphia, to include a New York office, Van Doren, Nowland, & Schladermundt, which he established in collaboration with designers Roger Nowland and Peter Schladermundt. Donald Dailey, of Van Doren's office in Philadelphia, designed the familiar Boston pencil sharpener, which would be used in millions of schoolrooms over the next 50 years. J. Gordon Lippincott opened a design office in New York with Walter P. Margulies, titled Lippincott & Margulies. He hired Read Viemeister as director of styling. Donald Deskey designed a bowling center ball-return system for the Brunswick-Balke-Collender Company, which would become an integral part of hundreds of modern recreational bowling alleys that attracted families after the war.

Harley Earl established an independent styling studio in Detroit, Harley Earl, Inc., to design nonautomotive products, but continued as vice president of styling at General Motors. John "Jack" Morgan had left Sears, Roebuck in 1943 and in 1944 opened his own office, Jack Morgan Associates, in Chicago, with a small staff. Jon Hauser (1916–1999) replaced Morgan at Sears. Peter Müller-Munk left his position as head of the industrial design program at the Carnegie Institute of Technology (CIT) in order to devote full time to his commercial design studio, Peter Müller-Munk Associates in Pittsburgh. Richard I. Felver, a Pittsburgh industrial designer, replaced him at CIT. John Gordon Rideout had closed his Cleveland office in 1941 but reopened it in 1943 in Chagrin Falls, Ohio, near Cleveland. In 1944, he joined with architect Ernest Payer to combine industrial design and architecture. That year, Edwin W. Fuerst, head of design at Libby Owens Glass Company, left Libby to open his own design office in Toledo, specializing in package design, which would operate until about 1952.

In February, 1944, the International Business Machines Corporation (IBM) Automatic Sequence Controlled Calculator (ASCC) was delivered to Harvard University by IBM, which had engineered, built and funded the project, but it had been conceptually designed by Harvard University's computing pioneer Howard H. Aiken (1900–1973). The ASCC used 765,000 components, 500 miles of wires, three million connections, and was 51 feet long, eight feet high, and two feet deep. It was the industry's largest electromechanical calculator and the first operating machine that could execute long computations automatically. The computer was later called the Harvard Mark I. Futuristic industrial designer Norman Bel Geddes, who had been hired by IBM for $50,000 to do the job, designed its enclosure. Aiken considered the elaborate case to be a waste of funds, which he thought could have been used to build additional calculator equipment. IBM was not happy because Aiken claimed himself as the sole "inventor" in the press release for the formal presentation to the university in August, ignoring several IBM employees who had helped build various components.

Raymond Loewy Associates (RLA) was the formal organization formed in 1944 by Loewy to recognize his partners: A. Baker Barnhardt (in charge of products, automotive, and packaging); Jean Thomson Loewy, his first wife; Jack Breen (business manager), William Snaith, considered by Loewy "a brilliant partner." Miss Helen Peters was Loewy's personal secretary. RLA designed an Electrolux B6 floor polisher as well as a 1,000 hp diesel engine yard switcher locomotive, the FM H-10–44, 195 units of which were produced by Fairbanks-Morse for the American Railroad. Gone was any evidence of streamlining. The engine was a simple black and orange rectangular box, distinguished mainly by a protruding roof visor mounted on the rear cab.

That same year, **Gilbert Rohde**, while still heading design at Herman Miller Furniture Company, passed away at age 50. In 1945, Herman Miller would appoint George Nelson as the company's new director of design. Nelson had become co-managing editor of *Architectural Forum* magazine in 1943 with Henry Wright and had that year written an article, "Grass on Main Street," in the *Saturday Evening Post*, which introduced the innovative idea of pedestrian malls. In 1945, Nelson and Wright developed a unique concept of storage wall systems, an outgrowth of their joint book, *Tomorrow's House*, published that year. The storage wall was featured in *Life* magazine, attracted national attention, and led to Nelson's position at Herman Miller. As his first project as director of design, Nelson designed a collection of 70 pieces. He would become an innovative and influential leader in postwar modern furnishings design.

Since 1941, the membership base of the American Designers' Institute (ADI) in Chicago had been shifting to New York. In 1942, its New York chapter questioned whether the organization should still be incorporated in Illinois. The ADI board apparently agreed, for in 1943, the Illinois secretary of state dissolved the corporation for failing to file an annual report as required. In May 1944, the American Designers Institute (without the apostrophe) was granted a new charter in Delaware based on an application filed by Alfons Bach, Ben Nash, and George Kosmak. ADI's administrative offices were promptly relocated to New York, and its new president for 1944 was Belle Kogan, who had been secretary/treasurer during the previous term of office.[5]

Most prominent industrial designers of the 1930s practiced in New York, and their well-publicized financial success over the decade had attracted the attention of New York State tax authorities, who decided that industrial design earnings were taxable under the unincorporated business tax statute. Accordingly, in 1940, Henry Dreyfuss, Walter Dorwin Teague, and Raymond Loewy were billed nearly half a million dollars in back taxes for the years 1935, 1936, and 1937. The three argued that their earnings were similar to architects, derived from services rendered rather than the sale of goods.

So the Big Three shared costs to engage attorneys to represent them in court, with Teague as their representative standing trial on April 30, 1941. The attorneys argued for exemption of Teague's income from the unincorporated business tax because "the vocation of industrial design is a 'profession,' rather than a 'trade' or 'business,' and section 386 of the tax law 'excludes the practice of law, medicine, dentistry, [and] architecture, which under existing law cannot be conducted under corporate structure, and in any other case in which more than eighty percent of the gross income is derived from personal services rendered by the individual.'" The case was dismissed.

This victorious result has been taken as legal evidence that industrial design is, in fact, a profession. From that time, subsequent attempts to tax industrial designers under the same law were unsuccessful. However, after about 1976, when taxes for "professional corporations" would be implemented, all professions, including architects and industrial designers, would be subject to taxation similar to business corporations. However, the tax challenge of 1941, and the intent to protect their legal identity as an individual service provider, probably explains the traditional obsession of designers such as Loewy, Dreyfuss, and Teague in requiring only their own name to be credited on all the work produced by individuals on their staff, a restrictive and resented practice that essentially disappeared in the 1970s, when tax codes were modified.

After the tax issue was resolved, New York's leading designers, including Dreyfuss, Loewy, Bel Geddes, Deskey, Arens and Wright, met informally to discuss how the profession could further its objectives of professionalism, influence the business world, and establish standards of practice. There were serious concerns regarding the American Designers Institute (ADI), the only such organization of national scale, as an appropriate organization for industrial design. First, the term "designers" in its title did not specifically refer to "industrial designers," the now commonly accepted specific term for the group of professionals who had successfully established the profession in the business and public worlds. That meant that any kind of "designer" could join ADI, not just industrial designers.

Second, Chicago furniture designers who were specialists in a single industry had formed ADI. Some still were, and yet, many of its members worked in many different, unrelated industries. The consensus of the leading designers mentioned above was that a national organization should specifically refer to "industrial designers," should be more exclusive, and should be restricted to those practitioners who had demonstrated proficiency in a broad range of products, not just one. At the time, there were about 2,000 industrial designers, but many were not well qualified in mass-production experience or skills.

On February 7, 1944, 15 industrial designers met to discuss and conclude the formation of a new organization in accordance with these objectives. They defined an industrial designer as "one who has successfully designed a diversity of products for machine and mass production." This was the fundamental philosophical difference between the two groups, as many members of ADI specialized in furniture or home furnishings. The Society of Industrial Designers (SID) was thus formed.

On August 20, 1944, at the first annual meeting of the SID, the group elected Walter Dorwin Teague as president, Henry Dreyfuss as vice president, Raymond Loewy as chairman of the executive committee, Harold Van Doren as treasurer, and Egmont Arens as secretary. Philip McConnell, formerly with the U.S. Treasury Department, and former secretary to the New York World's Fair Design Commission, was appointed executive secretary of the new organization.

In addition to those mentioned above, founding members included Donald Deskey, Norman Bel Geddes, Lurelle Guild, Ray Patten, Joseph B. Platt, John Gordon Rideout, George Sakier, Joseph Sinel, Brooks Stevens, and Russel Wright. Wright recalled that George Nelson, Charles Eames, and several other members of the ADI were considered for membership, but were thought to be not acceptable at the time, probably because their work was limited to furniture design.[6] Each of the SID founders invited one additional designer to join the next year, and after that, membership would be open to qualified designers. Some called the new organization "the Club," an exclusive organization led by the "Big Three," Teague, Loewy, and Dreyfuss.

The qualifications for membership in SID were spelled out in a 1945 flyer: "Membership ... shall be limited to practicing industrial designers ... [who have] practiced ... for several years [and] must have had three or more of his products manufactured and offered for sale. He must have ... served three or more manufacturers." These criteria specifically excluded businessmen, specialists, handicraft designers, and engineers. Each candidate for admission was required to submit a portfolio to the board of directors, where one negative vote could disqualify. Dues were levied according to individual income.[7]

So now, in 1944, there were two major design organizations, both of which had founding members who were pioneers of the profession. The ADI founders in 1938 were primarily furniture designers (or fabrics, ceramics, or other products in the traditional so-called art industries), but some ADI members later worked in a diversity of industries. SID founders were primarily designers in a diversity of industries (usually in the traditional so-called artless industries), but some later specialized. As the postwar era evolved, these distinctions would become blurred or irrelevant. They would all be industrial designers. But in 1944, with the solidification of this new and influential profession, there was considerable confusion and controversy about the formal definition of industrial design and industrial designers.

This confusion extended to educational institutions, which were faced with the new challenge of how to educate industrial designers. In 1944, Richard F. Bach, dean of education at the Metropolitan Museum, hosted a conference for administrators of art schools to discuss the matter. Invitees included Alexander Kostellow and James C. Boudreau of Pratt, as well as administrators from 44 other art schools. A summary of the conference stated that "the conference carefully avoided definition of the term 'industrial design,' fearful that the effort to arrive at a definition might constrain discussion."

Indeed it might. Most art schools' design programs were quite broad in scope and were not strictly industrial design programs. Only 15 of the 44 art schools had programs specializing in industrial design. Major changes in curriculum recommended by industrial designers included courses in the physical and behavioral sciences and the addition of workshops and laboratory courses similar to institutes of technology. Industrial design model building instruction required well-equipped workshops with heavy equipment such as lathes, band-saws, planers, drill presses and vented paint-spraying booths, as well as model instruction in soft and hard clay, plaster cast and forming, plastic heat forming, mold making, woodworking, metalworking, and high-quality finishing techniques, all requiring heavy expenditures. Specialized training in concept sketching and presentation rendering using pastels or gouache was also required to meet the standards of the profession.

These potential changes challenged the traditional "applied art" mentality of art schools and would require new specialized equipment and faculty. For art schools to recognize industrial design as a unique entity would be to separate it from existing curricula and require new faculty, or it could lead to the establishment of independent schools of industrial design, something art schools did not want to see.[8] Art school administrators argued that these sorts of changes would drive away some artistic students, which was probably true. As a result, many new industrial design courses in art school programs would not be as effective as in more technically adapted schools and would often be taught by existing art faculty with no experience in the field.

Walter Dorwin Teague, attending a later meeting of this same group, commented,

> We have created a profession, and some of [you educators] are delighted to teach it, with only the vaguest idea of what it is all about.... By our practice we will create a definition of industrial design and I trust the conference committee will be acute enough to discern it.[9]

What Teague was saying was that practitioners would define the profession as it evolved, and educators needed to figure out how to teach it. This is pretty much the way the issue was resolved over the next 50 years. In fact, the profession would constantly evolve and expand into all aspects of design (graphic, interior, architectural, etc.) as well as into dozens of related technical fields such as psychology, science, engineering, sales, and marketing. Every new technology would present new opportunities for industrial designers, who always sought to make technology understandable and user friendly. Industrial design would forever defy a fixed definition, and industrial design programs would continue to vary in quality, scope, and structure.

The invasion of Europe began in June 1944. On the day that the American troops entered liberated Paris in August of that year, Raymond Loewy, now a naturalized American citizen, was invited by the U.S. Department of State to speak on the radio to his former countrymen in French and was received in private audience by President Franklin D. Roosevelt.[10]

In May 1945, the war ended in Germany, and in August with the atomic bombs in Japan. Military personnel by the millions returned home to begin civilian life anew and needed housing, education, and jobs. Many married and went to school under the G.I. Bill. Those who had stayed at home had been deprived of new consumer products for four years. Aging cars, furniture, appliances, bathrooms, and kitchens, many from the 1930s, needed replacement or modernization.

Perhaps more significantly, the national mood was much deeper than simple relief that the war was over. Many had been economically deprived since the Depression began in 1929, unable to afford the many new products and advances of the 1930s, which were then unavailable during the war. To these people, they had endured 15 years of scrimping and making do without any of the wonderful modern goods of the "World of Tomorrow" that had been promised at the World's Fairs of both 1933 and 1939. Now, many had enjoyed good jobs and steady income during the war. Average weekly pay had doubled since 1940, from $24 to $44 in 1945. A large number of families had double incomes, because many women had worked on defense jobs. Millions had accumulated considerable savings, and they were eager to spend it.

The country struggled to make a rapid transition from wartime production back to domestic, which was, in a sense, almost as difficult as it was to make the opposite switch in 1941. New designs and tooling were required, both of which took time. But America was the wealthiest and most powerful nation in the world with the greatest manufacturing base and the largest consumer market. An economic boom began that would last for the next 15 years. From 1945 to 1950, annual housing starts increased from 200,000 to 1,154,000.

Developers were building new suburbs using crews of masons, carpenters, plumbers, electricians and painters who followed each other down streets and completed 30 homes per day. These mass-produced homes included a yard, a garage, shrubbery, shutters, and built-in appliances, including refrigerator, stove, washer and dryer—the perfect new owner package. All that was needed to start housekeeping was furniture, lamps, dinnerware, accessories, lawnmowers, ladders, and a car to drive to work. One of the popular accessories in such houses was a clothes dryer with a unique window in the front door that Brooks Stevens had designed in 1941 for the Hamilton Manufacturing Company.

The demand was high for the design of mass-produced housing, but most architects did not contribute because of their concern that it could reduce their commissions on individual houses. But a number of nonarchitect entrepreneurs stepped in with ideas, among them Buckminster Fuller, who in 1945 redesigned his original Dymaxion house of 1929, which had evolved into his 1942 wartime Dymaxion Deployment Unit.

Now, his final design used a central vertical stainless steel strut on a single foundation, which supported bicycle-wheel-like structures that hung down to support the roof, while beams radiated out to support the floor. Roof, ceiling, and floor were formed of wedge-shaped fans of sheet metal, creating a round floor plan. The entire exterior was aluminum, with many components built by the Beech Aircraft Company, which was seeking new postwar products. Several prototypes, including the Wichita House, were built, incorporating a prefabricated bathroom, water storage, heating, air conditioning and a convection ventilator built into the roof. Reactions to the prototypes were positive, but it was never produced industrially because Fuller considered the prototypes as "unfinished." Potential buyers found other options, and Fuller's company went bankrupt.

Industrial designers were again in high demand by manufacturers as they developed and tooled up for high-volume domestic production. The profession was attractive because of its reputation for high salaries or fees, because its leading practitioners were national celebrities, and because graduates were in high demand. In March 1946, the Society of Industrial Designers (SID) published an educational bulletin listing schools with programs in industrial design. There were 21 such programs, with another half dozen under development. But graduates from these programs would not reach the market before about 1949 or 1950.

At this point I need to remind you of something I mentioned in the preface. After 1934, and particularly after World War II, the number of industrial designers increased significantly. Since the subject of this book is the machine designers of the early 1930s, those who were the earliest pioneers prior to about 1935, these are the designers who are primarily followed throughout the book. To follow the careers and work of all industrial designers after 1935 would regretfully have been beyond the scope of this book. Although some are mentioned, information about them and their designs is limited. Accordingly, if you fail to find details of the many well-known and highly successful designers who have influenced the profession since 1935, or of their designs, this is the reason for their necessary omission.

Immediately after the war, of course, the supply of designers was limited to older prewar practitioners already at work, or to young prewar graduates who were returning from military service. This meant that the large design firms and companies that were in existence before and during the war, led by the machine designers of the 1930s, would be handling the lion's share of postwar demand for industrial design, and would be enjoying tremendous success.

Among those pioneer design firms was Raymond Loewy Associates (RLA) of New York, which in 1945 opened offices in Los Angeles, California, and Chicago, Illinois. For the latter, Loewy hired Richard S. Latham (1920–1991), a returning serviceman who had started before the war as a designer at Montgomery Ward. RLA designed a streamlined diesel locomotive, built by Fairbanks-Morse in General Electric's Erie, Pennsylvania, facility and introduced in 1945. The "A" units had an engineer cab above the engine with an automotive-style split windshield, similar to the 1937 Electro-Motive Super Chief, with which it was competing on the Atcheson, Topeka & Santa Fe Railway. The "B" units were cabless. Eighty-two A units and 29 B units would be built for American Railroads over the next four years.

RLA in 1945 was on retainer to no fewer than 79 American, British, and Swedish corporations as a design consultant. Britain was at the time considering its own renewal and regeneration after the war and invited Raymond Loewy to offer an American designer's view of British industry. He responded with a letter published in the *London Times* in November that expressed the typical philosophy of American industrial design, titled "Industrial Design: The Aesthetics of Salesmanship—an American View":

> In America the industrial designer has developed a design technique that has been successfully tested in the field during the last 15 years, and American products most generally carry with them the advantage of collaboration between the professional consulting designer and the industrialist. This partnership with industry and our philosophy of design are based on securing greater sales appeal—increased trade. I have heard much here in England about the Aesthetics of design. This to me is strange. We seem to find that the aesthetics of an industrial product will take care of themselves automatically after we have provided a balance between function, simplicity, and utility. In the average manufacturer's mind the word aesthetics has annotations of philanthropy and culture after office hours. Industrial design did not become standard practice in America because of aesthetics. There were too many other good reasons for using it—increased sales appeal of products and increased trade. And the final result is no less beautiful for having been designed on sound, unemotional business lines.[11]

This philosophy, which embraced the purpose of design as a business strategy to increase product sales by appealing to the tastes and desires of common consumers, was the antipathy of those at the Museum of Modern Art (MoMA) and other designers who saw the fundamental purpose of industrial design as an artistic, sculptural, and personal expression by an individual artist. This difference was the root cause of MoMA's criticism of streamlining, its general disregard of market value, and its dismissal of popular taste. It was why MoMA promoted product designs on the sole basis of aesthetic superiority, a quality judged by MoMA alone. This, of course, was the traditional mission of art museums for centuries—to educate the general public how to appreciate and understand the best of fine art. MoMA's acclaimed historical fountain of "good design," the Bauhaus, was founded with the legitimate objective of marrying art and industry. Now that this had been successfully accomplished in America, MoMA wanted to re-separate them into traditional, and unequal, categories of "good design" and "commercial design." Fortunately, the public would not recognize this elite distinction.

In 1944, Paul Block, Jr., publisher of the *Toledo* (Ohio) *Blade* newspaper, had commissioned Norman Bel Geddes to prepare a study suggesting future developments for the city of Toledo, which had lost population and appeared to be in decline. It was to serve as a model for urban planning, not just for Toledo, but for other cities in Ohio as well. Bel Geddes was well known for his work in city and traffic planning, for his City of Tomorrow for Shell Oil in 1937, for his Futurama at the 1939 World's Fair, and for his 1940 book, *Magic Motorways.* Geddes was offered a fee of $50,000 plus travel expenses to complete the exhibit within three to six months. The final cost would be closer to $75,000.

In response to his commission, Bel Geddes created Toledo Tomorrow, a giant model fifty-nine feet in diameter depicting Toledo transformed by a major reorganization of the city's transportation infrastructure. At a scale of one-inch equals 100 feet, the model showed an area of about 150 square miles, with limited-access freeways crisscrossing the metropolitan area. A new passenger depot located near city center served air, rail, and bus travelers in a single location. The model was created in Bel Geddes's New York office in 47 panels, each four by six feet, and shipped to Toledo for public unveiling on July 4, 1945.

Visitors walked around the model on an elevated walkway to simulate viewing the city from an altitude of 10,000 feet, and a special sound system provided audio descriptions at five points. The audio assured visitors that "if only a part of this master plan were brought to fruition, your Toledo would be a better city." As a dramatic effect toward the end of the visitors' walk, the lights were dimmed to suggest night, and special ultraviolet lamps were turned on to illuminate the model. Ultimately, while a few changes in the city resulted from the Bel Geddes model, they were not very noticeable, and the proposed central air-rail-bus terminal was never built.[12]

Also in 1945, Bel Geddes designed motorboats for Gar Wood Industries, luggage for Atlantic Products, and model-photography stories for *Encyclopaedia Britannia* on the Great Pyramid and the Panama Canal. In 1946 he would design a line of radios, washing machines, and company trademarks for the Federal Telephone and Telegraph Company, and for the U.S. Navy he would write, construct models, and advise in preparation of the official record of the Battle of Midway, the Naval battle that destroyed the Japanese Navy.[13]

At Ford, after the death of Edsel Ford in 1943, the man he had appointed head of Ford's styling studio in 1935, Bob Gregorie, left the company and established his own consulting business in Detroit. But at the request of Henry Ford II, he returned in 1944 to head the postwar styling section, which had been revised by management in an effort to blend engineering and design. However, in 1946, the new management (Henry Ford II) engaged industrial design consultant George Walker to work on Ford's first new postwar design, which would become the 1949 Ford. Walker, who had designed Nash cars as a consultant since 1937, had hired Joe Oros (1917–2012) in 1940, and just recently, Elwood P. Engel (1917–1986) in 1946. Henry Ford II had invited both Gregorie and Walker to submit competing design proposals, and Ford chose one of Walker's submissions.

Gregorie saw his days at Ford numbered, so he resigned, closed his consulting office, and returned to his other love, designing boats. He designed his own 42-foot trawler, *The Drifter*, and would sail for the next 15 years. Thomas L. Hibbard, an automotive designer since 1915, left General Motors to replace Gregorie as head of Ford styling and would also contribute to the design of the 1949 Ford.

In 1946, Raymond Loewy Associates' (RLA) Chicago office, largely through the work of Richard Latham and working with Greyhound president Orville S. Caesar, developed a series of concepts for a 40-foot Greyhound double-decker Scenicruiser bus to provide a high view of scenic landscapes from a second-level dome, similar to those popularized by railroads for cross-country travel. After some years of development by the General Motors Truck and Coach Division, the final design based on RLA's concepts would be introduced in 1954. Over a thousand would be built over the following two years and would prove to be very popular with the traveling public.

RLA also designed the most innovative and modern-looking postwar upright vacuum cleaner in 1946, the Magic Carpet S-2, for the Singer Company. Instead of the typical "broom"-type single handle, the S-2 had twin chrome handles that spread apart as they descended to the cleaner head, straddling a flat, closed, plastic housing for a cord reel. The motor housing was no higher than a pack of cigarettes, accomplished by turning the motor on its side, which enabled the cleaner to go easily under furniture. The cleaner could also be hung on a closet wall, saving considerable storage space. That same year, Richard Latham of RLA designed logical layouts for "ham" shortwave radio receivers I-20 and No. S-38 for the Hallicrafters Company.

Another diesel locomotive in 1946 came from Ray Patten, head of General Electric's Appearance Design division. Along with his assistant director, Gordon Florian, Patten designed passenger (PA) and freight (FA) versions of the American Locomotive Company (Alco) and GE diesel locomotives. Between 1940 and 1953, these companies jointly made and marketed large road locomotives. Patten described the design as "a locomotive so distinctive and powerful looking that it actually helps railroads sell their services to passengers and shippers." By 1946, Patten's design group at GE numbered 50 people, including 15 to 18 model makers.[14] Gordan Florian soon left GE to open his own design office in Bridgeport.

Otto Kuhler had designed more locomotives and railroad cars than Raymond Loewy, Paul Cret, and Henry Dreyfuss combined. But after the war began, his German accent became a liability, and he abandoned freelance work. In 1944 he worked for the American Car and Foundry, developing double-deck sleepers and subway cars, but was laid off in 1947. He retired to a ranch near Pine, Colorado, where he painted western scenes, exhibited them in museums, and raised cattle.

After the war in 1945, Teague, like Loewy, changed his sole proprietorship to a partnership, allowing senior staff, including C. Stowe Myers, with Teague since 1934, and Frank Del Guidice (1916–1993), who later became the company's president, to be partners in Walter Dorwin Teague Associates (WDTA). Del Guidice represented WDTA in seeking commissions from the Boeing Company, which met with success, and WDTA established a branch office in Seattle, Washington. Teague established an engineering division to handle the complexity of the projects at Boeing. Their first assignment was to convert Boeing's B-29 World War II bomber into a commercial airliner by transforming the raw unfinished interior of the plane's

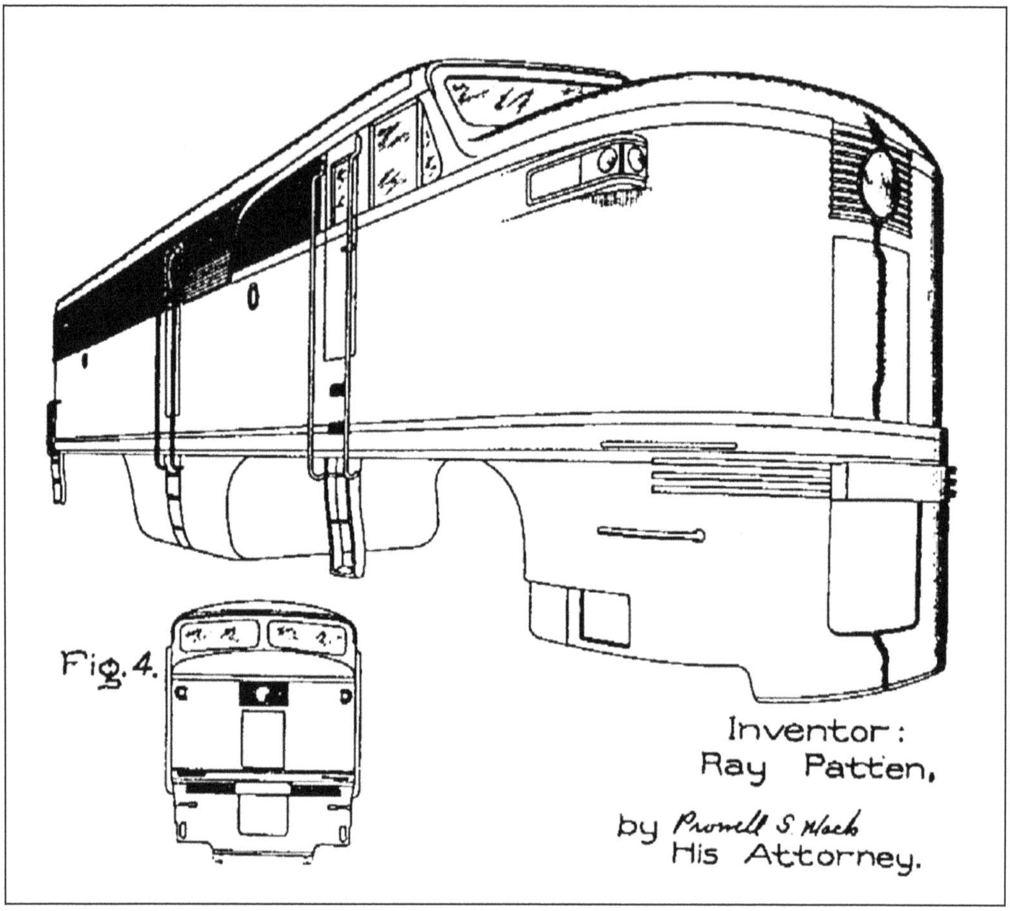

Patten locomotive design (Patent D153,896, U.S. Patent Office).

body into a more comfortable setting. They installed vinyl-clad panels for an easy-to-clean surface, introduced pleasing colors, and added cabin lighting as well as call buttons for flight attendants and overhead sections with individual reading lights and oxygen masks, all standard features of today's planes. The result was called the Stratocruiser, introduced to commercial service in 1946. WDTA would design interiors for a sequence of Boeing airliners far into the future, including the 707 (1958), the 737 (1963), the 747 (1969), the 767 (1982), the 777 (1995), the 777 Worldliner (2005), and the 787 Dreamliner (2011).

Donald Dailey, who had headed Harold Van Doren's office in Philadelphia since 1940, established his own firm in Philadelphia in 1946. That year, László Moholy-Nagy, who headed the School of Design in Chicago, wrote and published a book, *The New Vision,* but sadly passed away that same year. His wife, Sybil, would publish another of his books, *Vision in Motion,* posthumously, the next year. Moholy-Nagy's School of Design would be absorbed by the Illinois Institute of Technology as the Institute of Design after his death, and taken over by Serge Chermayeff (1900–1996). Brooks Stevens designed *Ben Hur,* the first home food freezer, which was blue, unlike all previous white home appliances. George A. Beck became manager of industrial design in General Electric's new electronics department in Syracuse, New York. Donald Deskey in New York expanded his firm, adding architectural, product design, and packaging departments.[15]

In 1945, Henry Dreyfuss had moved to California and opened a second office near his home in Pasadena. His office in New York, managed by a partner, William F.H. Purcell, and an associate, Robert H. Hose (1915–1977), both hired in 1946, designed RCA radios, televisions, and phonographs in 1946. In California, Dreyfuss's new client, Convair in San Diego, was seeking entry into the postwar aviation boom with a flying car developed by Theodore P. Hall, Convair's chief engineer. Styled by Henry Dreyfuss, it would be introduced in 1947. The Model 118 ConVairCar was a four-seat, two-door car with an aircraft engine and wings attached to its roof. Convair planned to build 160,000 of them, costing $1,500 each, and rent them at airports. The first prototype, Model 116, made 66 successful flights, but Model 118 crashed during a demonstration flight, injuring the pilot. The pilot had mistaken the auto fuel gauge for the aircraft's fuel gauge and ran out of fuel in midflight, which is what happened to Convair's plans.

Most of the automobile industry was flying, not to get in the air, but to get back to making cars for a market that was desperately in need. Millions of prewar cars were at the end of their life span, with rust, worn-out parts, or dated looks.

Chapter 11

Celebrity and "Good Design"

Raymond Loewy's fascination with automotive design extended into his private life. In 1946 he took two 1940 Lincoln Continentals and custom redesigned them for himself and a friend. The result was a modern version of the traditional brougham style car body in black with the front seat fully visible beneath a Plexiglas roof, as if for a chauffeur, and an enclosed cabin with round opera side windows for passengers in the rear seat. The custom hood ornament was a Frigidaire door handle.

Loewy often used the car for the showmanship that was his trademark. One of Loewy's student apprentices at the time, Budd Steinhilber, relates the time when he was asked to

> take Loewy's leather presentation portfolio down to a parking garage on 6th Avenue, and put it in a designated Ford convertible. Loewy was scheduled to make a presentation in Hartford, Connecticut. In the meantime, Loewy's personal chauffer was instructed to drive in from Long Island in his customized Lincoln Continental, and park it at curbside in front of the office building on Fifth Avenue. Large crowds began to gather, and soon Loewy, his wife, and her French poodle emerged ceremoniously from the building, stepped grandly into the Lincoln and drove off accompanied by the rumble of its 12-cylinder engine. However, they only drove around the corner and down 46th street to the parking garage where the Loewy's then transferred to the rental Ford convertible. The chauffer was dispatched back to Oyster Bay. Pure and simple theater![1]

Meanwhile, Loewy's office in South Bend had been busy during the war with the first newly designed postwar cars for the Studebaker Corporation, introduced in May 1946. Recall that all automotive companies were required to devote all their manufacturing capability to produce military equipment during the war and to suspend work on consumer goods. General Motors and Chrysler made M-4 Sherman tanks at the rate of 2,000 per month. Ford produced 18,000 B-24 Consolidated Aircraft Liberator bombers. Plymouth made Bofors antiaircraft guns and aircraft landing gears. Packard made V-12 engines for PT boats and aircraft. Willys-Overland, Bantam, and Ford made Jeeps. Studebaker made trucks, aircraft engines, and personnel carriers.

After the war, a few prewar-designed cars, such as Packard, Hudson, and Nash, were quickly introduced in 1946, using old tooling, and with only minor changes. Norman Bel Geddes designed a new grille and dashboard for the prewar Nash. Willys-Overland began production in 1946 of civilian jeeps using military Jeep tooling, the Model 463, 4WD, Jeep station wagon, designed by Milwaukee industrial designer Brooks Stevens to look like prewar "woodies."

But Loewy's office in South Bend, being independent of the Studebaker Corporation, was exempt from the war production restrictions of the automotive companies, and thus, starting in 1939, it was able to begin to develop Studebaker's postwar, 1947 models of the

Champion, Commander, and Landcruiser during the war. They were designed by Virgil Exner, head of Loewy's studio in South Bend, who was hired by Loewy in 1937, and Robert E. Bourke, who was hired by Exner in 1940, both members of Loewy's Studebaker team.

But Loewy had fired Exner in 1944 for conspiring with Roy Cole, Studebaker vice president of engineering, to bypass Loewy. Exner went to work as an employee of Studebaker under Cole, and Loewy replaced him with Gordon Buehrig, who had been hired as head of Loewy's Studebaker studio. Since he left Auburn in 1936, Buehrig had served with the Budd Company until 1938, after which he had worked as a freelance designer. Although an excellent designer, Buehrig was a shy and unassuming person and simply did not have the take-charge disposition needed to manage the studio. Bourke was soon put in charge, and both worked on the 1947 models, which were ready for tooling as soon as the war was over.

When they were introduced in late 1946, design professionals praised the innovative design, with its distinctive wraparound, "greenhouse" rear windows on two-door models; its low-slung, flush side styling; and the flat-back trunk instead of the traditional tapered look of the time. The trunk prompted a joke at the time that one could not tell if the car was coming or going. But it was new and quite different from any Detroit models. Eliot Noyes said it was "so refreshing it made other cars look out of date." *Life* magazine published a ten-page article on Loewy, Studebaker, and its new models. Over the next three years, Studebaker would break all its peacetime records for sales and profits.

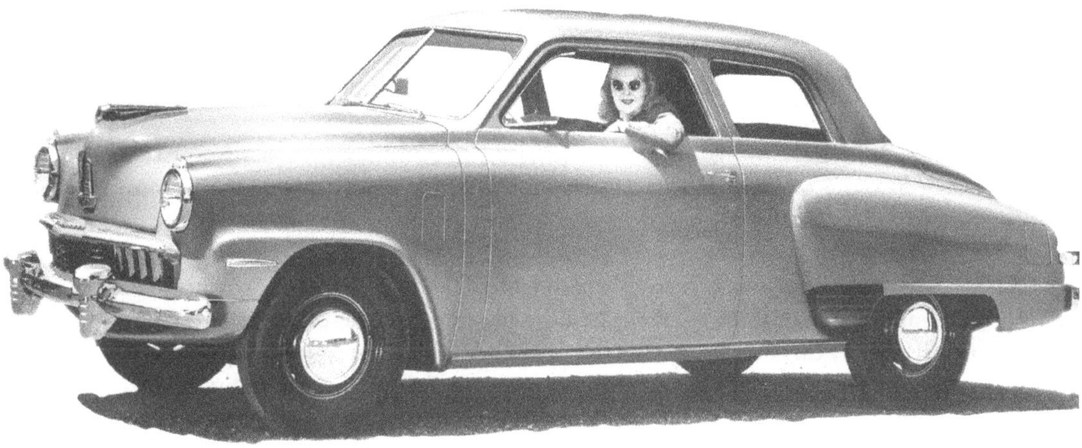

1947 Studebaker, by Raymond Loewy Associates.

Other design activities in 1947 included work by Raymond Loewy Associates (RLA) in its New York office. As consultant to Coca-Cola, RLA designed the red, streamlined dispensers made by Dole Deluxe that sat on soda fountain counters across the country, the work of Carl Otto and Clare Hodgman of Loewy's office. RLA also slimmed down the hourglass shape of Coca-Cola's iconic bottle and designed open-sided delivery trucks to expedite stock delivery and handling.

Tucker Madawick joined RLA in 1947 and with Carl Otto reopened Loewy's London office on Grosvenor Street, which had closed down in 1940 as the bombing of Britain began. Madawick handled the accounts of Electrolux of Sweden, Austin of England, Unilever, Gestetner, Lyon's Tea House, the Rootes Group, and Allied Ironfounders. In Chicago, John Morgan took on Mel Boldt as a partner in his consulting office, and Dave Chapman designed streamlined sewing machines for Montgomery Ward.

Lurelle Guild was engaged by Alcoa to design hardware/aluminum furniture, which included chairs, tables, consoles, vanity desks, bed headboards, nightstands, and chests. Alcoa sold over 400,000 Kensington and Wear-Ever chairs before discontinuing the line in 1952. Guild had a long list of clients and was so well known that many manufacturers featured his name and photograph on marketing materials.

In New York, Norman Bel Geddes had hired Elliot Noyes, who had returned from military service after the war, to be his design director. The office designed a secretarial desk for International Business Machines (IBM), a cigarette vending machine for Rowe Manufacturing Corporation, and a jukebox for the Lion Manufacturing Company. Bel Geddes was soon commissioned by IBM to design a new typewriter and had started working on it, but Bel Geddes's office closed in 1947 due to financial mismanagement. Noyes opened his own office that same year and completed the typewriter design started under Bel Geddes, which would be introduced in 1948 as IBM's Model A. Thomas J. Watson, Jr., president of IBM and a glider pilot friend of Noyes, retained him for product design at IBM.

Nathan George Horwitt designed a wristwatch in 1947 with a round, black face without numerals, symbols, or lines to mark hours or minutes, and with only a single gold dot marking the twelve-o-clock position, intended to be evocative of a sundial. Vacheron & Constantin–Le Coultre, a Switzerland firm, originally manufactured it. But Zenith Movado would proceed to produce the exact same design in 1948, without any credit or compensation to Horwitt. A prime example of design piracy, it would take 27 years in the courts until Movado finally settled with Horwitt for $29,000 in 1975. The watch would be included in the permanent design collection of the Museum of Modern Art in 1960 and become known as "the Museum Watch." A wall clock of the classic Horwitt design was produced in 1960 by the Howard Miller Clock Company and is still being produced today.

In June 1947, the Chicago, Milwaukee, St. Paul and Pacific Railroad introduced its streamlined flagship, the *Olympian Hiawatha,* which traveled from Chicago to Seattle/Tacoma, Washington. Brooks Stevens had designed it from stem to stern. The dramatic orange and maroon train was pulled by a massive Fairbanks-Morse diesel engine topped by a high, forward cab with automotive/aircraft-type split windshield, and a nose like a vertical cylinder with chrome horizontal bands. Many described the Skytop solarium observation cars as the most distinctive railway cars ever built. The rear lounge had glass windows, not just at eye level, but also in three tiers overhead like a greenhouse, to afford an uninterrupted view of scenery. In the forward area were sleeping compartments. The separate ladies' and men's lounges were designed with distinctly feminine and masculine character, respectively. That year, Stevens also designed a Trav-ler portable plastic tabletop radio, an Evinrude Sportwin outboard motor, a line of luggage for the Wheary Company, and an aviation plant in California for McCulloch Corporation's aviation division.

The Society of Industrial Designers (SID) in 1946 had initiated a massive effort to promote the work of its growing membership. Prominent designers were busy writing articles, lecturing, and granting interviews. The October 1946 issue of *Electrical Manufacturing* published an article by Walter Dorwin Teague titled "Industrial Design in Partnership with Engineering," which stressed coordination between the two disciplines to build better appearance *into* and not *applied to* a product. The design and arrangement of Teague's offices and studios were featured in a May 1947 issue of *Architectural Record.* The June 1947 issue of *Coronet* magazine featured a biography and personal interview with Henry Dreyfuss.

By 1947, the American Designers Institute (ADI) had several hundred members, while the Society of Industrial Designers (SID), a more exclusive organization, had only 78. For many years, SID was criticized by some as being an exclusive club, and compared to the ADI, it was. This was obviously the intention of the founders. Ten years after its founding, in 1954, SID membership would total 153, a tenfold increase over its original 15 founders. At that same time, membership in the original organization, the American Designers Institute (ADI), would be 200.

Unlike the ADI, the SID did not offer recommendations for a postwar industrial design education curriculum. Instead, in February 1947, it announced a program of evening courses at New York University for industrial designers, both professional and nonprofessional. Four courses consisting of 15 lectures each were scheduled for each semester of the 1948–1949 academic year. Various industrial designers would lecture on a variety of technical and philosophical subjects. Also in 1947, the SID organized a photo exhibition of 32 of its members, which opened in Philadelphia and traveled to 12 other cities.

Harley Earl, the dominant force in automotive design, started the 10-year Detroit battle of "tail fins" by adding modest but noticeable tail fins on the rear fenders of the 1948 Cadillac, which prominently elevated the brake light housings. Reportedly, Earl was inspired by the Lockheed P-38 warplane's twin tails. Already, Harley Earl's postwar use of excessive chrome and iconic streamlining details were coming under criticism by leading industrial designers, who generally favored simpler and cleaner designs after the war. "Streamlining" had become a derogatory term.

In his 1940 book, *Industrial Design,* Harold Van Doren had explained how the unfortunate term of "streamlining" had been associated with industrial design since about 1934:

> Streamlining has taken the modern world by storm. We live in a maelstrom of streamlined trains, refrigerators, and furnaces; streamlined bathing beauties, soda crackers, and facial massages. True streamlining is not nearly so important, nor so frequently met with in problems of industrial design, as the general public probably thinks.
>
> The manufacturer who wants his laundry tubs, his typewriters, or his furnaces "streamlined" is in reality asking you to modernize them, to find the means for substituting curvilinear forms for rectilinear forms ... [to] conform to the current taste, or fad if you will.... He expects, too, that unnecessary exposure of mechanical parts will be eliminated, that buttons will be substituted for levers, and control panels and dials will be organized into simple and easily read groups, tied in wherever possible with related elements of the principle forms.
>
> In 99 cases out of 100 this procedure is utterly unrelated to genuine streamlining. Then why does the client refer to it as such? ... Simply because, in the unbelievably rapid growth of the American language, words soon lose their specific and restricted meanings and assume a general significance embracing far wider fields than originally intended.[2]

The original term "streamlining" correctly referred to automobiles, aircraft and trains, which moved at high speed and therefore were made more efficient with streamlining. But when used on stationary objects, it was understandably criticized as unnecessary, and postwar designs generally abandoned the practice, except for Detroit and the automotive industry in general, and of course it is still used there today. But in the late 1940s, Detroit was the recipient of criticism, even though automobiles had reasonable justification to use streamlining under the original meaning. But streamlining was no longer the problem. Simply put, it was more a matter of an excessive amount of decorative chrome and extraneous details such as tail fins, elaborate hubcaps, and hood ornaments that Raymond Loewy would call "jukeboxes on wheels" in a 1955 speech he would deliver to the Society of Automotive Engineers.

The same critics also did not approve of the Detroit term "styling," which by now had become almost synonymous with decorative surface embellishment, rather than the ergonomic consideration and convenience of passengers. Cars were so low that entry was made difficult even without a hat. There were no seat belts or air bags. You may recall that Harley Earl had changed the name of his group at General Motors from the "Art and Colour Section" to the "Styling Division" in 1937. The term "styling," for many, suggested cosmetic decoration, excessive streamlining, and insincerity, rather than the more comprehensive practice of industrial design, which would have considered safety devices and user convenience. Had Earl used the term "industrial design" instead of "styling," he could have avoided such criticism, but he did not. It seemed that he and his colleagues envisioned their work as something different than designing toasters and refrigerators.

It was the beginning of a sustained split within the industrial design community, with three more or less independent groups: the single-product designers, who worked mainly, but separately, in the home furnishing industry (American Designers Institute); the automobile industry (Detroit); and the multiproduct/multiclient designers who had founded the Society of Industrial Designers in 1944. As the number of industrial designers began to increase exponentially after about 1948, when the first wave of new graduates arrived in the field, these three groups would slowly blend into the single identity of industrial design over the next 20 years.

In 1948, **Ray Patten** at General Electric passed away while heading GE's Appearance Design Division. Arthur BecVar, who had joined GE in 1944 after returning from military service, was appointed to fill his vacated position. BecVar changed the name of his group to the Industrial Design Division. Industrial designers like BecVar, who were employees and heads of corporate industrial design organizations, were in the minority at the time, representing only 15 percent of the members of the Society of Industrial Designers (SID). They were informally referred to as "in-house" designers. The remaining 85 percent of SID members were attached to independent consulting firms. And no, they were not referred to as "out-house" designers, but "consultants."

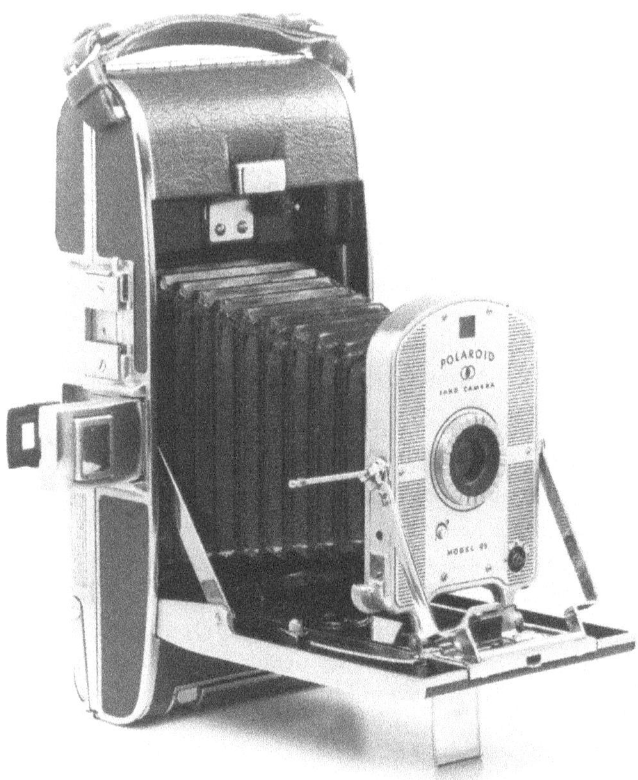

Polaroid camera Model 95, by Walter Dorwin Teague Associates.

As 1948 progressed, Alfonso Iannelli became a consultant for appliance manufacturer John Oster, designing mixers, hair dryers, knife sharpeners, and hair clippers. Walter Dorwin Teague Associates designed the first Model 95 Polaroid camera, produced by the Polaroid Land Company, which became an instant success. Brooks Stevens designed a refrigerator for the Coolerator Company and the iconic Willys-Overland Jeepster phaeton (a roadster with four seats rather than two), a sporty convertible that retained the familiar fenders and vertically slotted grille of the wartime Jeep. Ten thousand were sold the first year. In 1949, Stevens would design Harley Davidson's Hydra-Glide motorcycle and a Roadmaster bicycle for the Cleveland Welding Company.

By 1948, Barnes & Reinecke

had a huge staff of 375 and a shop of 50 machinists, but Reinecke sold his interest in the partnership and established his own office, J.O. Reinecke & Associates, specializing in industrial design and product planning, with offices on Ohio Street in Chicago, taking with him designers Jon Hauser, Harold Hart, Don Lowe, and Jack Knight, as well as a number of accounts. Clients included Caterpillar, Emerson, AT&T, Johnson & Johnson, 3M, McGraw-Edison, Westinghouse, and Zenith.

In Cleveland, Viktor Schreckengost had returned from military service to teach industrial design at the Cleveland School of Art (now the Cleveland Institute of Art). He also continued to design many consumer products, such as for his client Murray Bicycle Company, for whom he designed many children's transportation products, including pedal cars, tricycles, and banana-seat bikes well into the 1960s. Schreckengost also produced work in fine art, sculpture, and painting. In the early 1950s, he created monumental ceramic sculptures for the Cleveland Zoo (now Metroparks Zoo).

Ford introduced its first postwar car design in 1949 that would become an automotive classic. As mentioned earlier, Henry Ford II had in 1946 selected the design concept of industrial design consultant George W. Walker over that of Bob Gregorie, the head of Ford's styling studio, causing Gregorie to resign. Walker had actually submitted three quarter-scale clay concept models to Henry Ford II, one each by three of his staff members: Elwood Engle and Joseph Oros, Walker's key designers, and a model maker, Dick Caleal, just hired by Walker because of the clay model he had made at Walker's request. It was Caleal's model that got Walker the job with Ford.

How Caleal, a lowly model maker, created a concept that won out over two of the best automotive designers (three, if you include Gregorie) in the industry was unknown at the time. But a story of intrigue would come out, essentially because of Robert Bourke, then head of Loewy's Studebaker team in South Bend. Studebaker's chief engineer had ordered Bourke to cut Loewy's staff because of budgetary reasons. Bourke had to fire several people, including one of his model makers, Caleal. Now unemployed, Caleal had approached Walker for a job and was told by Walker that he would be hired if he built an acceptable quarter-scale clay model of a concept proposal for the 1949 Ford, for which Walker was trying to get a contract, within a week.

Caleal excitedly called Bourke with the good news and asked if he could help him plan the model. Bourke, wanting Caleal to get the job with Walker, gave him a wooden "buck" or

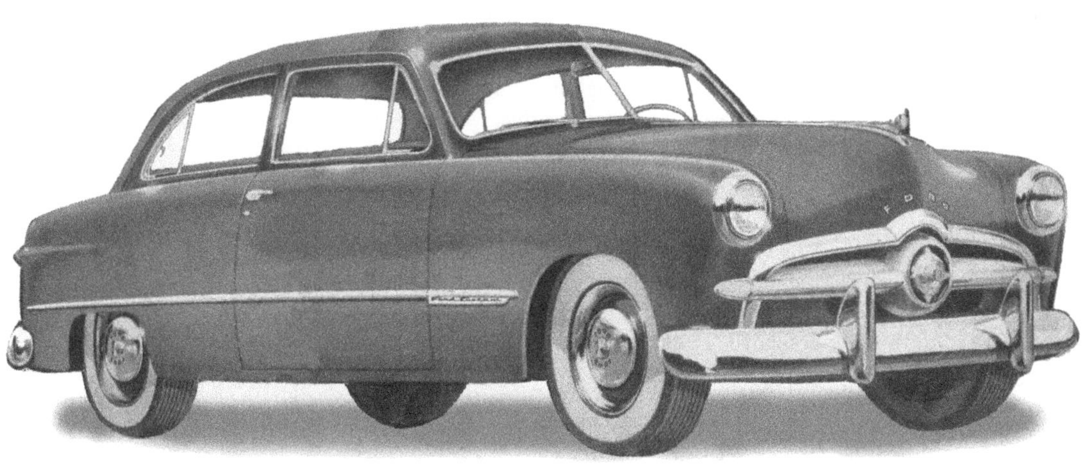

1949 Ford, by George Walker staff member Dick Caleal.

armature for the quarter-scale clay model, and some wheel models. Bourke and three other designers agreed to help Caleal with the model on evenings in Caleal's Mishawaka, Indiana, home when he had the model up to rough size, in order to help him meet the deadline. They did so on Caleal's kitchen table, heating the modeling clay in the kitchen oven. Based on Bourke's rough sketches, they all worked on the model at the same time, Bourke handling the front grille and motif.

Bourke had been inspired since 1940 by aircraft propeller hubs that looked like the tip of a bullet and had sketched numerous ideas with such a spinner hub on the nose of concept sketches (see also the 1950 Studebaker, described later). So it was not unexpected that this device would appear in the center of the clay model's grill. When it was finished, the model was 99 percent like the Ford that would be produced in 1949, with the spinner centered on the middle horizontal grille bar as its most distinctive feature.

Caleal took the model to Walker on the backseat of his car; Walker liked it, and Caleal got the job. Bourke was silent, since he had secretly used Loewy's staff and materials to help Caleal and didn't want to jeopardize Caleal's job. As a result, Walker, in the tradition of the day as head of his office, received full credit for the design, By 1978, the design would be hailed as "The Car That Saved an Empire—The 1949 Ford." But Bourke would spill the beans in 1985, and the story would be out.[3] In 2002, Ford would introduce a retro concept model called the Forty-Nine, which emulated the original, and in 2003, Ford would officially credit Caleal with the design of the 1949 Ford.

Gordon Buehrig left Loewy in 1949 and went to Ford, as head of the body development studio, and would work on the 1951 Victoria coupe and the 1952 Ranchwagon. George Nelson, who had opened his own office in 1947, was by now producing a flood of highly acclaimed furniture designs as design director of the Herman Miller Furniture Company, including his 1949 Basic Cabinet Series, his 1945 Storage Wall concept, and his 1949 Atomic Clock for the Howard Miller Clock Company, named because of individual balls marking the hours, which radiated from the center hub on thin sticks. It was the first of many of his clock designs.

Nelson, who had been the anonymous author of the influential 1934 *Fortune* magazine article "Both Fish and Fowl," now wrote another article in the July 1949 issue of *Fortune* titled "Business and the Industrial Designer," this time under his own name. He retraced the brief 20-year history of the profession and acknowledged that increasing sales is an integral objective of industrial design, but that following fashion trends or fads can be detrimental to small manufacturers. Using the example of automotive design, he explained,

> The general tendency of the little manufacturer is to tag along with the big three in the apparent hope that this course is the safest, although in point of fact it is the most risky because it means competing on the most unfavorable terms possible.

He then cited the 1947 Studebaker Champion designed by Raymond Loewy (with a lot of help by Bob Bourke),

> which demonstrated with great clarity the one way open to the small manufacturer: design leadership, [and which] ... established itself as a distinct entity in the public mind in the only effective way it could—through the appearance of the product itself. In other words, for most manufacturers in the lower brackets, the radical approach to design may be the safest business approach.

Nelson also challenged business to "get the most out of their designers" by suggesting that industrial designers of ability and integrity be established

> as a member of a company's policy-making group, with freedom to make his influence felt not only on product design but on all matters of general policy that affect design.[4]

In other words, he wanted designers to have an influential managerial role in industry, not simply function as artistic stylists. Nelson himself was a living example of design's future beyond "styling" as an influential director of design for Herman Miller. Over the next decades, many designers would climb the corporate ranks to become vice presidents with significant influence in business strategies.

About this same time, the Society of Industrial Designers (SID) published *U.S. Industrial Design, 1949–1950*, with 190 pages and 389 photographs, showing the work of members Egmont Arens, Robert Davol Budlong, Dave Chapman, Donald Dailey, Donald Deskey, Henry Dreyfuss, Gordon Florian, Lurelle Guild, Raymond Loewy, Peter Müller-Munk, Joseph Platt, George Sakier, Brooks Stevens, Viktor Schreckengost, Walter Dorwin Teague, Harold Van Doren, Russel Wright, and many others.

SID also conducted a Product Design Seminar at Lehigh University in Bethlehem, Pennsylvania, from July to August in 1949. Speakers included Walter Dorwin Teague, who spoke on "Rightness Resurgent"; Donald Dailey, who presented a case history; Carl Sundberg, who spoke on "Designing for Thermoplastic Materials"; Egmont Arens, who spoke on "Low Pressure Molding of Plastics"; J. Mcleod Little, who spoke on "Product Design"; Henry Dreyfuss, who spoke on "Design Analysis"; and Harold Van Doren and Gordon Florian, both of whom also spoke on "Design Analysis." The seminar fee was $500.

By 1949, Raymond Loewy Associates was incorporated and became known as the Raymond Loewy Corporation (RLC), which at that time had a staff of 200 working under partners Loewy; William Snaith (managed all department store work); A. Baker Barnhardt (in charge of all packaging, product, and transportation); and John Breen (business manager). Raymond Loewy Inc. also included many soon-to-be-famous designers who worked for Loewy without any public recognition for years, since only Loewy was allowed to receive design credit. Among them were Richard Latham, Peter Thompson, George Jensen, Clare Hodgman, Robert Bourke, Virgil Exner, Robert T. Tyler, Franz Wagner, Joseph Parriott, Jay Doblin, Gordon Buehrig, Carl Otto, Don Defano, Tucker Madawick, and countless others. In 1949, Loewy's organization designed a complete line of electrical home appliances, including refrigerators, ranges, automatic washers, ironers, and dryers for the Frigidaire Division of General Motors. The refrigerator was a simple white box, with only the handle end accented by a delicate chrome circle, a visual "target" to open the door.

But the most successful result of SID's publicity campaign was when Loewy pulled off the greatest postwar industrial design public relations coup by appearing on the cover of *Time* magazine on October 31, 1949, with the caption, "He Streamlines the Sales Curve." On the cover was a color drawing by Boris Artzybasheff (1899–1965), the famous artist who created many *Time* cover portraits of world personalities. This one featured a large image of Loewy's face, surrounded by ghostly images of Loewy's designs, including his 1950 Studebaker, S-1 locomotive, Greyhound bus, aircraft, boat, furniture, and appliance packages, all floating in a bright, blue sky background around Loewy's head.

"Blue sky," ironically, is a common design term describing imaginative but often impractical ideas, the opposite of Loewy's creations. The article, just over five pages in length, was in the "Business and Finance" section on page 68, titled "Modern Living: Up from the Egg" (Loewy had often described the egg as a perfect design), and reviewed a number of Loewy's current designs in progress, which included the new two-level Greyhound Scenicruiser, designed in 1946 but just being road tested, a face-lift for Manhattan's Gimbel Brothers department store, and models for three new ocean liners for American President Lines.

Included in the text was a brief history of industrial design and references to Loewy's pioneer colleagues who brought art to the Machine Age, including Teague, Dreyfuss, and Van Doren. Egmont Arens was cited as making chairs more comfortable by taking clay impressions of fat, skinny, and in-between posteriors to make a one-piece plastic chair; Dreyfuss for his form-fitting modern toilets for the Crane Company; Teague for Boeing Stratocruiser interiors, and Bel Geddes for transforming 10,000 square feet of lobby, dining, and storage space into a modern theater that extended the stage into the audience.

Loewy was the Steve Jobs of the day, and for similar reasons. They didn't invent the technology, but they took it and turned it into life-changing physical products that consumers loved and valued. There was a colorful description of Loewy as a person in the *Time* article:

> **The Shy Salesman.** Suave, grey-haired, medium sized (5 ft. 10 in.), Loewy talks in a subdued voice that is, at the same time, apologetic and compelling. His face is reposed, gentle, sad, and as inscrutable as a Monte Carlo croupier. Obsessively shy, he is always "Mr. Loewy" to his associates.... Despite his shyness, he is a crack salesman who throws no artistic tantrums. Far from turning out designs with offhand sureness, he works them over painstakingly until the client is satisfied. He also has an almost hypnotic power to impress, persuade and convince the toughest tycoon.[5]

The *Time* article also described Loewy's extravagant lifestyle, and what he did with his annual gross income that exceeded $3 million (equivalent to $29 million in 2013 dollars), compared to $1 billion in yearly sales of Loewy-designed products (equivalent to $9.77 billion in 2013). In addition to his Manhattan apartment atop a building on 58th Street, in 1946 he had designed and built a $100,000 ($977,000 in 2013) dream house, Tierra Caliente, in the desert near Palm Springs, California, complete with an outdoor swimming pool that curved into the living room. Summer found him in France, where he had three homes, one a red-tiled villa, Uriane, in St. Tropez, purchased in 1930; another a 16th-century manor, La Cense, outside Paris, built by Henry IV, and purchased by Loewy in 1933; and an apartment on Paris's Quai d'Orsay.[6] Not mentioned was the Loewy home in Tetelpan, Mexico, built in the 1940s and also called Tierra Caliente. Loewy had just married his second wife, Viola, 30 years his junior, in 1948.

The use of the term "streamline" on the cover of *Time* magazine with Loewy's picture illustrated the persistent popularity of the term in 1949 to describe industrial designed products. By this time, the surviving holdout for 1930s "streamlining" was the automotive industry, led by General Motors. Most design professionals were abandoning the term, as well as the practice, on stationary products. Nevertheless, by coincidence, that same month, Harold Van Doren, then president of SID, who was also part of SID's promotion campaign, wrote a feature article, "Streamlining: Fad or Function?" in the October 1949 issue of *Design,* the monthly journal of the British Council of Industrial Design (CoID, established in London in 1944), which had just begun publication of its professional journal. Van Doren defended streamlining, which had been under fire by both the Museum of Modern Art (MoMA) and Europeans: "Much so-called streamlining is imposed on the designer by the necessity of obtaining low cost through high-speed production." Using the refrigerator as an example, he compared the traditional costly process of bending sheet metal at sharp angles and assembling multiple components, versus the large metal stamping presses adopted by most manufacturers since about 1936:

> Since steel stampings techniques make it impossible to produce hard, 90 degree edges of pointed corners, any designer engaged by a refrigerator company was already committed to these large radii and softer contours.

He concluded his defense: "Is it fair to damn the designer to eternity because his designs echo the one really new visual phenomenon of his age?"[7]

Van Doren essentially was saying that most of what critics were calling streamlining was the result of manufacturing requirements for economy, which was exactly the explanation he gave in his 1940 book. He was correct, but critics of industrial design had a tendency to be more interested in artistic criticism than facts.

The AT&T 500 telephone designed by Henry Dreyfuss Associates (HDA) was introduced in 1949, a sleeker and more compact version of the original 1937 model. The unsung designer was Robert Hose, an associate of HDA. In 1954, the 500 would be the first telephone to be offered in colors other than black, and would still be the most commonly used phone in the United States in 1995. In December 1949, contributing to the Society of Industrial Designers' (SID) ongoing campaign to promote industrial design, Dreyfuss made an address to the American Society of Engineering Educators. He suggested a revision to the traditional three elements manufacturing relied upon for success: "utility, quality, and price," which he said could be easily duplicated by competitors. He then spelled out the five "vital factors the industrial designer must consider":

1. **Convenience of use** (including utility and safety).
2. **Maintenance.**
3. **Costs** (consideration of tooling, general manufacturing costs, and distribution problems).
4. **Sales** (merchandising knowledge, public acceptance, and awareness of competition).
5. **Appearance** (last but not least, the designer must add his knowledge of form, line, proportion and color to integrate the product into a pleasing whole).

These principles would become the credo of HDA, repeated in magazine interviews and lectures, and were picked up by many industrial designers well into the future. In his address, Dreyfuss then criticized the traditional use of extraneous decoration, imitation of materials other than those used in manufacture, and sentimental icons:

> With the application of this formula by the industrial designer, the extraneous decoration which the manufacturer had previously thought sufficient attraction to sell his product, began to disappear. Sullivan, that great forerunner of modern architecture, said that form should follow function. The industrial designer picked up that watchword, and rose buds, griffen's heads, and other doo-dads became increasingly conspicuous by their absence. In short, the industrial designer permitted the function of the product to dictate its appearance.

Mindful that he was speaking to engineers, Dreyfuss tactfully added: "I should like to underscore here the fact that no industrial designer worthy of the name could contemplate this work without the close cooperation of engineers."[8]

In the fall of 1949, the 1950 Studebaker was on the market, priced at $1,487 ($14,500 in 2013) and designed by Loewy's South Bend Studebaker studio, with a staff of 30 under Robert Bourke. The body was basically the same as the successful 1947 model for cost-reduction purposes, so the design needed a dramatic front end to make it obviously different than the 1947. Bourke was competing for its design with Virgil Exner, who was in 1947 working directly for Studebaker's vice president of engineering, Roy Cole. Bourke won the competition and secured the Studebaker account for Loewy, Inc., over the next seven years with his famous aircraft nose spinner (see the 1949 Ford, described earlier) to create the now classic "bullet-nose" Studebaker. The first month, Studebaker sold more cars than any in its history. During its production life of three years, over 500,000 would be sold. Studebaker would become the largest independent carmaker, fourth under the Big Three, and the only serious challenger to Detroit styling. In 1948, Exner had left Studebaker and returned to Detroit to work for Chrysler, where he would become director of styling.[9]

Ever since 1940, when it created its own separate department of industrial design, the Museum of Modern Art, led by architects, had been trying to establish its influence on the field of industrial design, which had been monopolized by nonarchitect Machine Age pioneers. MoMA had consistently criticized the designs on the market as "modernistic and streamlined aberrations." In 1941 it had conducted its Organic Design in Home Furnishing Competition, won by architects, and promoted "organic" design (no "vain ornamentation or superfluity"; read: no streamlining) as its aesthetic standard. In 1946, the new director of MoMA's industrial design department, Edgar Kaufmann, Jr., had invited members of the American Designers Institute (ADI) and the Society of Industrial Designers (SID) to attend a Conference on Industrial Design: A New Profession. The discussion was largely philosophical and resulted in no specific actions. If MoMA's objective was to gain a voice in the new profession, it failed to do so. So MoMA decided to go it alone and jump into the consumer market, acting as an arbiter of public taste.

In January 1950, the first Good Design exhibition opened in the marketing center of the Chicago Merchandise Mart. It was conceived and launched by Kaufmann, in collaboration with MoMA's director, René D'Harnoncourt, and Wallace O. Ollman, general manager of the Mart, who issued this joint statement:

> It is the first time an art museum and [a] wholesale merchandising center have cooperated to present the best examples of modern design in home furnishings.... [We] believe and hope that in combining their resources they will stimulate the appreciation and creation of the best design among manufacturers, designers and retailers.[10]

There would be three Good Design exhibitions annually over the next five years: at the winter market in January and the summer market in July, both of these at the Merchandise Mart, and in the fall, selections were transported to New York for display at MoMA during the pre–Christmas period. Selection of home furnishings for the exhibition was by a three-person committee consisting permanently of Kaufmann, and a designer and retailer appointed for each show. Public opinion was studiously avoided. Selections of 150 to 250 products were chosen from the showrooms of various manufacturers at the Mart and relocated to the exhibition on the 17th floor. At the winter market, the exhibition was open only to wholesale buyers, but after that it would be open to the public, with a guided tour offered for a fee. A well-known designer or architect designed each exhibition. A "Good Design" tag was attached to each selected design, all of which were available in the open market, and each manufacturer could purchase a kit of promotional materials, including "Good Design" tags, labels, suggested window displays, and literature to promote sales of their product as a MoMA-selected "Good Design."

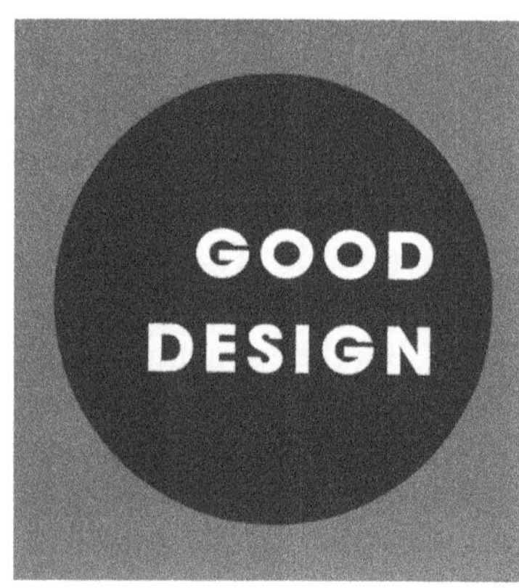

"Good Design" tag.

What exactly was "Good Design"? Kaufmann defined it like this:

> A frequent misconception is that the principle purpose of good modern design is to facilitate trade, and that big sales are a proof of excellence. Not so. Sales are episodes in the careers of designed objects. Use is its first consideration.

Thus sales, the fundamental purpose of industrial design, was tossed aside. This seemed somewhat hypocritical from a man who conceived the entire "Good Design" program in a commercial setting, which promoted the sale of products branded with the MoMA "Good Design" labels. Further, the first consideration for selection seemed to not be use, but simply aesthetic, abstract, and artistic presentation of form. Elliot Noyes in 1941 had said "Good Design shows the taste and good sense of the designers. On none is there arbitrarily applied decoration."[11]

"Arbitrarily applied decoration," to MoMA, included any commercial message, such as an obvious brand name or decal, or any element that interrupted the purity of geometric form. This definition was "art" masquerading as "design," with the only criteria being the personal aesthetic choice of the individual designer. Genuine designers design not for themselves, but to meet the limitations and sales needs of manufacturers, to adapt to the physical requirements of users, and to appeal to consumers at the point of sale. Eva Zeisel (1906–2011), revered ceramic tableware designer who won countless "Good Design" awards, at age 90 said, "Good Design [as defined by MoMA] was never good business, it only appealed to a very narrow elite. 'Good Design' was nothing more than lack of applied decoration."[12]

Actually, it was more than that. The MoMA standard of "Good Design" were designs devoid of any emotional, representational, or playful characteristics, the subtle qualities that make designs more interesting, more personal, and more meaningful to average consumers. Faber Birren (1900–1988), a famous color consultant of the 1950s who monitored and published color trends, once remarked that any product recognized by MoMA as art was "the kiss of death" commercially. The bottom line was this: "Good Design" depended entirely on how it was defined and judged, and by whom. It is interesting to compare MoMA definitions of "Good Design" with the pragmatic business definitions of industrial design by Henry Dreyfuss. He was not impressed with the museum's idea of good design. When asked by visitors from an English museum in which museums he would like to have his designs exhibited, his reply was, "They are called Macy's, Marshall Field's, and the May Company—similar to your Eaton's, Simpson's, and Morgan's."[13]

A number of "Good Design" awards went to imported European products. Europe had largely recovered from World War II, found distribution for their products in the United States, and promoted them through trade exhibitions. Germany, Japan, and Italy had rebuilt their manufacturing capabilities, opening the gates for exported automobiles, furniture, and electronic products to the American market, and stiffening the competition for all U.S. manufacturers. Among these imports were Scandinavian products in 1948; the Volkswagen Beetle from Germany, which entered the U.S. market in 1949; and the French line of porcelain-enameled cast-iron cookware, Le Creuset, in 1950.

The Museum of Modern Art and other U.S. museums promoted European design aggressively through exhibitions of Italian designs and designers such as Marcello Nizzoli (1887–1969) of Olivetti (typewriters), and Corradino D'Ascanio (1891–1981) of Lambretta (motor scooters) and Giovanni Ponti (1913–1993). It seemed as if art museums were clinging to their age-old assumption that European design was always superior to American design and were urging U.S. designers and industry to imitate.

It never seemed to occur to museums that U.S. design had a unique personality of its own, or that it was U.S. design that was inspiring and leading the rest of the world to seek American markets. If other countries wanted to access the enormous U.S. market, they had to design for the U.S. consumer, and to accomplish this they often needed American industrial designers to advise them, or to hire American designers to work overseas.

But now, another war began, only 10 years after the last, this time in western Asia.

CHAPTER 12

Competition and Controversy

The Korean War began in June 1950, threatening industry with restrictions of critical materials. Peter Müller-Munk used the threat of austerity to urge manufacturers to reduce excessive decoration of products:

> The electrical and automotive industries ... have reveled in ... superfluous shiny decoration ... to catch the customer's eye and attention.... The effect of this has been deplored by every serious designer.... Chrome and fancy trim used freely without a design need now can be easily taken off.[1]

In 1950 the New York and Philadelphia design offices of Van Doren, Nowland, & Schladermundt announced they would operate as two independent firms. Van Doren remained in control of his Philadelphia office as Harold Van Doren & Associates, and the New York firm became known as Nowland & Schladermundt.

Brooks Stevens designed the Skylark pattern for Formica, the postwar surface of choice for kitchen and bathroom countertop surfaces. Skylark was called the "boomerang" pattern, because of the multicolor, thin line shapes resembled boomerangs, as well as alluding to supersonic airplane wings and modernity. Henry Dreyfuss Associates designed an aluminum outdoor telephone booth for AT&T, which became ubiquitous across the country until cell phones became the norm in the 1990s. C. Stowe Myers was recruited from Teague to join Loewy, Inc., in New York. He would manage Loewy's Chicago office from 1952 to 1954. Arthur BecVar, head of design at General Electric in Bridgeport, Connecticut, relocated in 1950 to Louisville, Kentucky, to become GE's manager of industrial design for major appliances. Donald McFarland would become manager of product planning at GE Bridgeport and would later replace BecVar as head of small appliances.

In Chicago, Robert Budlong designed a transverse, two-slot Radiant Control T-20 toaster for Sunbeam that lowered bread automatically and raised itself silently when done. It would remain in production until 1995. Also in 1950, Budlong designed, with Sunbeam staff designer Ivar Jepson, the compact Sunbeam Shavemaster electric shaver with a smooth rounded head, the first compact, ergonomic shape held in the palm of the hand, rather than the previous elongated shape held like a hammer. He continued to design radio cabinets for Zenith.

Five years after the war, the U.S. economy was back to normal, and businesses were humming along, despite the Korean War, which had reinitiated the draft for single young men. The first generation of postwar industrial designers had graduated and was expanding the field into the smallest of manufacturing companies. Annual salaries for designers started at about $4,500 ($43,500 in 2013 dollars). There was no longer a question of the value of industrial design to successful products, but there remained a residual resentment by some engineering managers, who were still in charge of most product development, and who guarded their traditional authority in matters of product design. Proposals by industrial designers, even the most famous ones, were not always accepted.

But designers persevered and integrated themselves into sometime hostile environments. Americans were in the mood to spend, manufacturers still had little international competition, and design organizations were expanding and promoting their profession. They were generally good times.

By 1951, many Americans were living comfortable middle-class lives, with all the latest appliances, cars, and furnishings. Russel and Mary Wright were designing paper Christmas ornaments, paper table service, table linens, and draperies, among a range of accessories and pottery. They also wrote and published *A Guide to Easier Living,* which placed them in the position of being the Martha Stewart of their time. The book, in the manner of Stewart, suggested a variety of ways that young families could better organize their homes and make housekeeping easier, such as buying furniture that was easy to clean, how to organize a room for maximum living space, and simple ways to maintain a pleasant and attractive living environment.

The American Designers Institute (ADI) changed its name to the Industrial Designers Institute (IDI) in 1951 in order to address the continuing criticism by the Society of Industrial Designers (SID) that "industrial design" was absent in its title, thus accepting members from any specialized product industry, or even one-of-a-kind craftsmen. SID, of course, only accepted designers who designed products in a variety of mass-production industries. This remained as a fundamental difference in the two organizations, as well as their geographical separation of headquarters (ADI in Chicago, SID in New York). The subtle change in title was enough to convince the Chicago Society of Industrial Designers (CSID), founded by Dave Chapman and others in 1938, to merge with IDI. Chapman had been president of SID in 1950, and Russel Wright in 1951. SID had chapters in New York, Chicago, Detroit, and Los Angeles. The competing organizations were growing closer. By 1955, IDI would relocate its headquarters from Chicago to New York, where many of its 200 members resided.

In 1951, IDI, possibly inspired by MoMA's "Good Design" exhibitions, initiated the first national industrial design awards program, which recognized the best designs and credited the actual designers by name. This did not sit well with their rival society, SID, where large offices like Loewy, Dreyfuss and Teague *never* credited anyone but their namesake. But over the next 15 years of these awards programs, most SID members began to credit individual designers, except Dreyfuss, who remained adamant about the issue. No designs from Henry Dreyfuss Associates (HDA) ever won IDI awards, probably because none were submitted. Nevertheless, the awards broke the ice, and many individual designers finally received public design credit for their work, which greatly enhanced their reputations and careers.

Dreyfuss appeared on the cover of *Forbes* magazine in 1951, part of the Society of Industrial Designers' publicity campaign. His firm that year designed an innovative round wall thermostat for Honeywell, the interiors for American Export Lines' SS *Independence* and SS *Constitution* ocean liners, and a lavatory sink for Crane.

Carl Otto resigned from Raymond Loewy in 1950 to establish his own firm, Carl Otto Design, with offices in New York, Stockholm, and London. His New York office was in the Squib building on Fifth Avenue. He hired Harold Vanderhyde; Jay Doblin (1920–1989), recently director of the night school at the Pratt Institute; Albrecht Graf von Goertz, the equivalent of an Earl (all ex–Loewy men); and Alban Gifford Jackson (b. 1922), a New Zealand boat designer who had just arrived in America in 1949, and who in 2013, at age 91, would receive New Zealand's Order of Merit for his distinguished career in industrial design.

Otto and his staff designed the 1951 Schick Model 20 electric shaver in collaboration with Norman Gray, chief engineer at Schick, Inc., reducing the bulk and length of the shaver by rearranging the internal components from the traditional "in-line" configuration to a more compact, wider shape, more easily held. The design won an IDI design award in 1951 and became the standard typeform for shavers thereafter, reaching six million in sales.

Otto's firm would also design the 1952 Edison Voicewriter portable recorder (competitor to Dictaphone), which would win an IDI award in 1953. That year, Otto worked on an account in England from the Standard Motor Company to design the Triumph Vanguard. He had hired Tucker Madawick, an automotive design expert with Loewy, to help with clay models and the final design. Otto's international firm would close in 1954, and Madawick would return to the Loewy office as an account executive. Clare Hodgman had also resigned from Loewy in 1951.

John Gordon Rideout, one of the founders of the Society of Industrial Designers in 1944, had closed his Chagrin Falls, Ohio, office in 1946 and passed away at age 53 in 1951. Greta Von Nessen designed the Anywhere lamp, which could be used on a table, mounted on the wall, or suspended from the ceiling. Franz Wagner, Richard Latham, and Don Defano, of the Raymond Loewy Corporation, designed the elegant Flight bathroom scale that the Borg-Erikson Corporation would introduce in 1953. It would be selected by *Fortune* magazine as one of the top 500 designs of all time and would win an IDI national design award in 1954. In Chicago, John Morgan patented a deep fryer.

In 1951, Walter Paepcke, head of the Container Corporation of America, initiated the International Design Conference in Aspen (IDCA), which would become an annual event, bringing top designers and philosophers to the lovely natural beauty of the Colorado ski resort. Among the founding members of IDCA was H. Creston Doner, director of the Libby-Owens-Ford design department.

The machine design founders were still suffering under their unfortunate "streamlining" reputation. In 1951, Walter Dorwin Teague, in an article promoting American design in Britain's *Art & Industry* magazine, chided his British colleagues for

> a certain amount of "ribbing" ... on what they consider our penchant for "streamlining" ... [which has become] a popular metaphor for anything smooth and svelte ... and it is carefully avoided by designers except when it can be used with technical accuracy [as in aerodynamics].[2]

Of course, while the use of the term was fading as a general criticism of American industrial design in general, the automotive industry continued to utilize streamlining "with technical accuracy." Harley Earl and General Motors (GM) led the automotive design field. In 1951 Earl exhibited a concept convertible called LeSabre, named after the North American F-86 Sabre jet fighter plane used in the Korean War, then still in progress. It featured the first genuine curved, wraparound windshield, which had been in development by Libby-Owens-Ford Glass Company since 1946. Tail fins and a grille simulating a jet plane air intake completed the aircraft theme. LeSabre would set GM style trends for the next ten years, as cars became lower, sportier, and costlier.

Few American car designs were included in the Museum of Modern Art's exhibition 8 Automobiles in 1951, MoMA's attempt to subtly criticize the current style trends of Detroit, and to suggest what Detroit should be designing. The eight examples were chosen as historical works of art, MoMA's usual definition of "Good Design." Five of the eight were European; the only U.S. designs were the 1937 Cord, the 1940 Lincoln Continental, and the 1941 wartime military Jeep.

In 1951, Loewy wrote and published his book, *Never Leave Well Enough Alone,* which was a flamboyant, romanticized, and self-promotional autobiography that influenced scores of young men into the profession that seemed to offer the wealth, influence, and celebrity that Loewy enjoyed. As a hobby in 1952, Raymond Loewy established an office in Paris, Compagnie de L'Esthetique Industrielle (CEI, Company of Industrial Esthetics), headed by Harold Barnett. For the next 30 years, the office would design packaging, architecture, and products throughout Europe.

By this time, U.S. industrial design students by the hundreds were graduating from about 30 institutions of higher learning with industrial design programs, over 20 new ones since the end of World War II. New firms were being started by recent graduates across the country to provide local, and less expensive, industrial design services than the large, urban offices in Chicago, New York, or the West Coast. Most would not be as fortunate as Loewy, as industrial design salaries and fees were being reduced because demand no longer exceeded supply, as it had in the 1930s and 1940s.

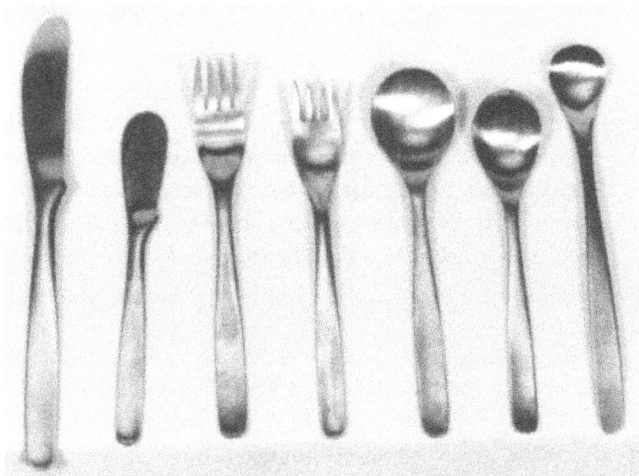

Hall Cutlery Highlight flatware, 1952, by Russel Wright (from *Russel Wright Designs,* by Kerr).

In 1952, Hall Cutlery introduced a stainless steel flatware pattern called Highlight, designed by Russel Wright, to match his design of Highlight dinnerware, introduced by Hall in 1951. The dinnerware had won MoMA's Home Furnishing Award. The Highlight flatware was the first modern and contemporary flatware designed in the United States since World War II and would be imitated by many competitors. Wright thought it was his best and informally called it "Pinch" because of the way the narrow part of the handle just before the business end was "pinched" into a small but strong diameter. In 1952, Wright's wife Mary died; their only adopted daughter, Annie, was only two years old. Wright

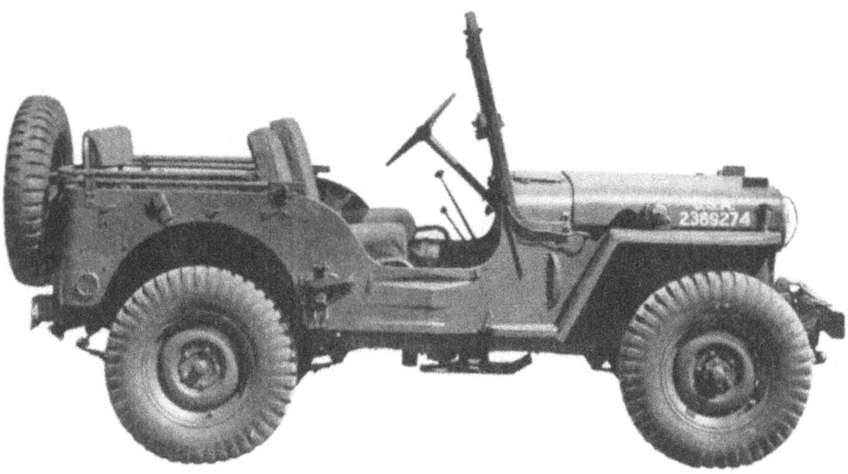

Jeep, 1941.

closed his studio and retired to his retreat, Manitoga, a wooded mountainside in Garrison, New York, where he would build his modernist home and Dragon Rock, his studio. There, he would raise and care for his daughter as a single parent.

In 1953, the Northern Industrial Chemical Company would introduce Wright's innovative and inexpensive plastic dinnerware called Residential. It would be the first-ever plastic tableware, which was injection molded in high quantities using Melmac, the trade name of a melamine plastic compound. With Wright's name, known for his informal casual living designs and styles over the previous 15 years, Residential was highly popular. The design would receive MoMA "Good Design" awards that year and the next.

The Korean War ended with an armistice in July 1953, allowing a return to peacetime activities. Veterans were eligible for benefits under the G.I. Bill, including state and federal for education and unemployment compensation. Returning veterans would result in another surge of graduates by 1957, including additions to the growing number of industrial designers.

Walter Dorwin Teague Associates (WDTA) had grown to about 200 designers, one of the largest consultancies in the country. Its office, at 444 Madison Avenue in New York, was spread out over three floors. Teague's son, W. Dorwin, had left Bendix and rejoined the organization in 1953 as an associate. A year later, he was made a full partner, with 8 percent shares of the profit, the same as all the other partners, of which there were six (his father received 42 percent). Dorwin designed a .22 rifle for Remington Arms, and WDTA soon designed a Steinway piano and delivery trucks for UPS.

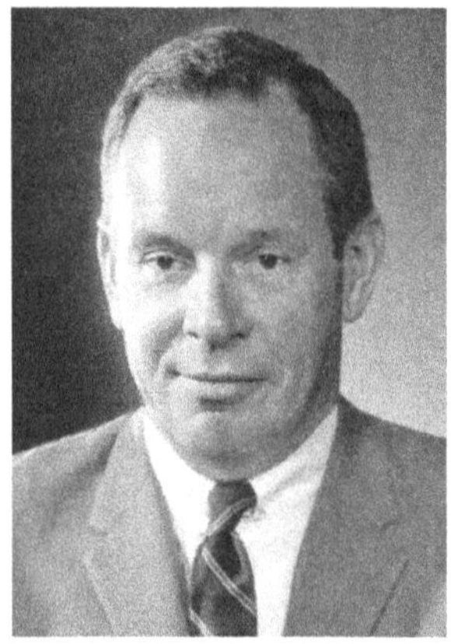

W. Dorwin Teague, Jr., ca. 1960 (courtesy Harry Teague).

The year 1953 saw several innovative automobile designs. One was the Chevrolet Corvette, designed by Harley Earl and his GM staff. It was the first U.S. sports car produced in quantity and the first with an all-fiberglass body. It was an instant commercial success and became an immediate classic. Another classic was introduced the same year, the Studebaker Starliner, designed by Robert E. Bourke of Raymond Loewy's group in South Bend, including Tucker Madawick. The Starliner blended European and U.S. design styles into a low, sleek, and graceful masterpiece, acclaimed by the Museum of Modern Art as a work of art, an extremely rare compliment to any American design by MoMA. The Raymond Loewy Corporation was also responsible for Greyhound's double-deck PD-4501 Scenicruiser bus, largely designed by Richard Latham, that would be introduced in 1954. Belle Kogan won an IDI award for excellence in glassware.

Dave Chapman, of Dave Chapman, Inc., in Chicago, won an Industrial Designers Institute (IDI) national design award in 1954 for his innovative School Furniture Line, manufactured by the Brunswick-Balke-Collender Company. Chapman that year organized a subsidiary company, Design Research, Inc., to provide long-range planning for clients.

Gustav Jensen, one of the ten most prominent machine designers of the 1930s, who continued as a one-man studio after World War II making designs and sculpture, passed away at age 56 in 1954, largely forgotten.

Alexander Kostellow had headed the Pratt Institute industrial design program since 1944 and had expanded the program to four years. In 1952 he had created an Experimental Design Laboratory at Pratt, providing professional company designers office space adjacent to student drafting and shop areas. They conducted experimental work, lectured, and advised students. Participating companies included General Motors, Shell Oil, and Owens-Corning Fiberglass

Corporation. In 1954, both Kostellow and his wife, Rowena, designed a Kitchen of Tomorrow for GM's Frigidaire division that was exhibited at GM's annual Motorama. The kitchen had stainless-steel appliances that retracted into cabinets. While engaged in special summer consultation work for the styling division at General Motors despite poor health, **Alexander Kostellow** passed away at the age of 58. Robert A. Kölli would replace him at Pratt until 1962, when Rowena Reed Kostellow (1900–1988), Alexander's wife, who had been associate professor of design, would chair the department until 1966.

The United States' 20-year monopoly in industrial design was coming to an end, with many international events devoted to industrial design and designers as part of the development of an international design community. New professional design organizations and international consulting groups were formed in a number of European countries and Japan. Japan established industrial design programs with the assistance of a number of U.S. designers, including Russel Wright, George Nelson, Freda Diamond and Jean Reinecke, who traveled to Japan after 1950 to assist in this process.

Germany's Rosenthal China engaged the Raymond Loewy Corporation (RLC) as consultants. In 1954, urged by Richard Latham, head of Loewy's Chicago office, Rosenthal began using international designers for its special division of Studio-Linie products, starting with Coffee Service 2000 by Latham for RLC. These products were distributed internationally in special showrooms called Studio Houses, and over the next 30 years, Rosenthal would employ more than 100 designers from 11 nations for this exclusive line. Industrial design was in the process of becoming an international and global business. Latham in 1955 would team together with Bob Tyler and George Jensen, all previously on Loewy's Chicago design staff, to form the new firm of Latham-Tyler-Jensen, with a staff of 14.

In 1954, The Tenth Triennale was held in Milan, Italy, with the theme "Industrial Aesthetics." Peter Müller-Munk, then president of the Society of Industrial Designers (SID), sent

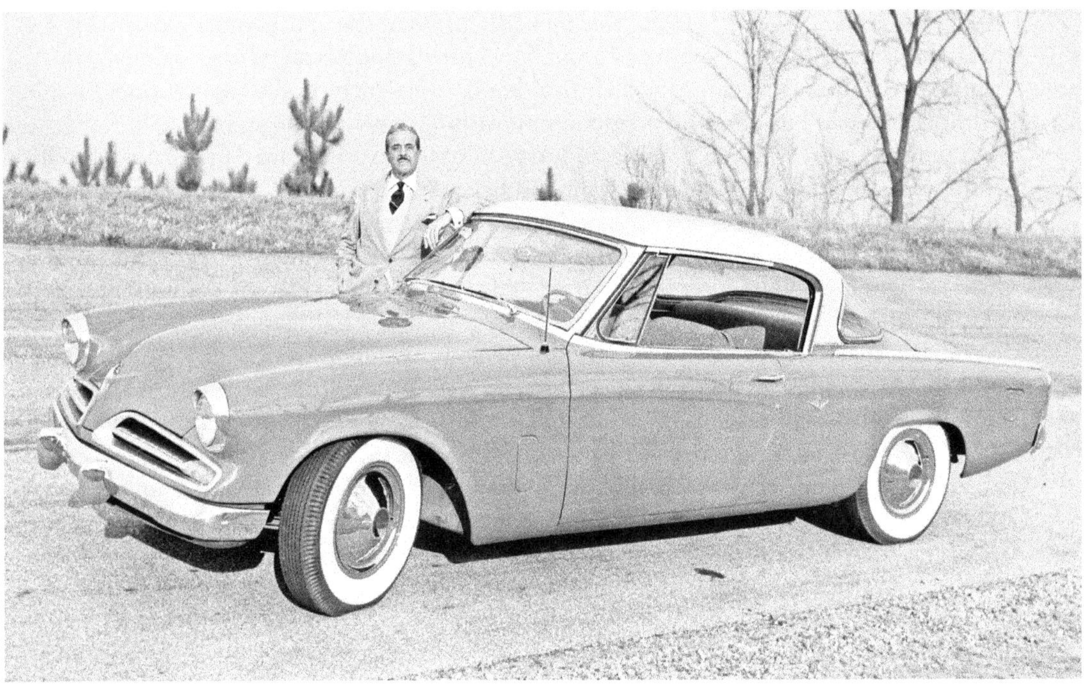

Raymond Loewy with Studebaker Starliner, 1953.

exhibits of 40 U.S. products designed by its members for exhibition in Milan, including two Buckminster Fuller geodesic domes, the innovative architectural structures that would make Fuller famous. It was also the tenth anniversary of the founding of SID, and to celebrate its anniversary, SID published *Industrial Design in America,* a 224-page hardcover book with color illustrations and case histories, publicizing the work of its members. In addition, because industrial design was becoming a worldwide phenomenon, the name of the organization was changed to the American Society of Industrial Designers (ASID). At the time, there were 153 members with chapters in New York, Chicago, Cleveland, and Los Angeles.

Also in 1954, *Industrial Design,* the first magazine dedicated solely to the profession, founded and edited by Jane Fiske Mitarachi (later, Jane Thompson), published its first issue. Thompson had been editor of *Interiors* magazine in which a regular "Industrial Design" section had been initiated by Donald Dohner in 1943, and which had been continued after his death by his partner J. Gordon Lippincott in 1944. The first issue of *Industrial Design* magazine initiated its Annual Design Review, which featured the magazine's selections of the "best designs" of the year, which would become a promotional tool for designers to submit entries and be recognized for their work. Recognition by *ID* magazine was a highly coveted honor valued by designers, clients, and employers.

Stowe Myers opened his own design office, Stowe Myers Design, in Evanston, Illinois, in 1954. He would receive awards for his design of buildings and exhibits at the U.S. exhibit for the 1959 International Trade Fair in Casablanca, Morocco, as well as designs for Tinkertoy packaging, a CENCO Instruments laboratory meter, and a Rheem Manufacturing Company water heater drain valve.

By 1954, interest in MoMA's annual Good Design exhibitions was slowing down. During that year's winter market, in addition to the Good Design exhibit, the Merchandise Mart staged an exhibition called Good Companions: Today in Tradition, which showed "what the manufacturers of traditional furniture, fabrics and lamps are providing for the homes of today." The reality was that "Good Design," as defined by MoMA, was not popular either with the general public, or with the majority of home furnishing manufacturers, who, unlike MoMA, depended on public acceptance for their livelihood. Even today, traditional (antique) home furnishing styles far outsell so-called modern furnishings by a wide margin.

So much for the home furnishings industry, but MoMA still wanted to expand its architectural idea of "Good Design" into what had become the exclusive territory of industrial designers: electromechanical products, major appliances, and automobiles (the latter at its 1951 8 Automobiles show). Accordingly, the jury for the 1955 winter market exhibition of "Good Design" included Arthur BecVar, head of design at General Electric in Louisville, Kentucky, and then president of the American Society of Industrial Designers (ASID). It was MoMA's first token recognition of organized design on any "Good Design" jury. BecVar, Edgar Kaufmann, Jr., and Just Lunning of Georg Jensen's New York store selected 400 products for this final Good Design exhibition, which *Interiors* magazine noted with surprise included a number of "industrial designs that hardly qualified as home furnishings," such as cameras, telephones, kitchen equipment, and household appliances.[3] The historical dividing line between household furnishings (the former art industries) and mechanical equipment (the former artless industries) was gradually being erased, but the Good Design exhibitions were concluded with the resignation of Edgar Kaufmann, Jr., from MoMA in 1955. "Good Design" had run its course.

However, in Germany, the legacy of the Bauhaus lived on. In 1953, it was resurrected by the Hochschule für Gestaltung (HfG, Institute for Form), which was founded in Ulm as the "New Bauhaus" and funded by the German government and the American High Commission

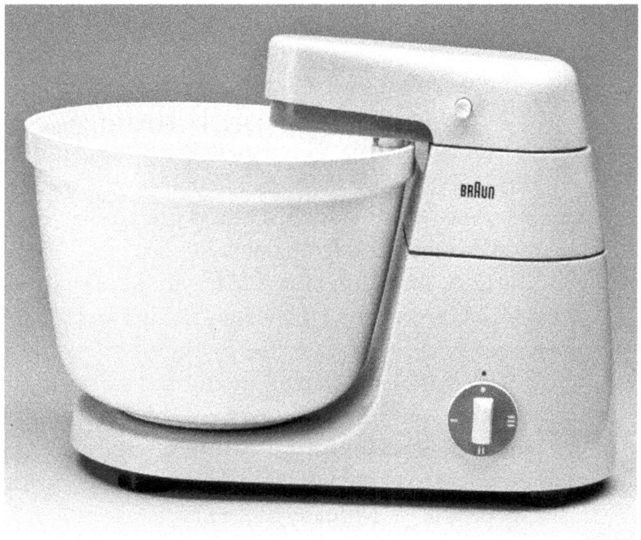

Braun Kitchen Machine, 1957, by Gerd Alfred Müller.

to foster national economic recovery for Germany. Max Bill (1908–1994), a former Bauhaus student, was named director. The school initially focused on product design, but by 1956 it would abandon this in favor of pure theory.

At the same time, Max Braun's sons, Artur and Erwin, were reorganizing the German appliance company Max had founded in 1921. They were inspired by the HfG in Ulm and initiated a corporate design program emulating the Bauhaus tradition, headed by Dieter Rams. The company's products—spare, undecorated, and geometric in form and neutral in color—soon began to be introduced internationally.

In 1954, Henry Dreyfuss Associates had designed the interiors of *Columbine*, President Eisenhower's presidential plane, and in 1955 his design of the unique, spherical-shaped Model 84 canister vacuum cleaner (that floated on its own exhaust air) for the Hoover Company debuted. But Hoover balked at his high fees and, according to some engineers at Hoover, at his arrogance in insisting on impractical design changes. Hoover dropped him for a newer, younger, less costly, and local consulting firm, Smith, Scherr, & McDermott of Akron, Ohio, which had been founded in 1947. In Philadelphia, Philco introduced Harold Van Doren's design for the first refrigerator with a two-way opening door in 1955, with a V handle in the center that pulled left or right. Detroit firm Sundberg-Ferar designed IBM's 702 computer.

Chicago industrial designer **Robert Budlong** passed away in 1955 at age 53, and his business was taken over by one of his associates, Ken Shory, Sr. (1921–1991); the business was renamed Ken Schory Associates. Schory would later partner with Tom Steinbach, and their firm would be named Schory-Steinbach Associates. Also in Chicago, Waltman Associates, headed by Chauncey "Chick" Waltman, was designing a sewing machine for Sears, Roebuck, and a three-way mixer for Nutone.

Ann Swainson, the lady who initiated the Bureau of Design at Montgomery Ward in 1931, died in 1955 at the age of 55. Swainson remained head of the Bureau as Sewell Avery was decimating the company. Avery had led the firm through its phenomenal growth in the 1930s, but during and after the war, he was convinced there was going to be another Depression. He pulled back on any modernization or expansion. In 1953, as Sears was approaching its second billion-dollar mark, Montgomery Ward fell below one billion. By 1955, the Bureau was reduced to Swainson and an assistant, confined to a desk in a corridor of the warehouse building. That year, Avery would be ousted, but it was too late. The loss of her Bureau was too much for Swainson.[4] In 1955, Dave Chapman would return to Montgomery Ward as a consultant, where he had worked from 1934 to 1936 under Swainson, to reorganize and reform the Bureau of Design, which would now be headed by Frederick W. Priess, one of Swainson's former designers. The department would survive until 1976, when Mobil acquired Sears.

The U.S. Department of State in 1955 retained Russel Wright to survey possibilities for developing cottage industries in Southeast Asia. He explored native handicrafts and established centers of manufacturing for these products. George Nelson incorporated his office in New York as George Nelson Associates, Inc., and relocated his studio to 251 Park Avenue South. He had opened his initial design studio in 1947.

Clare Hodgman and Robert E. Bourke, colleagues from both Sears and Loewy organizations, joined to form an independent office, Hodgman and Bourke, Industrial Design Consultants, in New York in 1955. Their accounts would include Coca-Cola, Mack Trucks, GAF cameras, Volvo, Superior Coach, and Electrolux, SA. Hodgman lived in Mamaroneck and was awarded dozens of design patents.

In 1955, the first retractable hardtop car by Brooks Stevens, the Gaylord Grand Prix, was introduced, as was the Ford Thunderbird, designed by George Walker and his styling staff. That year, Walker became vice president of styling at Ford, a position he would hold until 1961. But criticism of Detroit design was still rising.

Henry Dreyfuss published his book *Designing for People* in 1955, which encapsulates his personal design philosophy on the title page: "We bear in mind that the object being worked upon is going to be ridden in, looked at, talked into, activated, operated, or in some other way used by people individually or en masse."[5]

His book included recollections of his career and illustrations of his most important work. Perhaps most importantly, it introduced "Joe and Josephine," diagrammatic charts of typical anthropometric males and females, in typical standing, sitting and reaching positions, each with appropriate average body dimensions indicated as if on a mechanical drawing, as well as comparable measurements of the largest (97.5) and smallest (2.5) percentiles of people. Joe and Josephine's children were also diagrammed. Dreyfuss had accumulated most of these measurements during World War II, when he measured large cohorts of military personnel in order to design equipment and control panels. The Joe and Josephine charts would be invaluable to the profession as scientific measurement tools.

The book established Dreyfuss as a leader in what was at the time called "human factors," later better known as ergonomics: the specialty area of adapting product designs to human sizes and functional physical capabilities. Henry Dreyfuss Associates would later expand and improve the Joe and Josephine measurements into numerous detailed charts that would become essential industrial design tools. The next year, Bell Telephone Laboratories debuted a wall-mounted telephone designed by HDA, which became popular in kitchens where counter space was at a premium.

The postwar period was over. The United States became involved in the Vietnam War in 1956 and would see an escalation of that war over the next 20 years, complicated by the Cold War with the Soviet Union that would last for 35 years. These events not only dominated national priorities but also contributed to vast cultural, social, political, and generational changes in the country over this period. Despite this, the industrial design profession advanced and expanded, following the examples and standards set by the machine designers of the 1930s.

The Cold War was raging in 1956 when Nikita Krushchev made his "We will bury you" speech and the Soviet Union broke the back of the Hungarian revolt. The United States was deeply invested in Central European problems, and European exhibits were the means for promoting U.S. interests. That same year, Walter Dorwin Teague Associates (WDTA) was selected by the Department of Commerce to design the permanent U.S. pavilion and exhibit in Moscow at Gorky Park. At the time, WDTA was also working on the interiors and furnishings for the new Air Force Academy and a new museum at West Point.

Eliot Noyes was retained to develop a new IBM corporate design style to improve the quality of IBM products, graphics, exhibitions, packaging and architecture. The result was the first U.S. "house style" program, a comprehensive and consistent public visual image of a cor-

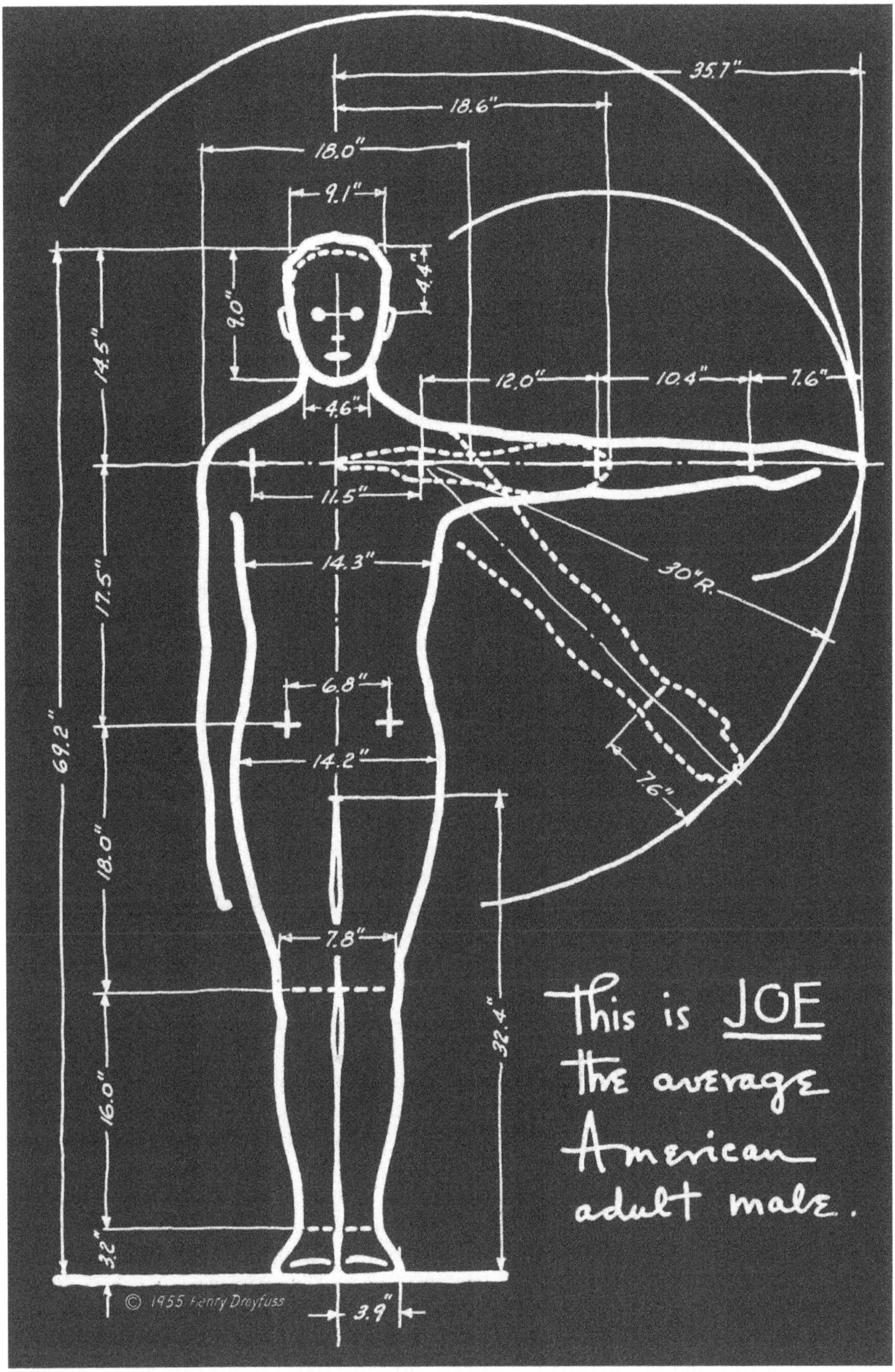

Dreyfuss's ergonomic chart, 1955 (reprinted from *Designing for People* by Henry Dreyfuss, by permission of Allworth Press, an imprint of Skyhorse Publishing, Inc.).

poration, much of which is still being used by IBM today. Also in 1956, the Raymond Loewy Corporation was amended as Raymond Loewy and William Snaith, Inc., with William Snaith as managing partner.

In 1957, the year before *Sputnik* ignited the Cold War space race to the moon, Walter Dorwin Teague Associates (WDTA) designed the U.S. pavilions at both the Zagreb Fair in Zagreb, Croatia, and the Vienna Fair, both funded by the U.S. Information Agency (USIA), and both led and supervised by W. Dorwin Teague, Jr. Henry Dreyfuss Associates designed the "deer" logo for John Deere; Sundberg-Ferar designed the Miracle Kitchen for Whirlpool with a robotic vacuum cleaner; the Raymond Loewy Corporation designed the Alouette utility helicopter for the French firm, SNCASE; and Donald Deskey designed packaging for some of the most common and popular everyday household items, including Cheer detergent (1958) and Gleem toothpaste (1954).

Sadly, **Harold Van Doren**, one of the founders of the American Society of Industrial Designers, passed away in 1957 at age 62, and his Philadelphia firm was taken over by Harper Landell (1918–1997), a Van Doren employee of ten years. The firm would be known as Harper Landell & Associates until it would dissolve in 1975. Van Doren's papers can be found at Syracuse University Library, Special Collections.

During the decade after the war, John Vassos became involved with Greek war relief efforts through the American Relief for Greek Democracy (ARGD), whose honorary chairperson was Eleanor Roosevelt. Vassos became one of the most honored Greek Americans and developed close friendships with other well-known Greeks such as Aristotle Onassis, who would marry Jackie Kennedy in 1968. Besides his work for RCA and other companies, Vassos designed murals for theaters and studios, including the Riviera, Syosset, and Rivoli theaters in New York; the Egyptian Theatre in Los Angeles; the Condado Beach Hotel in Puerto Rico; the WCAY studio in Philadelphia; and the Radio and TV Center in Washington, D.C. Vassos also designed trade and exhibition pavilions in New York, India, and Pakistan.

Brooks Stevens Design Associates designed a range of product prototypes in 1957 for the Evinrude division of the Outboard Marine Company, including the 14-foot-diameter Fisherman circular boat; the 16-foot Lark runabout; the Starflight cabin cruiser; the Housefloat houseboat; the Heli-bout, a combination helicopter/boat; and the Showboat, a miniature version of a 19th-century Mississippi river packet. They were all built to attract attention at boat shows, and they accomplished that in spades, with considerable publicity, but they were never produced in quantity.

In 1957, Jean Reinecke relocated to Los Angeles and opened a second office in Pasadena, California, where he designed the 3M Décor Dispenser Model C-15, a swooping plastic desktop tape dispenser in numerous colors and textures, that would be introduced in 1961, and which would be popular for the remainder of the century. He would lecture at a number of universities, including MIT, Cal Tech, UCLA, Stanford, ITT, the Chicago Art Institute, and the University of Illinois.

Arthur BecVar, head of GE's industrial design group of 45 for major appliances in Louisville, Kentucky, designed a new kitchen center with Robert W. Blee. The nine-foot-long grouping included a combination washer-dryer, a DisposAll equipped sink, a dishwasher, and a range with oven, all under a single, seamless, stainless-steel countertop.

The Industrial Design Educators' Association (IDEA) was formed in 1957 for industrial design educators, who were denied membership in both the American Society of Industrial Designers (ASID) and the Industrial Designers Institute (IDI) because they were not active commercial design practitioners. The first president of IDEA was Joseph Carriero (1920–1978), chair of the industrial design department at the Philadelphia Museum School of Art.

In London, the International Council of Industrial Designers (ICSID) was formed, comprised of 23 professional and promotional design organizations from 17 countries. Peter Müller-Munk, of the ASID, was elected president of the new international organization, which still exists today. Over many years, ICSID would hold design seminars in many countries to promote the use of industrial design.

At a 1954 talk to the local advertising club in Minneapolis, Brooks Stevens had described the mission of industrial design as "planned obsolescence," a phrase that would become almost as controversial in industrial design as "streamlining" had been. By 1956 Stevens was using the phrase frequently and defining it as "instilling in the buyer the desire to own something a little newer, a little better, a little sooner than is necessary."[6] Actually, to us in the 21st century, this sounds like exactly what happens with the latest and newest version of Apple's iPhone, or any new version of a popular product. Although Stevens admitted to originating the phrase, he did not, and he may not have known it, because he was a student at Cornell at the time when Elmo Calkins, the "Dean of Advertising" partner of Calkins & Holden advertising agency, in his 1932 book, *What Consumer Engineering Really Is*, wrote that

> obsoletism is another device for stimulating consumption. The consideration of style is a consideration in buying many things. Clothes go out of style and are replaced long before they wear out. That principle extends to other products—motor cars, bathrooms, radios, foods, refrigerators, furniture.[7]

Calkins went on to say the principle of obsoletism was the solution for solving the Depression because buying things increases prosperity for business, consumers, and the country. In fact, that is precisely what happened.

But in Vance Packard's 1957 book, *The Hidden Persuaders,* Packard (1914–1996) accused industry for "manipulating" consumers and "inducing" desire. He pointed to Stevens' "planned obsolescence," and by implication, all industrial designers, as the leading culprit in this "immoral" practice. In fact, he called Stevens "the crown prince of planned obsolescence." Packard's book reached a wide audience and was hailed by many as a noble cause, because the idea of secretly and deliberately manipulating consumers was considered underhanded in some way. Packard viewed the practice as wasteful, but producing only products that are functionally "necessary" because "old ones" were no longer usable might be morally "frugal," but would be totally unacceptable to modern consumers wanting something newer and better. Consumers have not just functional "needs," but also upward mobility "wants and aspirations." Packard, a reporter and writer, obviously knew little about business, design, or consumers, but it was a time when the public had become suspicious of industry in general, partly because anti–Vietnam War activists referred derisively to the "wartime military-industrial complex."

Nevertheless, Packard's book put industrial design on the defensive, and in 1958, Dave Chapman, Walter Dorwin Teague, and Henry Dreyfuss mounted a campaign to publicly address critics. They openly spoke out against what they called "phony obsolescence" that added gimmicks and superfluous changes to products, but they agreed with "planned obsolescence" that resulted in genuine product improvements. Stevens himself argued that the item that is replaced by a new product often finds its way into a lower-priced secondary market and is purchased by someone who otherwise would not have been able to afford a new one. He saw it as essentially recycling of material goods that benefited the lower class at the same time as it produced a healthy economy.

The controversy faded in a year or so, but it was an early warning that social issues were becoming more relevant to the public than technological or business issues. The generation of Depression survivors was aging. Many of them had since 1930 been without a decent standard of living and had struggled for five years after 1945 to acquire the material wealth and comfort of the middle class.

Now their children, the so-called baby boomers, blessed since birth with all the comforts and security their parents provided, had the luxury of concerning themselves with social, political, and moral issues such as civil rights, clean air, clean water, and environmental concerns. These would soon become mainstream preoccupations with the boomer generation, along with the nuclear age. The first nuclear power plant, in Shippingport, Pennsylvania, went into service in 1957, and in 1958, the first nuclear-powered merchant vessel was launched. In 1958, Walter Dorwin Teague Associates designed the U.S. Atoms for Peace exhibit for the Atomic Energy Commission, held in the Palais des Nations in Geneva, Switzerland, and Brooks Stevens Design Associates designed a futuristic Oscar Meyer Wienermobile, about which he declared, "There's nothing more aerodynamic than a wiener."

Machine Designer giant **Norman Bel Geddes** passed away in 1958 at age 65. Since his office closed in 1947, he had mostly returned to his first love, the theater, by designing stages and theaters, including a television stage for NBC. This shift in priority may have been his tribute to his daughter, who had become a famous stage and screen actress. His autobiography, *Miracle in the Evening,* which he wrote starting in 1957, would be edited by William Kelly and published posthumously in 1960. It is a wonderfully written story of his early life but is primarily about his personal experiences, with very little about his professional industrial design career, other than a listing of his designs and clients at the end of the book.[8]

By the time of Bel Geddes's death, his daughter, Barbara Bel Geddes (1922–2005), had become a famous actress in films such as *I Remember Mama* (1948), for which she won an Academy Award for the Best Supporting Actress, and Broadway productions such as *The Moon Is Blue* (1951–1953). Before his death, Bel Geddes had the satisfaction of seeing the Federal Interstate Highway System under construction, initiated by President Eisenhower in 1956. It was the full-scale embodiment of Bel Geddes's 1939 miniature Futurama concepts, and at $25 billion over ten years, it was the largest public works project in American history throughout that time. The archives of Bel Geddes's theater and industrial design papers reside at the Harry Ransom Center at the University of Texas in Austin.

The automotive industry was being threatened by foreign competition. From 1955 to 1959, as imports of Volkswagens, Karmann Ghias, Porsches and Mercedes from Germany; Toyotas and Nissans from Japan; Morris Mini Minors, Triumphs and MGs from England; and Saabs from Sweden increased from 1 percent to 10 percent of the U.S. market. Traditional Detroit brand names such as DeSoto disappeared, or merged to form Studebaker-Packard (1954) or American Motors (1954, Hudson/Nash).

But Detroit continued with the same styling approach it had followed the previous ten years. In 1957, the first vice president of styling at Chrysler, Virgil Exner, had introduced a new Forward Look line of cars with tail fins even higher than General Motors. There was one exception to this pattern. In 1956, under George Walker, Ford staff, including Gordon Buehrig, chief body designer, had sought to create the elegance of its classic 1940 Lincoln Continental with the new Lincoln Continental Mark II. It had no excessive tail fins, and in fact it seemed to be part of the reason Ford outsold GM for the first time in history in 1957. In 1961, the design would win a national design award from the Industrial Designers Institute.

On the other hand, in late 1957, Ford introduced the least successful car design since the 1934 Chrysler Airflow, this time the 1958 Edsel, a new brand name honoring Henry Ford's son, designed by vice president of styling George Walker and his staff after extensive market research. The vertical air intake in the Edsel's front nose, resembling an open-mouth human scream, seemed to be the offending feature, as the public had become more accustomed to horizontal grills and air intakes. The market failure would cost Ford $350 million and force cancellation of the Edsel brand in 1959. In June of that year, Walter Dorwin Teague expressed the general consensus of most industrial designers when he spoke at a dinner meeting of the Detroit chapter of the Industrial Designers Institute:

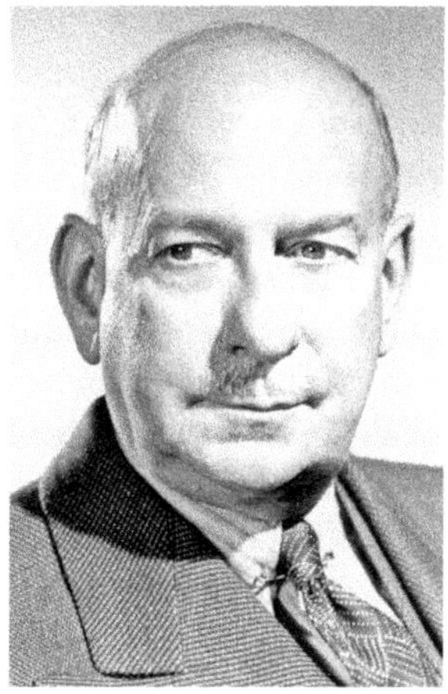
Walter Dorwin Teague, ca. 1950s.

> I agree heartily with millions of others that our characteristic car design has in recent years become lousy, ugly, ostentatious, impractical, and excessively costly; and what is worse from management's point of view, this kind of design has come a spectacular cropper in the market.[9]

The driving force behind this Detroit design direction since 1927, Harley Earl, retired in December 1958, naming William Mitchell (1912–1988) his successor as vice president of styling at General Motors (GM). Earl's garish chrome-trimmed and tail-finned 20-year era had run its course. The last of Earl's creations would be the 1959 Cadillac El Dorado with tail fins 42 inches above the ground (about chest high). Fortunately, this would be the last gasp of tail fins. Mitchell favored smoothly sculptured, unadorned shapes, and GM's cars would be transformed into more sophisticated designs.

However, Harley Earl, Inc., Earl's independent studio, continued operations, winning IDI national design awards in 1955 with the Sweepmaster for the Bissell Carpet Sweeper Company, in 1958 for the Secretary Model thermo-fax copying machine for 3M Company, and in 1959 for the Cinetronic Electric Eye 8 mm movie camera for Argus, a division of Sylvania Electric Products, Inc.

Braun products in Germany, which had been transformed by its introduction of Bauhaus-style products in 1955, had attracted international attention in 1957 with the introduction of its Model KM 3/KM 31 Kitchen Machine. The Museum of Modern Art (MoMA) enthusiastically welcomed this resurgence of its favorite international style by creating a permanent display of Braun products in 1958. In fact, many designers in the United States also praised and imitated the style because of its simplicity, cleanness, and clarity. The products were obviously of high quality, but they cost more than similar American products, which limited their acceptance in the mass market. Perhaps more significantly, such characterless, colorless, spare design simply did not appeal to the U.S. mass market.

In 1958, the Aluminum Company of America (Alcoa) had initiated a series of advertisements promoting concept designs of well-known designers. In 1959, Alcoa presented the first industrial design awards, which included one for Peter Müller-Munk Associates (PMMA). PMMA also consulted with U.S. Steel regarding its newly introduced vinyl-coated steel sheets with decorative patterns embossed into the vinyl, and which could be drawn or stamped like any sheet metal.

By this time, according to a February 1959 *Fortune* magazine survey, there were 300 total industrial design organizations, and 180 total members of IDI and ASID combined (a significant decline in membership). One hundred twenty corporations had internal design departments, compared to 180 independent consulting offices. The obvious result was an increase in competition in the market and the lowering of the exorbitant fees of the big-name designers. The day when a Raymond Loewy could get $50,000 for designing a cigarette package was gone forever. Retainer fees now ranged from $500 to $5,000 per month, depending on a firm's reputation. The 20 largest industrial design firms named by *Fortune*, half of which were led by original machine designers of the 1930s, included the number of employees where known: Raymond Loewy Corporation (200); Walter Dorwin Teague Associates (140), Henry Dreyfuss Associates (50); Dave Chapman, Inc. (50); Peter Müller-Munk Associates (30); Russel Wright (25); George Nelson (24); Harley Earl, Inc.; Brooks Stevens; and Donald Deskey. The remaining 10 were younger and smaller firms.

Industrial designers were becoming part of the popular culture. The 1954 film *Executive Suite* starred William Holden as vice president of design and development Don Walling, who wins the presidency by vowing to replace the shoddy furniture produced by the company with quality products. In the 1959 film *North by Northwest,* Cary Grant meets a pretty girl, Eve Kendall, played by Eva Marie Saint, who is posing as an industrial designer.

AT&T introduced its Princess phone in 1959, designed by Henry Dreyfuss Associates (HDA), with a petite size for teenage girls and an illuminated dial to facilitate bedroom usage. AT&T was the iPhone of the 1950s. HDA also designed the NASA/AT&T Bell Labs Mercury control room for the fledgling space program, interiors for the Lockheed Constellation aircraft, and a Strobonar photoflash for Honeywell. That year, HDA published Measure of Man, a book featuring Joe and Josephine that expanded the applications of ergonomic data to many more design situations and circumstances.

Now, the increasingly chilly Cold War with the Soviet Union spawned a new, political market for industrial design. Some would call it propaganda.

CHAPTER 13

Cold War, Mergers, and Computers

In the summer of 1959, the U.S. Information Agency (USIA) mounted an exhibition in Moscow, the American National Exhibition, as the first of many such cultural exchanges between the Soviet Union and the United States. The exhibit, designed by George Nelson and Charles Eames, turned into the site of a vigorous Cold War debate between Soviet premier Nikita Krushchev and then-vice president Richard Nixon. Featured at the exhibition was an innovative geodesic dome by R. Buckminster Fuller, which attracted worldwide fascination bordering on a space-age fad. Nineteen similar such exhibits in the USSR by American designers over the next 32 years would be seen in a total of 87 cities across 12 time zones by an estimated 20 million Soviet citizens to promote U.S. national political objectives, and to win the hearts and minds of the public with personal cultural diplomacy. Eventually, by 1991, Soviet citizens would be convinced that Soviet promises of free consumer goods would never be realized, and they rejected communism for a more democratic government.

Also in 1959, the Museum of Modern Art displayed three Structures by Buckminster Fuller, including his geodesic dome, plus other designs of his from 1927, which included a circular house on a central pole called a 4-D Utility Unit, and the development of a half plane, half car called the 4-D Zoomobile.

In Los Angeles, the famous "Skyscraper" furniture designer, **Paul T. Frankl**, passed away in 1959 at age 73. He had left New York in 1934 to establish a gallery in Los Angeles, where his clients included many Hollywood stars of the late 1930s. He had applied his talents to many projects, from interior design, posters, greeting cards and theater sets, to selecting objects from Europe, the Orient, and Mexico for sale in his shops. His fortunes had shifted often in his career from prosperity to poverty, and his personal life ranged from depression over the death of his first wife, bitterness over divorce from his second wife, and emotional security with his third.

A month after President John F. Kennedy was elected president in 1960 came the death of the most highly esteemed and famous pioneer of industrial design, **Walter Dorwin Teague**, who died on December 5 in Flemington, New Jersey, just two weeks shy of his 77th birthday. Just the year before, Teague had addressed the Royal Society of Arts in London, and in June of that same year, he had spoken to the Detroit chapter of the Industrial Designers Institute. In 1963, the American Society of Industrial Designers (ASID) would honor Teague, one of its founders, with the organization's first scholarship program, the Walter Dorwin Teague Scholarship, eligible to selected junior students majoring in industrial design. Teague's papers are archived at Syracuse University Library, Special Collections, and at his surviving original organization, now called Teague.

His son, W. Dorwin Teague, who had worked with his father from 1934 to 1942 and had later rejoined the organization in 1953 as partner, now was number two in Teague's professional office, Walter Dorwin Teague Associates (WDTA), with Robert Harper as CEO and president. Dorwin Teague and other WDTA associates in 1960 received a national design award from the Industrial Designers Institute (IDI) for the design of the Euphorian dental chair for the Ritter Company. A functional patent for it is in Teague, Jr.'s name. That same year, WDTA also designed the new Polaroid Electric-Eye 900 camera and a landmark mathematical model for the U.S. Navy's Missiles and Quality Control System, still in use today.

Another IDI award in 1960 went to Dave Chapman, Inc., in Chicago for its research report and recommendations on the design of facilities for U.S. schools, produced for the Educational Facilities Laboratories, Inc., and to William L. Mitchell and the General Motors Styling Staff for the Chevrolet Corvair. The latter, a so-called compact car, was a significant improvement in styling, with no tail fins, but its oversteering tendency earned it the wrath of Ralph Nader, who dubbed it "one of the nastiest-handling cars ever built." His 1965 book, *Unsafe at Any Speed,* would prod Congress, which in 1966 would pass the National Traffic and Motor Vehicle Act, empowering the government to set safety standards for new cars after 1968, which would soon be enacted.

Designs of 1961 included the Tower Capri portable typewriter for Sears, Roebuck by the Sundberg-Ferar design firm in Detroit, which won an IDI national award; the IBM Selectric typewriter by Eliot Noyes, which by 1975 accounted for 75 percent of the U.S. electric typewriter market; and the Singer 660 sewing machine by Henry Dreyfuss Associates. HDA also designed the Civil War Information Building in Richmond, Virginia. George A. Beck, head of GE's industrial design light military electronics department, received one of Alcoa's annual industrial design awards. In a major corporate move, Eliot Noyes was retained by Westinghouse to dramatize its image, including logo and design standards, and to reorganize its internal design departments in order to better compete with General Electric. Also in 1961, Raymond Loewy established Raymond Loewy/William Snaith, Inc., and Raymond Loewy/William Snaith Associates, Inc., with William Snaith as president and Raymond Loewy as chairman of the board.

In 1954, the Studebaker Corporation had merged with Packard to form Studebaker-Packard, and with its combined internal design staffs, as well as its financial difficulties, it felt that it no longer needed to pay Raymond Loewy's extravagant design fees, so it canceled his contract in 1956, even though Loewy had designed Studebaker cars for the previous 20 years. Loewy's Studebaker studio in South Bend shut down, and Robert Bourke relocated to Loewy's office in Chicago. But now, in 1961, Studebaker was in serious trouble, and the company engaged a new president, Sherwood E. Egbert, from the McCullough Company, where Egbert had worked with Brooks Stevens on a number of design projects.

Egbert engaged Stevens to design the 1962 models of the company's mainstays, the Lark sedan and the Gran Turismo Hawk, with a meager budget of $7 million to tool both cars. Stevens accomplished the task with minimal changes to the previous year's models, which had been designed by Loewy. But Egbert knew such minor changes were not enough to save the company and wanted to demonstrate that Studebaker could still produce an innovative design, so as a last ditch effort he rehired Raymond Loewy to design a sensational new sports car, which would become the 1962 Avanti. Loewy retreated to his Palm Springs home with designers Bob Andrews, Tom Kellogg, and John W. Ebstein. The resulting design, sleek, graceful, and unlike any car from Detroit, was an instant success, but it was too late for the company, which would fold in 1963. Still, the Avanti would be brought to life again and again by sequential private interests until 1991.

John Tjaarda, the exceptional automotive designer of the 1936 Lincoln Zephyr, passed away in 1962 at age 65. In 1961, he had pursued an interest in solid-state rocketry and joined the Aerojet-General Corporation in Sacramento as a technical advisor. His son Tom Tjaarda (b. 1934) would follow in his dad's footsteps as an automotive designer and would establish his own automotive design office in Italy, Dimensione Design, in 1984.[1]

In 1962, **Chauncey "Chick" Waltman** also died at age 66. A well-known Chicago furniture and lamp designer starting in 1920, he joined the industrial design movement in the 1930s practicing as Waltman Associates at 165 Chicago Avenue. The office later moved to 310 E. Chicago Avenue. His son, Charles T. "Chubb" Waltman (1919–2006), joined him after World War II, and the office operated until about 1960.

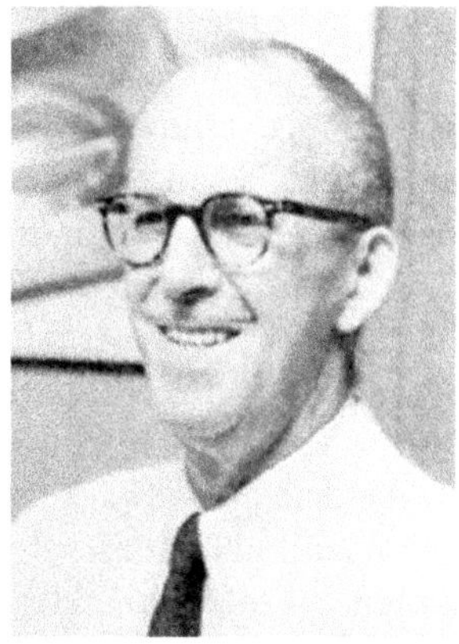

Chauncey "Chick" Waltman, ca. 1950s (courtesy William Waltman).

Also in 1962, H. Creston Doner chaired the Advisory Board of the Showcase National Corporation for the United States and Canada, and over a number of years, he was a consultant to the architects and designers of the 1964 World's Fair in New York, Expo 67 in Canada; HemisFair 68 in San Antonio; Tomorrowland in Disneyland, California; and the Bay Area Regional Transit (BART) system in 1969.

In the world of expositions, Walter Dorwin Teague Associates designed the U.S. Science Center at the 1962 Century 21 Exposition in Seattle, which was called by some "a virtual cathedral of science." The Spacearium held 750 people for a simulated voyage first through the solar system, then through the Milky Way Galaxy and beyond. In 1963, Walter Dorwin Teague Associates would design the Maxwell House Automatic Coffee Machine for the General Foods Corporation.

The new International Council of Societies of Industrial Design (ICSID) convened its Third Congress in Paris, and the Industrial Designers Institute (IDI) held an exhibition at the Louvre, organized by Alfons Bach, who had moved his practice to Palm Beach, Florida, in 1959. Among the large selection of American designs were the 1962 Avanti by Loewy and the 1956 Lincoln Continental Mark II by George Walker and his staff at Ford.

By 1963, international design was becoming a common practice. Sixteen American industrial design firms had offices abroad, and manufacturers from 22 foreign countries had contracts with American designers. The Cold War pitted the United States against the Soviet Union in providing a better standard of living to their respective citizens. After Russia's *Luna 1* flight to within 3,700 miles of the moon, Soviet premier Nikita Krushchev had boldly predicted in 1959 that by 1966 the Soviet Union would "catch up" to America and would be distributing basic consumer goods to Russian citizens "free of charge." Industrial designers were well aware that they were now competing globally. At ASID's annual conference in 1962, Raymond Loewy had spoken of such competition:

> Nothing great has ever been achieved—whether by a nation, a business, an individual—without quality.... Now, we are at a point where that belief is being put to the test. Americans are in a massive race to maintain our leadership against other nations in trade, training, goods, and taste.[2]

Raymond Loewy had just designed the interior and external livery for Air Force One when President Kennedy was assassinated in Dallas, Texas, in November 1963. The Boeing 707 was a specially modified Boeing C-137 Stratoliner. Loewy had met with Kennedy and conducted research in the National Archives for the assignment, for which he did not charge a fee. He found the uppercase typeface called Caslon, which was used widely spaced in the country's name in the first printed copy of the U.S. Declaration of Independence, and used it for the large lettering on each side of Air Force One's fuselage, reading, "United States of America." A presidential seal was on both sides of the fuselage, and an American flag on the tail. The plane took the president to Dallas and returned his body to Washington. The following year, at the request of Jacqueline Kennedy, Loewy designed the 1964 five-cent John Kennedy memorial postage stamp for the U.S. Postal Service. That same year, Loewy established Market Concepts, Inc., a subsidiary for confidential research operations. Now reorganized as Raymond Loewy/William Snaith, Inc., Loewy's office won a national Industrial Designers Institute (IDI) Award for the design of the Time-Master, made by the Dictaphone Corporation, as did William Mitchell of the General Motors Styling staff for the Pontiac Grand Prix.

Kem Weber, the pioneer who began designing modern furniture in 1926 and was inspired by the Paris Exposition of 1925, passed away in California in 1963 at age 74. After the war, Weber had moved back to Santa Barbara, California, and worked mostly as an architect, building a handful of houses there and in nearby Montecito. By the late 1950s, his health had begun to fail. Also in 1963, Henry Dreyfuss Associates designed the first touch-button phone, the Touchtone 2500 for AT&T, and the Automatic 100 camera for Polaroid.

In 1964, Ford's sporty Mustang won the Industrial Designers Institute's national design award. It was Ford's answer to GM's Corvette and was a huge success with the 16- to 24-year-old baby boomer crowd. George Walker had resigned in 1961, so the Mustang was designed by the new vice president of styling, Eugene Bordinat (1920–1987), and his staff, driven by Lee Iacocca, who had been named Ford vice president in 1960.

Even as Studebaker was dying, Brooks Stevens, without Studebaker's agreement, had prototyped a promotional concept car for Studebaker, which was initially called the Mercebaker, a "contemporary classic" sports car based on the antique Mercedes SSK of the 1927–1929 period. Stevens' sons, Steve and David, built the prototype in six weeks during 1964 and were in the process of towing it to the auto show to create a sensation for Studebaker when Studebaker executives told Stevens they would not be showing it. But Stevens knew the director of the auto show, who agreed to display the prototype as a product of Brooks Stevens Associates. The car was so well received that Stevens funded the production of the car by his sons as a business enterprise, renaming it the Excalibur SSK Roadster, which enjoyed success as the first "replicar," a design based on a historically classic car.[3]

In April 1964, the New York World's Fair opened at Flushing Meadows in Queens, with the theme of "Peace through Understanding," and dedicated to "Man's Achievement on a Shrinking Globe in an Expanding Universe." The fair would run from April to October in 1964 and 1965. The grounds were dominated by a 12-story high, stainless steel, skeletonized model of the earth, appearing to float on a ring of fountains in the center of a circular lagoon. It was designed by Peter Müller-Munk Associates for U.S. Steel and was called the Unisphere, which still remains in Flushing Meadows today.

Attendees traveled the fairgrounds on a monorail designed by Walter Dorwin Teague Associates, who also designed the Festival of Gas, including the House of the Future, featuring a gas-powered refrigerator and a gas oven designed for Norge. Most regarded Futurama II as the most popular exhibit, designed by Henry Dreyfuss Associates (HDA) for General Motors. Like its 1939 fair predecessor, designed by Norman Bel Geddes, it was a show in which visitors seated in moving chairs glided past elaborately detailed miniature 3-D model scenery, showing what life might be like 60 years in the "near future" of 2024. Bel Geddes, who had died six years earlier, would have been proud, but he may have wondered why, after 25 years, Dreyfuss hadn't come up with something better.

Also in 1964, HDA designed the symbol for American Airlines, the Swinger camera for Polaroid, and the Trimline desk phone for AT&T, which would be accepted into the permanent collection of the Museum of Modern Art. George Nelson won a prestigious Alcoa award for Herman Miller's Action Office I, an innovative system of office furniture, but later it became known that Robert Propst had actually designed it. Like other major designers, Nelson claimed credit for all designs in his organization. Following in the footsteps of Raymond Loewy, Buckminster Fuller, now famous for his geodesic domes, was featured on the cover of *Time* magazine.

Now it was time for the three separate industrial design organizations to get together.

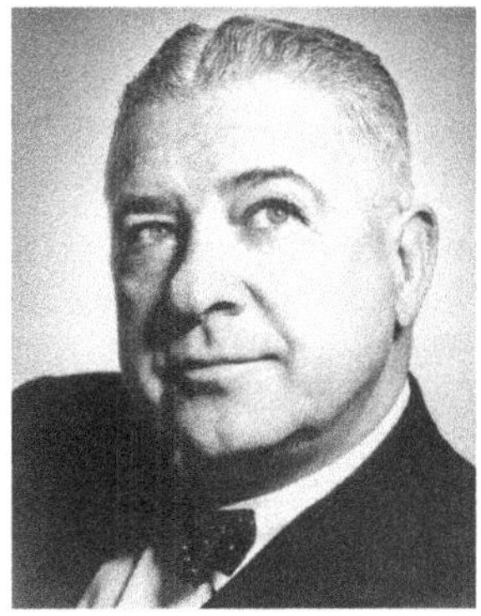

John Vassos, ca. 1950s (courtesy Industrial Designers Society of America).

After much discussion and negotiation since 1958, the three industrial design organizations, the Industrial Designers Institute (IDI), the American Society of Industrial Designers (ASID), and the Industrial Design Education Association (IDEA), agreed to merge and form the Industrial Designers Society of America (IDSA) in 1965 to provide a single voice for industrial design in America. The first president of IDSA was Henry Dreyfuss, one of the founders of the SID in 1944, and the first chairman of the board was John Vassos, one of the founders of the ADI in 1938 (in 1951, it had become the IDI). Vassos had retired in 1964 as a consultant of RCA design, where Tucker Madawick had taken over as manager of radio, phonograph, tape and television design in 1959.

IDSA membership at the time was estimated at 650. The first casualty was the IDI national design awards, which had been initiated in 1951. As we know, Dreyfuss did not like the idea of public awards because they insisted on identifying and crediting the specific designer by name. Big-name designers required that any design made in their firm must be credited to the principal's name, no matter who actually designed it. There would be no further national design awards after 1965 until 1980, when Dreyfuss would be gone.

Russel Wright was invited to Japan in 1965 to work for more than 100 companies. Later, he would work with the Johnson administration to establish the Park Beautiful Program, forming an association with public land and its reclamation for public use.

Alfonso Iannelli passed away in 1965 at the age of 77 in Chicago. In 1955, he had sculpted the large-scale Rock of Gibraltar relief for the Prudential building (now called Prudential Plaza) in Chicago, which was his last major work. In 1982, the entire contents of Iannelli's studio would be purchased by Kelms Cott Gallery in Chicago. That same year, **Egmont Arens** passed away from a heart attack at age 78. He was one of the founders of both the Society of Industrial Designers (SID) in 1944 and the Package Design Council (PDC) in 1952. A month after his death, his widow, Mathilde, published a collection of his early poetry. She wrote that he would often say, "You know, I came to New York to be a poet." His papers are archived at Syracuse University Library, Special Collections.

Winners of the 1966 Design in Housewares Award Program included Hamilton Beach's "hole-in-handle" electric knife and its Gourmet Center by Chapman, Goldsmith & Yamasaki, and West Bend's coffee percolator, broiler rotisserie, and Teflon-coated cookware by David Painter Industrial Designers.[4] That same year, Raymond Loewy/William Snaith designed the Exxon logotype.

George Nelson's office designed an exhibition, Hand Tools USA, for travel within the

Soviet Union by the U.S. Information Agency (USIA) as a cultural exchange between the two Cold War countries in 1966. Another USIA exhibition, Industrial Design USA, would also be designed by Nelson's office in 1967. These exhibits, mentioned earlier, demonstrated the power of design as public diplomacy to change the hearts and minds of people and helped win the Cold War.

In 1967, Raymond Loewy/William Snaith Associates, Inc., designed the trademark for the Shell Oil Company. That year, Raymond Loewy/William Snaith Associates, Inc., in Chicago was merged with Raymond Loewy/William Snaith Associates, Inc., in New York[5]Henry Dreyfuss Associates designed a prototype videophone for AT&T. The HDA organization included James Conner, Niels Diffrient, and Donald Genaro. W. Dorwin Teague resigned from Walter Dorwin Teague Associates in 1967, which had been headed by Bob Harper since Walter Dorwin Teague passed away in 1960. Milton Immermann soon took over as president and CEO of WDTA, and Dorwin resigned to start his own business, Dorwin Teague, Inc., in Englewood Cliffs, New Jersey, which specialized in both industrial design and engineering.[6]

Peter Müller-Munk, head of Peter Müller-Munk Associates in Pittsburgh, died unexpectedly and tragically in 1967 at the age of 60. Müller-Munk had been instrumental in the formation of the International Council of Societies of Industrial Design. **Joseph Platt**, one of the founders of the Society of Industrial Designers in 1944, passed away at age 73 in 1968. That year, Henry Dreyfuss retired at age 64, the same year as the tragic assassinations of Martin Luther King, Jr., and Robert Kennedy. These events, the never-ending Vietnam War, antiwar student demonstrations, and desegregation struggles created a sense of national unrest and helplessness. They were emotional and traumatic times that would be lifted only temporarily in 1969, when U.S. astronauts first landed on the moon.

The industrial design profession also suffered from mistrust by virtue of its prominent role in the corporate business world, which was now seen by many as being responsible for environmental damage, obscene profits, and neglect of global social responsibility. In 1968, R. Buckminster Fuller was awarded the World Medal of Architecture for his 250-foot-diameter, twenty-story-high geodesic dome that had housed the entire U.S. pavilion at Expo 67 in Montreal, Canada.

Fortune magazine, which had introduced industrial design to the business world in 1934, now in 1968, exactly 34 years later, published an article, "The Decline of Industrial Designers," in February, claiming lack of innovation. This seemed to be more a matter of perception than reality. The profession continued as normal, even though many of its famous pioneers were aging or passing away, and design news was dominated by younger personalities, who were as innovative as ever.

More likely, the headlines of civil unrest, campus rebellions, and other pressing social issues had pushed industrial designers and design off of the front pages. Many in IDSA were grappling with the organization's declining membership, financial problems, and its diminishing reputation among young designers, who were criticizing IDSA as an "old boy's club." There was a kernel of truth in this, since many of the "old guard" were longing to return to a time in their youth when leading industrial designers commanded top dollar, and were celebrated as national personalities. But this time had passed.

Still, some of the "old guard" were still in the thick of things. In 1969, Detroit design firm Sundberg-Ferar designed the Bay Area Rapid Transit (BART) system in San Francisco, the first U.S. regional transit system since 1907. Frank Del Guidice now headed Walter Dorwin Teague Associates, and Raymond Loewy opened his third London, England, office. Nathan Horwitt designed wall clock No. 4601 for the Howard Miller Clock Company. Henry Dreyfuss

Associates designed American Standard fixtures and fittings, and the model 500 helicopter for Hughes Aircraft.

In Philadelphia, **Clyde Shuler**, designer of "tombstone" wooden radios of the early 1930s, passed away at age 77 in 1969. While regarded as passé at the time, the gothic shapes of his radios in the early 1930s were simply a logical extension of "skyscraper" furniture design of the 1920s. Shuler had taught industrial design at the Philadelphia College of Art since 1933 and had retired from there in 1952. He had established his own office, Clyde Schuler Associates, at 1717 Sansom Street, working on residential architecture and commercial design. By 1961, he was a partner in the office of his old friend, William Heyl Thompson.

Boeing 747 "jumbo jets" made their first flights. Walter Dorwin Teague Associates (WDTA) designed Boeing interiors, and Henry Dreyfuss Associates (HDA) designed American Airlines interiors. Design in Steel design awards, initiated in 1963 by U.S. Steel, were

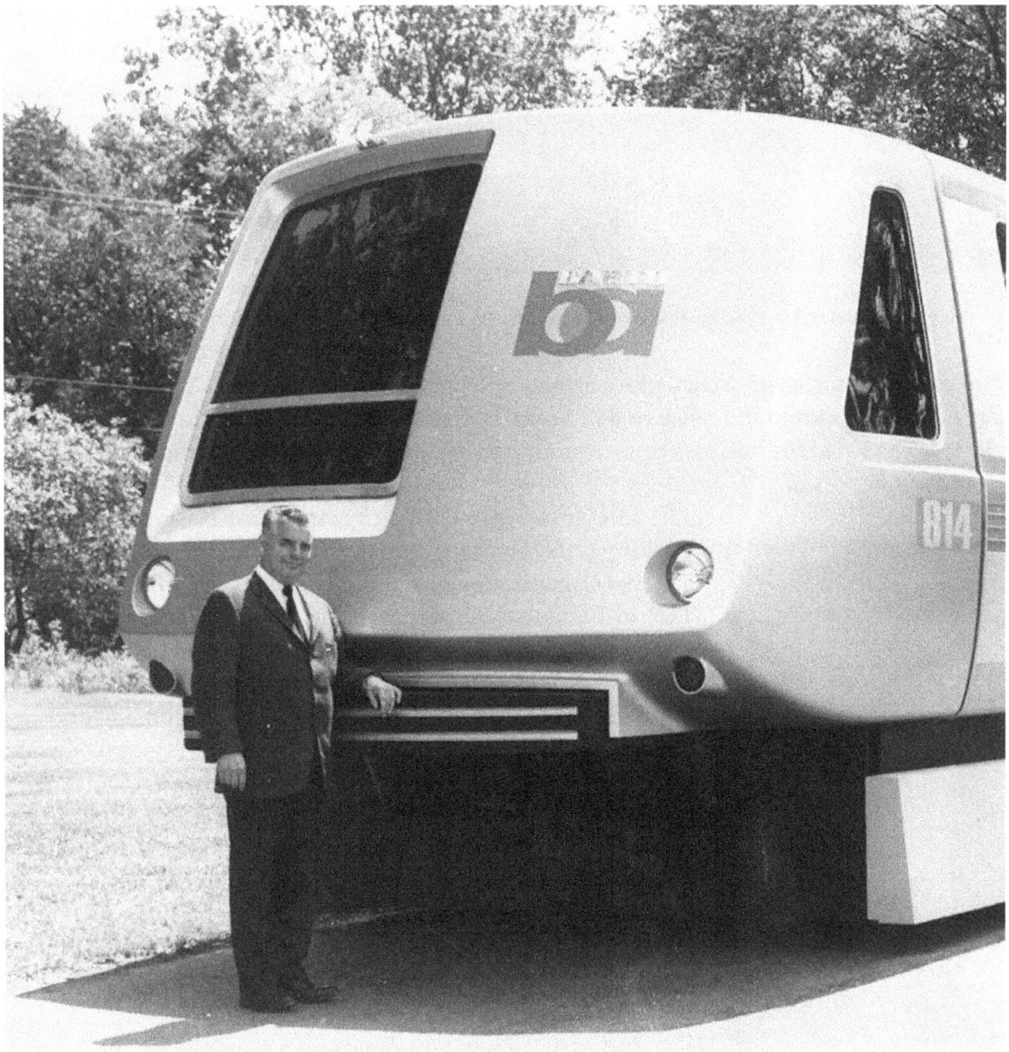

Carl Sundberg with Bay Area Regional Transit System, 1969 (courtesy Sundberg-Ferar, www.sundbergferar.com).

awarded in 1969 to Henry Dreyfuss Associates for a transmission tower for the Southern California Edison Company, and to Sundberg-Ferar for a subway transit vehicle designed for the New York Metro Transportation Authority.

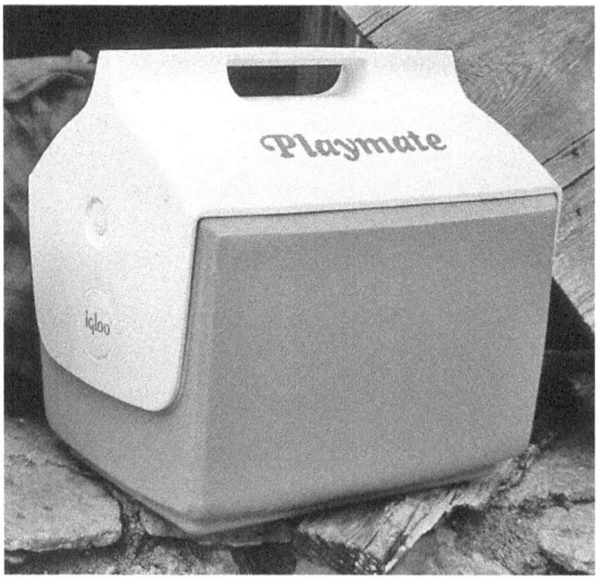

Playmate cooler, by Marlan Polhemus of GYS.

In Detroit, **Harley Earl**, the man who had most influenced automotive design over the previous 30 years, passed away at age 76 in 1969. They say he is still remembered at General Motors with this prayer: "Our Father, who art in Styling, Harley be thy Name." Amen to that. His private consulting firm had merged with Walter B. Ford II (1920–1991) Design Associates in 1964 to form Ford & Earl Design Associates. Ford had been in GM Styling under Earl. In 1970, Dave Chapman, Inc., of Chicago was reorganized with Paul Specht as president and partner in Goldsmith Yamasaki Specht, Inc. (GYS), which that year designed the ubiquitous Playmate plastic picnic cooler for the Igloo Corporation.

Dreyfuss closed his Pasadena office in 1970, and his staff was relocated to 4 West 58th Street in New York, where Bill Purcell and Don Genaro were senior partners and Niels Diffrient and Jim Conner were junior partners. They moved the office to 888 Seventh Avenue with more space. In the early 1970s, Dreyfuss had sold his business to these four partners, and focused on international symbols by himself.[7]

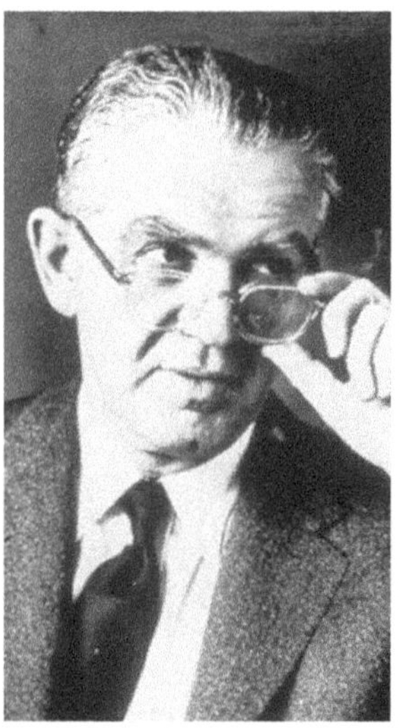

Henry Dreyfuss, ca. 1955 (courtesy Henry Dreyfuss Associates).

Dreyfuss, representing the American National Standards Institute (ANSI), chaired the first meeting in 1971 of the International Organization of Standards' Technical Committee (ISO/TC) in Berlin, which set international standards for 145 signs and symbols. In 1972, Dreyfuss would use these in his publication, the *Symbol Sourcebook,* a virtual encyclopedia of international symbols. HDA designed the Hyster lift equipment and the Flicker, a woman's razor for the American Safety Razor Company. Sadly, that same year, **Henry Dreyfuss** and his wife, Doris, both terminally ill, designed their own death by carbon monoxide at their home in South Pasadena. Until his death, Dreyfuss had been a faculty member in the engineering department at Caltech in Pasadena. The personal archives of Henry Dreyfuss would be donated to the Smithsonian in 1973. Henry Dreyfuss Associates, however, remained active and profitable. New associates Jim Ryan and Bill Crookes were hired.

Industrial design now came under attack by Victor Papanek (1925–1998), chair of design at the Kansas City (Missouri) Art Institute, whose 1972 book, *Design for the Real World,* the book jacket of which characterized Papanek as "an ingenious designer who charges the industrial design establishment with criminal negligence of almost everybody," featured such declarations and accusations as, "It is about time that industrial design, as we have come to know it, should cease to exist," and, "Today, industrial design has put murder on a mass production basis."

Papanek claimed the moral high ground in his book using his students' designs, which addressed child disability problems, created radio receivers made of juice cans and cottage industries for Third World countries, designed human-powered vehicles, and created packaging made of pea pods. Papanek's idealistic, socially conscious, and antibusiness philosophy appealed strongly to 20-year-old young adults who optimistically aspired to change the world, which to them appeared to be failing in many respects. They felt that their unique role, as a generation, was to challenge authority and the establishment to achieve change of any sort. Within a year, the Vietnam War would be over, conscription into the military would end, and the volunteer army began. The focus of youth would then be more on social and environmental issues.

Over the years, change did occur in industrial design as well. Public design emphasis on profits and sales was reduced (although this still occurred), museum preoccupation with aesthetic purity was reduced (although this still occurred), and public commitment to solving social and environmental problems increased.

The government increasingly became involved with legislation to address these social and environmental problems, and the design profession participated where it was relevant. In January 1973, the Industrial Designers Society of America (IDSA) relocated its national office from New York to Washington, D.C., where an expanding government was becoming activist in nature. IDSA was receiving grants for working with government programs, such as the Product Safety Commission and Barrier Free Design. IDSA at the time had a total membership of 723 and 10 regional chapters. New York, Chicago, and the West Coast were no longer the only centers of design. Manufacturers were moving south for lower wages, and small, local design offices were popping up all over the country.

A 1973 study, "Industrial Design Education in the U.S.A.: A Factual Summary," by Joseph Carriero, chair of the Department of Design and Environmental Analysis at Cornell University, revealed that there were 41 higher education programs of industrial design, and approximately 2,000 students pursuing a career in industrial design, of which 363 were female and 1,737 male. There were also 211 graduate students pursuing master's degrees in 29 programs. Faculty members totaling 1882 were devoted to full-time teaching of industrial design, plus 129 part time, and 200 support people.

That same year, Raymond Loewy/William Snaith Associates, Inc., designed the *Skylab 2* space station for NASA, launched on May 25, 1973, a project Loewy regarded as the most important of his career. It was the first manned mission to *Skylab,* and the first U.S. orbital space station. Loewy emphasized habitability and comfort for the astronauts, with a wardroom for meals and relaxation, and a window to view the Earth and space. Skylab missions could last for months. Each astronaut had a private sleeping area the size of a walk-in closet, with a curtain, sleeping bag, and locker. That same year, Loewy signed a contract with the Soviet Union to be its consultant for mass production and over the next few years collaborated on the design of Russian automobiles, farm tractors, cameras, and hydrofoil vessels.

Henry Dreyfuss Associates designed an open-sided street telephone booth for AT&T, a Bard Parker surgeon's scalpel, and the Polaroid SX-70 camera in 1973, and in 1974 published *Humanscale 1/2/3*, extensive ergonomic data and diagrams laminated in plastic for durability and frequent use for industrial designers to determine accurate dimensions for human use of products and equipment. President Nixon resigned as a result of the Watergate scandal, and Gerald Ford, vice president, took the office of president.

It was in 1973 that **Virgil Exner**, who started in automotive design in 1933, passed away at age 64. He had a heart attack in 1956 but resumed work in 1957. In 1962 the president of Chrysler ordered him to reduce the size of their cars. The standard-size models Exner had already designed were downsized by associates over the objections of Exner. When they failed on the market, Exner was fired as the scapegoat. He retired in 1964 and was replaced by Elwood Engle (1917–1986). Exner continued consulting for various car companies from his office in Birmingham, Michigan, teaming up with his son, Virgil Exner, Jr., to design watercraft for the Buehler Corporation. He also designed "revival cars," including the revival of Stutz in the 1970s.

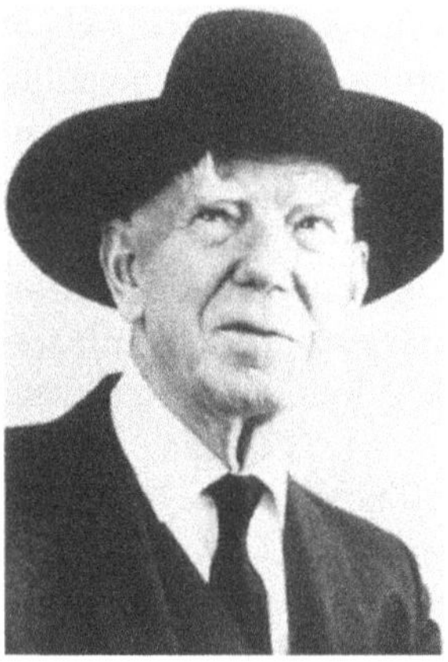

Joseph "Jo" Sinel, ca. 1970s (*Designscape* magazine, February 1977).

The year 1975 saw the passing of **Greta von Nessen** in Murraysville, Pennsylvania, at age 75. She and her husband, Walter von Nessen (1889–1943), founded Nessen Studios in 1927 to design lamps. She continued the family studio, which survives to this day as Nessen Lamps, Inc.

That same year, **Jo Sinel**, one of the founders of the Society of Industrial Designers, passed away at age 86. A New Zealander who was designing products in 1923 for an advertising agency in New York, and eventually worked for 55 such agencies, he became a noted industrial designer and educator. His papers are archived at the California College of Arts, where he taught for so many years, and where he received an honorary doctorate of fine arts.

The Smithsonian's Renwick Gallery in Washington, D.C., in 1975 mounted a retrospective exhibition of the work of Raymond Loewy, which included his Studebaker Avanti. In 1974, Loewy's partner, William Snaith, had died, and the Loewy firm was nearing bankruptcy. By 1975, Loewy had already established Raymond Loewy International by merging all his coexisting organizations.

That year, Henry Dreyfuss Associates designed the Athena 2000 sewing machine, the International Women's Year symbol for the United Nations, and in 1976 Polaroid's Pronto camera. The Concorde supersonic jet began service in 1976 with interiors designed by Raymond Loewy/William Snaith, and the Cooper-Hewitt Museum in the Andrew Carnegie mansion in New York became the National Design Museum, a branch of the Smithsonian.

Russel Wright, one of the founders of the Society of Industrial Designers in 1944, passed away in 1976 at the age of 72. He had transformed his 79 acres of quarry land in Garrison, New York, into a natural scenic wonder called Manitoga. His papers are at Syracuse University Library, Special Collections, and at Manitoga, operated by the Russel Wright Design Center. Manitoga is designated as a National Historic Landmark by the U.S. Department of the Interior and is open to the public, with tours and hiking trails.

Also in 1976, Walter Dorwin Teague Associates designed the Kenworth Sleeper truck, the first to house a sleeper cabin. The design enabled more effective and efficient long-haul driving.

In 1977, the Raymond Loewy/William Snaith, Inc., office in New York closed under Chapter 11 bankruptcy, and industrial design operations were sold to former employee David Butler to form Lister and Butler, Inc.

Loewy's offices in Paris, London, and Sweden remained open. In 1976, Raymond Loewy International had been sold to Sheridan Funding Corporation, James Sheridan, president. He established Museum Administration and Planning Services, Inc. (MAP), as a subsidiary and opened an office in Atlanta, Georgia. Function and Vision, Inc., was formed as "debtor in possession" to handle liquidation of Raymond Loewy International, with James Sheridan as president. Sheridan Associates was also established, specializing in interior design.[8]

George A. Beck, who began his career in 1931, passed away at age 69 in 1977 after retiring from General Electric; Eliot Noyes, the head of the Museum of Modern Art's first industrial design department in 1940, also passed away at age 67; and **Otto Kuhler**, designer of early streamlined 1930s steam locomotives, who had relocated in 1969 to Santa Fe, New Mexico, and in 1976 to Denver, Colorado, died at age 83.

By 1977, the International Style architecture of Walter Gropius and Mies van der Rohe (both had died in 1969), which had been promoted by the Museum of Modern Art from 1932 throughout the 1960s, was coming under increased criticism as sterile, bland, and characterless, or worse, hostile to the natural environment and humanity. That year, the groundbreaking design of the Pompidou Center in Paris, a complex of public library, museum of modern art, and music center, violated all the International Style rules by its dramatic "inside out," high-tech look, with exposed plumbing; electrical, ventilation and safety systems in bright color codes; and with exterior staircases, somewhat like antique urban fire escapes.

Architects began abandoning the International Style by designing buildings that blended in with the surrounding buildings, rather than standing alone as rectangular, glass-box monuments to modern design. The term "postmodern" began to be applied to such architecture, which was generally described as witty, hybrid, eclectic, and concerned with issues of history, local context, metaphor, and ambiguity. It was a sea change in architectural style that would soon find its way into product design.

Henry Dreyfuss Associates designed the Singer Touchtronic sewing machine in 1978. That year, **Dave Chapman**, who had worked as a designer at Chicago's Century of Progress Exhibition in 1933–1934 and founded Chicago's most successful industrial design firm in 1936, passed away at age 69, but his firm would survive under new management as Goldsmith Yamasaki Specht. Chapman's papers are at Syracuse University Library, Special Collections. Charles Eames, who dominated postwar furniture design, also died that year at age 71. Deskey, Inc., was founded in Cincinnati, Ohio, in 1978 as a division of Donald Deskey Associates of New York.

During the 1970s, the computer was becoming the center of a technological and informational revolution. The first prototype personal computer, the Altos desktop, appeared in 1974; in 1975, 19-year-old William Gates III and 22-year-old Paul Allen founded Microsoft; in 1977, Steven G. Wosniac and 21-year-old Steven Jobs founded Apple Computer in a California garage; in 1978, the first electronic typewriters with memory appeared; and in 1979, Sony introduced its Walkman. By 1983, the first cell phones would appear. These new technologies would generate hundreds of new consumer products, and when they entered mass production, industrial designers would design most of them. Technology continued to provide the raw material that industrial designers transformed into products for humans.

By 1979, membership in the Industrial Designers Society of America (IDSA) would top 2,000. It had taken organizational steps over the previous few years to replace its leadership with younger members, and to address some of their concerns. IDSA was no longer a sort of small private club governed autocratically by an older generation, including many of the pioneers, to be sure, but by a more democratic representation of younger chapter chairmen on the board. Membership requirements were relaxed, automotive stylists were included, and it became professionally managed by specialists in association management, rather than by a closed-room board of past presidents. Chapters had doubled to 20, as had membership, and financial health was restored. There were 675 consulting offices nationwide, and over 6,000 industrial designers working within corporations. At the time, it was estimated that there were 10,000 industrial designers worldwide.

Also in 1979, Raymond Loewy published *Industrial Design,* a comprehensive, illustrated book about his design career. He was the surviving member of the original Big Three founders of the profession, and therefore still a public figure who was featured in any historical review of industrial design.

The Industrial Designers Society of America initiated annual national design awards in 1980, known as the Industrial Design Excellence Awards (IDEA), which were presented at IDSA's annual meeting that year, and which continue to this day. They were the first such awards since 1965, when the first IDSA president, Henry Dreyfuss, shut down the previous IDI awards. Also in 1980, Henry Dreyfuss Associates designed the Polaroid 600 Sun Camera and published *Humanscale 4/5/6*, additional ergonomic reference tools for industrial designers, following *Humanscale 1/2/3*, published in 1974. The final publication in this valuable series would be *Humanscale 7/8/9*, published in 1991.

Helen Dryden, who was the "highest-paid woman artist of 1927" and a famous cover illustrator of *Vogue* magazine in the 1920s, passed away in 1981 at age 94. Since 1937, when she designed the interior of the Studebaker as Raymond Loewy designed the exterior, the only thing known about her was that in 1956 she was living in a $10-a-week hotel room paid for by New York City's Welfare Department. At the time, she referred nostalgically to her (1925) $200-a-month Tenth Street apartment in New York.

Other pioneer designers were passing. Both **Montgomery Ferar** and his partner at Sundberg-Ferar since its founding in 1934, **Carl W. Sundberg**, died in 1982, both at age 72. Sundberg had retired in 1975. From 1970 to 1980, Ferar had taught industrial design at the Center for Creative Studies in Detroit. The firm they established continued active practice after their death and is still in business today.

Raymond Dietrich, the well-known automotive designer, passed away at age 88 in 1982. He had left Chrysler in 1940; founded Ray Dietrich, Inc., in 1949; worked for Checker Cab in 1953; and for the Lincoln and Mercury divisions at Ford until his retirement to Albuquerque, New Mexico, in 1969. His wife, Marion, donated his papers and records in 1995 to the Gilmore Classic Car Club of America Museum at Hickory Corners, Michigan.

In 1983, George Nelson installed an exhibit, Design since 1945, organized by Katherine Hiesinger, at the Philadelphia Museum of Art, along with a 247-page, illustrated catalog of the exhibition by Hiesinger. The exhibition and catalog were a sort of obituary of modern design or "Good Design" as defined by the Museum of Modern Art (MoMA) from 1950 to 1955. At the time of the exhibit, museums were no longer trying to dictate good design but were focused on its history, a sure sign of age.

As early as 1979, architectural and design critic Ada Louise Huxtable (1921–2013) had written about the difference between the Bauhaus/Museum of Modern Art concept of industrial design in the 1950s versus the pragmatic and successful reality of industrial design as developed and implemented by the designers of the Machine Age:

> Those who think they know how it was in those days [the 1950s] are in for a surprise. What was admired then is discarded now, and what was rejected is being revived.... Industrial design, that curious amalgam of promotion, marketing, technology, and taste invented by the 20th century to serve and shape a consumer economy ... [whose practitioners] shared the optimism of the architectural theorists ... who believed that the marriage of art and industry could provide answers for most of the ills of society, or at least put beauty and utility within the reach of all. But while the tastemakers were in hot pursuit of a mythical "machine art" realized largely in handmade

> prototypes or limited production for educated users, the industrial designers were successfully creating mass-produced objects for a universal market.[9]

However, the pragmatism, simplicity and repetition of so-called Good Design had its limits. The design community was searching, as it often had, for new styles, new looks, and new directions.

CHAPTER 14

Postmodern and Passages

Finally, after 30 years, designers were tiring of the black boxes and white cubes of Bauhaus-style modern design. It was Italians who broke the ice in 1981, when an avant-garde Milanese group founded an organization called Memphis. It created the design media event of the year with garish furniture designs with bold colors and laminates that emphasized surface decoration more than function. Presented to the beat of rock music in the manner of a fashion show, the designs were deliberately designed to be antifunctional, preposterous, and faddish. They were colorful, cheerful, and playful, the direct opposite of Bauhaus traditions. One of the designs, the Casablanca sideboard, would become the icon of Memphis.

By 1984, some U.S. educators were seeking new theories to free design from the formal dictates of so-called modern design. They came up with likening design to language, which they called "product semantics," defined as, "The study of the symbolic qualities of man-made forms in the context of their use and the application of this knowledge to industrial design. It takes into account not only the physical and physiological functions, but the psychological, social, and cultural context, which we call the symbolic environment." Got that? Although this was a highly academic and theoretical theory, many designers began to abandon the formal "good design" of the 1950s.

Many new designs were being described as "postmodern," which meant that they had departed from the formal, Bauhaus-style, simple, geometric style to more informal, expressive, humanistic, and playful styles. These included cameras that suggested humanoid facial features, modern furniture that imitated 18th-century styles such as Sheraton or Chippendale, teakettles with a little bird whistle on the spout, a bathroom shower squeegee with S-shaped forms suggesting waves, and travel irons that looked like miniature steamships. In 1990, Alessi would introduce a lemon juicer called the Juicy Salif that looked like an insect and would become a cult object.

In effect, design now offered an infinite variety of styles that could just as easily be based on traditional historical forms as those of future science fiction. The "ideal" style was that which best fit the specific individual product's function and usage and its primary audience, the consumer. Ironically, these stylistic modifications were little different conceptually from the highly criticized (by elite modernists of the 1950s) streamlining or skyscraper styles of the 1930s, which were simply 50 years ahead of their time!

One can only wonder what the remaining pioneer machine designers of the 1930s thought of this deja-vu development. Unfortunately, many of them were disappearing. **R. Buckminster Fuller**, inventor of the geodesic dome, passed away at age 88 in 1983. In 1985, he would be posthumously honored with the naming of a newly created form of carbon molecule (C-60) as "buckminsterfullerene" (nicknamed "buckyballs") and others in its class (fullerenes), such as "buckytubes," discovered in 1991, because the carbon molecule C-60 resembled a geodesic dome.

Gordon W. Florian, who had started with General Electric in 1934, passed away in 1984 at age 75. He had designed a locomotive in 1940 for GE when he was assistant director to Ray Patten, but he left GE in 1946 to open his own office. He taught design at the Junior College of Connecticut, which became the University of Bridgeport after the war, and in 1949 he established a four-year industrial design program there. From 1950 to 1954 he lectured to engineering and business students at MIT. A lifelong artist of landscapes and a member of the Connecticut Watercolor Society, he exhibited all over the East Coast, including Silvermine. In 1984, he was named Fairfield (Connecticut) "Artist of the Year."

Lurelle Guild, a prolific industrial designer since 1924, passed away at age 88 in his home in Darien, Connecticut, in 1985 after a long career. He had practiced design until the 1960s. His papers are at the Syracuse University Library, Special Collections, and his products are in various museums. He had been born in Syracuse and studied at the university.

In the mid–1980s, Henry Dreyfuss Associates was coming apart. The office had moved to 423 West 54th Street. Partners were Don Genaro, Jim Conner, Bill Crookes, Gordon Sylvester, Jim Ryan, and Jack McGarvey. Several associates started a separate firm and took with them other employees and most of the client base. Bill Crookes bought out the remaining partners and what clients were left, mostly John Deere Agricultural Products, and moved the office to Ann Arbor, Michigan.[1]

In 1985, Worldesign 85, the first Congress of the International Council of Societies of Industrial Design (ICSID), was held in the United States, hosted by the Industrial Designers Society of America (IDSA). International registrants of the conference voted Charles Eames the "Most Influential Designer of the 20th Century."

John Vassos, one of the founders of the American Designers' Institute in 1938, and RCA design head from 1931, also passed away in 1985 at age 93. He had been president of the Silvermine Guild of Artists and Writers in Connecticut in 1936, 1940–1947, and 1949–1955, and had retired from heading RCA design in about 1959. In 1977, the last retrospective of his work was held at the Silvermine Guild in Connecticut. In poor health and nearly blind, Vassos remained a feisty old man until the end. Vassos's papers are also at the Syracuse University Library, Special Collections.

The year 1986 saw the passing of more pioneer designers. **John Morgan**, Sears, Roebuck's first designer in 1934 with dozens of design patents, passed away at age 83 in McHenry, Illinois. After his retirement, he had continued his successful career in fine art. **Ray Sandin**, Hotpoint's first designer in 1935, died at age 76 after serving Hotpoint for 51 years. **George Nelson**, the man who alerted the business world to industrial design in 1934, passed away at age 78. He revolutionized furniture design at Herman Miller after the war and invented the concept of pedestrian malls. His papers are at the Syracuse University Library, Special Collections. Carl Otto is estimated to have died around this time, but there is virtually no information about him since he closed his office in 1956. He was last known to be in England.

Raymond Loewy, one of the Big Three industrial designers, died in 1986 at age 93. He had retired in 1980 at the age of 87 and had returned to his native France, where he passed away at his Monte Carlo residence. In 1992, his second wife, Viola, and the British American Tobacco Company would establish the Raymond Loewy Foundation in Hamburg, Germany, to promote industrial design internationally. An annual award of €50,000 was granted annually to outstanding designers. Loewy's daughter, Laurence, in 1998 would establish Loewy Design in Atlanta, Georgia, to manage her father's interests in the United States. Laurence would die in 2008, and her husband, David Hagerman, would manage Loewy Design and the Loewy estate, which would catalog the Loewy archives and raise funds to open the Raymond Loewy Museum of Industrial Design. Collections of Loewy archives are in the Raymond Loewy Collection at the Hagley Museum in Wilmington, Delaware, and in the Manuscript Division of the Library of Congress in Washington, D.C.

The Brooklyn Museum in 1986 published *The Machine Age in America, 1918–1941* and organized a related exhibition in Brooklyn from October 17, 1986, to February 16, 1987. The exhibit would continue at the Museum of Art, Carnegie Institute, in Pittsburgh from April 4 to June 28, 1987; at the Los Angeles County Museum of Art from August 16 to October 18, 1987; and would conclude at the High Museum of Art, Atlanta, Georgia, from December 1, 1987, to February 14, 1988. The exhibition and book was a masterful retrospective of this important era of American modern design.

In 1987, the National Endowment for the Arts presented a five-part television series on PBS, *America by Design,* exploring the history of American architecture and design. Raymond Loewy dominated industrial design history in the series. That same year, **Jean Reinecke**, who had opened his industrial design office in Chicago in 1930, died in Pasadena, California, at the age of 78. He had closed his Chicago office in 1974 and had continued designing until 1986 in his Pasadena office, where he was also on the Council of Advisors of the Art Center College of Design.

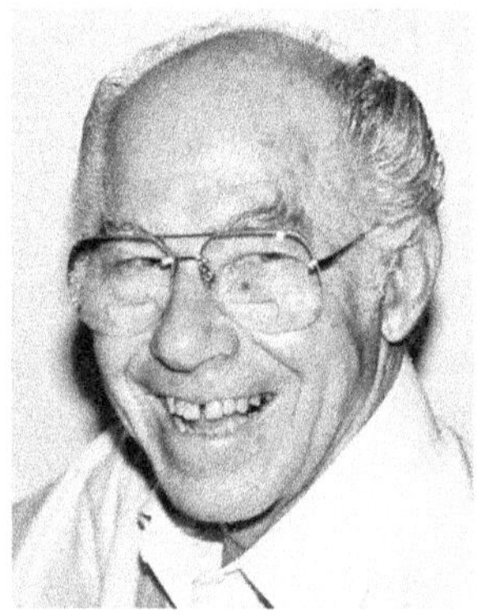

Jean Reinecke, ca. 1970s (courtesy Barbra Reinecke Sedory).

In 1988, **Edwin W. Fuerst**, who established his first Toledo design studio in 1929, passed away. He had taught industrial design at Ohio State University from 1948 to 1949. In the mid-1950s he became a freelance painter and moved to New England, becoming special assistant to the vice president of sales for the Plax Corporation in Hartford, Connecticut, from which he retired in 1966. **George Sakier**, one of the founders of the Society of Industrial Designers (SID), died at the age of 91, also in 1988. He had designed glassware for Fostoria Glass from 1929 to 1979, when he retired to continue his lifelong career as a painter of abstract landscapes, with natural elements reduced to geometric forms. His paintings included Red Mesa, #011, and Collage C008. Yale University initiated a George Sakier Prize for excellence in photography.

Donald Deskey, a prominent industrial designer since 1926, and one of the founders of the Society of Industrial Designers, died in 1989 at the age of 94 in Vero Beach, Florida, the town to which he retired in 1975. In 1988, Deskey had exhibited a one-man show, Donald Deskey, American Modern Design, an exhibit of his furniture designs from the 1920s and 1930s, at Fifty/50, a Manhattan gallery. After the exhibit he donated his papers to the Cooper-Hewitt National Design Museum in New York.

Several pioneer industrial designers passed away in 1990. **Nathan Horwitt**, the designer of the classic Museum Watch in 1947, passed away at his farm in Stockbridge, Massachusetts, at the age of 92. Between 1945 and 1948, his farm was used for training exercises of the Jewish underground force, the Irgun. Horwitt supported the creation of the state of Israel and served as a member of the American League for a Free Palestine. For these efforts, in 1980, he had been awarded the Jabotinsky Medal by the prime minister of Israel, Menachem Begin.

Leo Jiranek, a prominent furniture designer since 1924, passed away at age 90. In the

1960s he had founded and was president of the Jiranek School of Furniture Design and Technology in New York City, and in 1984 he was awarded a doctor of arts degree by the Kendall College of Art and Design.

Gordon Buehrig, noted automotive designer, died at age 86. He had retired from Ford in 1965 and taught for five years at the Art Center College of Design in Pasadena, California. In 1979, he designed the Buehrig motorcar, a limited-production carriage-roof coupe. In 1981, he was named by the Society of Automotive Historians as one of the most significant contributors to the development of motorcars. In the decade before his death, he directed precision car models for the Franklin Mint, and in 1990 he was inducted into the Automotive Hall of Fame.

Robert Lepper, a faculty member in Carnegie Tech's first degree program in industrial design in 1934, died in Pittsburgh in 1991 at the age of 85. Lepper had remained on the faculty in the program until his retirement in 1975. Andy Warhol had been one of his students from 1945 to 1949. In 1986, Lepper was prominently included in the Brooklyn Museum's exhibition and book, *The Machine Age in America, 1918–1941,* which regarded his murals of the 1930s and 1940s as rivaling those of Diego Rivera. In 1989, Lepper had received the Industrial Design Society of America's (IDSA) educator award.

In 1991, *Business Week* magazine sponsored the annual IDSA Industrial Design Excellence Awards (IDEA), then celebrating ten years of operation, and featured the awards in a major, illustrated article in the magazine. This annual recognition of industrial design in a major business magazine benefited both the manufacturers, the 800 to 1,000 design offices in the country, and the individual designers. The number of awards would vary from year to year, but by 1997 IDEA would confer a total of 141 gold, silver, and bronze awards.

Graduate students at the Pratt Institute in Brooklyn formed the Association of Women Industrial Designers (AWID) in 1991. Their key collaborative project would be a 1994 exhibition at the Pratt Manhattan Gallery in New York and a related catalog with the title, Goddess is in the Details: Product Design by Women.

H. Creston Doner, who opened a design office in Akron, Ohio, in 1932, and was a longtime design director at Libby-Owens-Ford Glass Company, passed away in 1991 in Toledo at age 88. He had retired from LOF after 30 years in 1968, and from 1970 he was a member of the Toledo Society of Communicating Arts. At the end of 1991, the Soviet Union also came to an end and collapsed to become a republic of individual states, ending the Cold War and leading to a better life for Russian citizens.

Clare Hodgman, who started his career with General Motors in the early 1930s, died in 1992 at the age of 81. He had retired from Hodgman & Bourke in 1969 and moved to Daytona, Florida, where he excelled in contemporary watercolors. **George Walker**, who began his career in the automotive industry in 1928, died at the age of 97 in 1993. In 1959, he was a guest challenger on the TV panel show *To Tell the Truth*. After retiring from Ford in 1961, he moved to Gulf Stream, Florida, where he became mayor in 1976. By 1985, he was in Tucson, Arizona, where he was interviewed by the Ford Design History Center.

In 1994, President Bill Clinton presented eight Presidential Awards for Design Excellence to designers, including industrial designers. The automobile industry rebounded in 1994 to capture 75 percent of the U.S. market share, and for the first time in 14 years led the world in car production. But many other corporations were in serious trouble. One hundred fifty-one Fortune 500 companies, or 30 percent, lost money in 1993. The previous decade had been a litany of mergers and acquisitions, as many companies sought to convert failing markets into cash.

By this time, computer aided design (CAD) was driving the manufacturing, engineering, and industrial design of products. Installation of CAD technology was revolutionizing in-house and consulting design firms across the industry. The historical restrictions of machine tools that controlled design were now replaced by computer-aided tooling that enabled designers to specify complex shapes without regard to those earlier physical restrictions. CAD proficiency was a prerequisite job requirement for all industrial designers and for anyone in the product development field. The days of hand-drawn renderings of product concepts were over, replaced by realistic computer-generated illustrations.

In 1995, **Stowe Myers**, who had been hired by Norman Bel Geddes in 1932 to illustrate Bel Geddes's book, *Horizons,* died in 1995 at age 89 in New Orleans. He had been a member of the faculty at the Institute of Design of the Illinois Institute of Technology, where he taught a class entitled "Design Counsel." Myers retired in 1976 to paint and was named official artist of the U.S. Air Force in 1978. In 1988, he moved to New Orleans, Louisiana. His papers are at the Syracuse University Library, Special Collections.

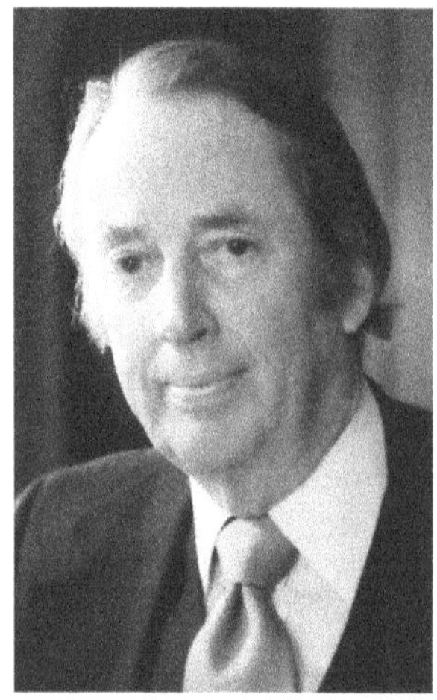

Brooks Stevens (Brooks Stevens Archive, Milwaukee Art Museum, gift of the Brooks Stevens Family and the Milwaukee Institute of Art and Design).

In 1995, Walter Dorwin Teague Associates designed Power Rangers toys. **Brooks Stevens**, one of the founders of the Society of Industrial Designers in 1944, passed away that same year at age 84. His son, Kipp K. Stevens (b. 1951), had joined Brooks Stevens Design Associates in 1978 as vice president and in 1979 had become president and CEO. From 1983 until his death, Brooks taught design at the Milwaukee Institute of Art and Design.

In 1996, **John M. Little**, who had opened an industrial design office in Minneapolis in 1931, passed away in 1996 at the age of 90. He had joined Harold Van Doren's Toledo office in 1936 and managed it after the war. He was a lecturer at such universities as Harvard, Lehigh, Cincinnati, Toledo, and Wisconsin. **Robert E. Bourke**, designer of Studebakers for Loewy, passed away that same year at the age of 80. After his partner, Clare Hodgman, retired from their partnership, Hodgman & Bourke, in 1969, Bourke had moved his office to Westport, Connecticut, where his home was located.

In Chicago, the firm of Goldsmith Yamasaki Specht (GYS), with Paul Specht as the last principal, was sold in 1997 to Sundberg-Ferar, for whom Specht served as senior design council until 2000. Sundberg-Ferar maintained four offices in the United States and one in the Far East. Also in 1997, the Cooper-Hewitt National Design Museum opened a retrospective, Henry Dreyfuss Directing Design: The Industrial Designer and His Work. Concurrently, Rizzoli International published *Henry Dreyfuss, Industrial Designer: The Man in the Brown Suit,* by Russell Flinchum, of the museum.

A 1997 *Time* magazine article, Where the Jobs Are, predicted 76,000 new product design jobs by 2005 and called industrial design "one of the hottest professions in terms of job growth and compensation." The forecast was based on figures from the Bureau of Labor Statistics, which stated that industrial designers, then totaling 9,000, were among those related professions and were expected to number 21,000 by 2005.

In 1999, **Alfons Bach**, who had established his furniture design firm in New York City in 1932, passed away at age 95. He had developed the Ridgeway Shopping Center in Stamford, Connecticut, in 1945, and after relocating to Palm Beach, Florida, in 1959, he developed the Palm Trail Plaza in Delray Beach. He retired to Pensacola, Florida, in 1992, where he continued his career as a watercolor artist. In 1997, he was honored at a reception for an exhibition of his work at the Visual Arts Gallery of Pensacola Junior College. An extensive archive of his drawings and papers is at the Cooper-Hewitt National Design Museum in New York.

Belle Kogan, a successful female industrial designer who opened her own office in New York in 1932, died in 2000 in Israel at the age of 98. She had closed her office in 1970; moved to Israel under contract with K.V. Design, a subsidiary of the Koor Corporation; and set up a studio offering comprehensive design services. In 1972 she left the studio to work as a consultant.

In the March 20, 2000, issue of *Time* magazine, an article, "The Rebirth of Design," subtitled, "The Redesigning of America," claimed that form was the dominant characteristic of many product designs. Forms were soft, voluptuous, highly ergonomic and even exotic. The "black boxes" of the 1970s were definitely out. Bauhaus geometric shapes were out. The color and form of functional details such as keys, levers, and switches had become the new decorative ornamentation of the era.

The turn of the century was a big year for retrospectives. The Metropolitan Museum of Art opened an exhibition, American Modern, 1925–1940: Design for a New Age, which traveled the country for several years. The Bard Graduate Center for Studies in the Decorative Arts mounted another exhibit, Women Designers in the U.S.A., 1900–2000: Diversity and Difference, along with a related book, which included work by a number of female designers.

NASA launched the International Space Station in 2000, with the crew quarters designed by Walter Dorwin Teague Associates (WDTA). The station is the area of a football field, weighs 861,804 pounds, and has been in continuous operation since 2000. In 2001, WDTA would work with Microsoft in the design of their first gaming console, the Xbox.

By 2000, the Industrial Designers Society of America (IDSA) had 3,300 members in 28 chapters around the country, and by 2001, it had relocated to new headquarters near Dulles airport in Washington, D.C. In 2002, IDSA and the China Industrial Design Association (CIDA) sponsored a China-America education conference in Beijing. U.S. companies had been using Asian manufacturers for years to produce products at lower costs, and the number of U.S. designers traveling to China, South Korea, and Taiwan was increasing dramatically. China was producing thousands of industrial design graduates and competing with American practitioners.

Arthur BecVar, who began his career in industrial design in Cleveland in 1935, died in 2003, at the age of 91. He had retired from General Electric as head of design in Louisville, Kentucky, in 1976, after more than 30 years of service. His papers are at the Syracuse University Library, Special Collections.

Also in April 2003, the Cooper-Hewitt National Design Museum in New York tried to capture the current state of design with an exhibition, National Design Triennial: Inside Design Now. The museum found it difficult to categorize a dominant style in the eclectic menu of what many called "postmodern," so the catalog simply referred to the body of work as being "by contemporary American designers." Some designs strayed so far from the geometric Bauhaus style of the 1950s that a magazine editor remarked, "Ornament is back and we've got to relearn how to use it."

W. Dorwin Teague, Jr., son of Walter Dorwin Teague, who started his career with his father in 1931, passed away at the age of 94 in 2004. He had relocated his Englewood Cliffs office to his home in the early 1970s and continued to design ski equipment, a water-powered toothbrush, a bicycle brake, and a snowmaking apparatus, among many others, until 1993, when he arranged for his papers to be donated to the Henry Ford Museum. Renowned designer John Barratt joined the WDTA team as CEO. **Eugene "Bob" Gregorie**, head of design at

Ford from 1935 to 1946, also died in 2004 at the age of 96. In 1943 he had retired to design and sail his boat, and in the mid–1970s he had relocated to St. Augustine, Florida.

In 2004, the annual Industrial Design Excellence Awards (IDEA) became international in scope, open to all designers around the world. Of the 1,300 entries, 305 were from other countries, clear recognition that industrial design had become an integrated global profession. A 2004 IDSA survey of 183 corporate design offices reported that design increased product sales by 47 percent and increased sales prices by 32 percent, while only increasing product cost by 24 percent. The numbers suggested that industrial design was still an essential business tool for manufacturing profitability.

By 2006, the Industrial Design Excellence Awards ceremony at IDSA's annual conference was becoming quite sophisticated, with black ties optional, and with an ambience not unlike Hollywood's Oscar Awards. It was the number of gold awards to design teams from Asia that was a surprise: 25 percent of the total. This was compared to 8 percent in 2004 and 2005.

Early in 2006, China's president, Hu Jintau, at a national conference on technology, had confirmed the country's new emphasis on design and innovation, stating, "Innovation is the core of the nation's competitiveness. Only with strong capacity for innovation can a country win the initiative in the international competition."[2] American educators teaching design in China confirmed that there was excitement in the air in China, reminiscent of the postwar U.S. attitude of "can do" in the 1950s and 1960s. Over the next few years, Asian design would increase in quality and quantity, and American designers would often be working in China, Taiwan, and South Korea.[3]

A permanent Raymond Loewy Gallery, emphasizing Loewy's transportation designs, opened in 2006 at the O. Winston Link Museum in Roanoke, Virginia. The museum was housed in the restored Norfolk & Western Railway Station, originally designed by Loewy's firm in 1949. Loewy's daughter, Laurence Loewy, was instrumental in establishing the gallery. That same year, Loewy was celebrated as "the Father of American Design" in a symposium held by the Hagley Museum in Wilmington, Delaware, where Loewy is prominent in the Pennsylvania Railroad Corporation records held by the museum.

In 2007, the name of IDSA's Industrial Design Excellence Awards competition was changed to the International Design Excellence Awards to recognize that the competition, and industrial design in general, was now global in scope.

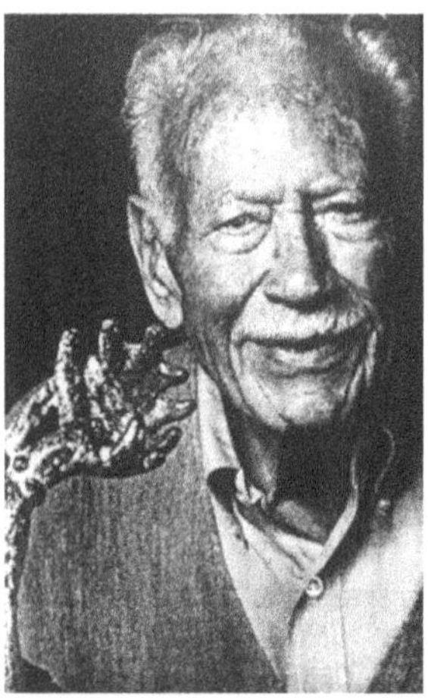

Viktor Schreckengost, ca. 2000 (photograph by Herb Ascherman).

In 2008, **Adolph "Al" Lambach, Sr.**, who had begun his career as a furniture designer in Grand Rapids, Michigan, in 1932, died at the age of 96 in California. As a founding member of the American Designers' Institute (ADI) in 1938, he was a member of organized design, including IDSA, for a total of 70 years, probably a record. He had worked as a freelance furniture designer until his retirement in 1982, when he carved replicas of famous ships for his collection and taught at the Art Center College of Design. He donated his many drawings to the industrial design collections at the Hagley Library in Wilmington, Delaware, and to Syracuse University Library, Special Collections.

Also in 2008, **Viktor Schreckengost**, who developed an early industrial design program at the Cleveland Institute of Art in 1933, passed away at the age of 101. In 2000, he had been honored with a retrospective exhibition and catalog at the Cleveland Museum of Art, called Viktor Schreckengost and 20th Century Design. Soon after, the Viktor Schreckengost Foundation was established in Cleveland. In 2006, a Viktor Schreckengost National Centennial Exhibition celebrated his 100th birthday by highlighting his work in 140 simultaneous nationwide exhibitions. That year, he was also honored at the White House as a recipient of the National Medal of Arts for "enduring contributions to the artistic life" of the United States.

By 2008, *Business Week*'s annual publication of the IDEA international design awards noted that a strong sense of social responsibility ran through many of the winning entries, including a wheelchair that freed hands for disabled athletes; an inexpensive computer laptop for children in Third World countries; and a book that helped governments to design clearer, more useful ballots and polling place signage. Design buzzwords such as "green," "sustainability," "environmental," and "disability" permeated design competitions, conferences, education, and debates. In 2010, sponsorship of IDEA design awards would be taken over by *Fast Company* magazine.

As Barack Obama was elected the first African-American president in 2008, professor Eric Anderson of Carnegie Mellon University was elected IDSA's first African-American president, and African-American industrial designer Charles Harrison (b. 1931) received a Lifetime Achievement Award at the White House as part of the Cooper-Hewitt National Design Museum awards program. Harrison, the first African-American executive hired at the company's Chicago headquarters, was hired by Sears in 1961 and worked there until his retirement in 1993.

By 2009, there were about 50,000 industrial designers, of which about 3,500 were members of IDSA. Median annual salaries were around $55,000. This was at the depth of the severe financial recession, when unemployment reached 10 percent and Chrysler and General Motors received bailout funds from the government. But industrial design is traditionally a recession-proof profession, and in 2009 there were over 20 position advertisements on the IDSA website, and thousands more posted on the Internet. IDSA had a staff of 12, an annual budget of about $5 million, and 27 chapters in five national districts. By 2009, in the annual International Design Excellence Awards, there would be 1,631 entries, and 66 of 150 awards would go to 15 countries other than the United States.

In 2011, the U.S. Postal Service issued stamps honoring 12 of the nation's most important and influential industrial designers with an image of a design created by each designer. They were Frederick Hurten Rhead (1936 Fiesta dinnerware), Walter Dorwin Teague (1934 Baby Brownie camera); Norman Bel Geddes (1940 Patriot radio), Raymond Loewy (1934 pencil sharpener), Donald Deskey (1927 table lamp), Gilbert Rohde (1933 clock), Greta Von Nessen (1951 Anywhere lamp), Russel Wright (1950 Highlight/Pinchware tableware), Henry Dreyfuss (1937 Model 302 AT&T telephone), Peter Müller-Munk (1935 Normandie pitcher), Dave Chapman (1947 streamlined sewing machines), and Elliot Noyes (1961 Selectric typewriter).

They are all gone now, those extraordinary designers of the Machine Age, and many are forgotten. Their proud legacy is today's prospering profession of industrial design, which they invented in the 1930s, practically out of thin air, by transforming our everyday environment of 19th-century, unwieldy, unattractive, mechanical devices, from automobiles to toasters, office equipment, and kitchens, into sleek and attractive works of art that enhance our daily experiences emotionally and visually, as well as functionally. Today, we regard our man-made physical environment of mechanical, electronic, and digital devices as an unremarkable and a natural result of modern technology, as if they magically appeared out of nowhere.

But in 1930, neither technology nor engineering knew how to shape the products it produced by the millions into forms that were designed to address human convenience, utilization, or emotional needs that made them attractive and desirable to consumers. Ordinary artists, the Machine Age designers of the 1930s, bravely ventured from their comfortable lives into a new world of complex, dirty factories, business offices, and brutal sales competition in order to transform everyday products into modern design, which encouraged people in the depths of the Depression to visualize a better future and lifted a failing manufacturing industry into profitability. A recovered economy led to the highest standard of living in the world after World War II, and the machine age designers established an educational system and business models for the new profession that became integrated into the modern world of manufacturing, production, and sales.

Today's industrial designers perform the same services for manufacturers and consumers that the machine designers of the 1930s did, only better and faster. The process remains the same, except that instead of hand-drawn illustrations of proposed designs and manually prepared mechanical drawings, computers now perform these tasks with the instructions of designers. Prototype models are often not built by hand but generated into three dimensions by computers. Contrary to many pessimists in the 1960s, who concluded that all consumer products had already been "designed" to perfection, technology continues to generate new products that need to be shaped and modified for human use and understanding, and this is likely to continue indefinitely.

Many of the organizations and designers mentioned in this book ceased operations over the years due to death or business closure, but, surprisingly, quite a few of the design organizations of the founders survive to this day, a tribute to the tenacity of their founders and a source of pride to their heirs. They continue to design the world about us.

In 2006, Walter Dorwin Teague Associates, founded in 1926, now called Teague, in Seattle, Washington, celebrated its 80th anniversary as a leading force in industrial design and its 60-year working relationship with Boeing, with a Frank Gehry-designed Experience Music Project held in downtown Seattle, and an elegant publication of Teague history and designs, *Design This Day,* the same title as his 1940 book. In 2011, Teague's relationship continued with Boeing with the interior design of the 787. That same year, Teague acquired a Munich-based design studio to expand into the European market. Teague staff in 2012 totaled 250 designers and support staff.

In 2012, the Art and Colour Section of General Motors that Harley Earl founded in 1927 celebrated its 85th anniversary, and it continues to design new cars each year, most of them still "streamlined," but no longer called that. Similar early corporate design organizations, such as at Ford, Chrysler, General Electric, Westinghouse, the Radio Corporation of America, Herman Miller, Owens-Illinois, American Standard, among others, are still operational, but many have changed names through mergers or acquisitions, and still others have gone out of business.

Sundberg-Ferar, founded in 1934 by its namesakes, is a still a highly successful design firm in Walled Lake, Michigan, with a 15,000-square-foot studio and a 7,000-square-foot prototype shop, with a talented, multidisciplinary team led by seven key personnel. A recent design was a Precedent golf cart for Ingersoll Rand's Club Car division. Clients since 2003 include Coca-Cola, General Electric, General Motors, Procter & Gamble, Sears, and Whirlpool.

Peter Müller-Munk Associates (PMMA), founded by its namesake in 1938, continues as an active design firm in Pittsburgh, well known for way-finding signage design at major airports and public spaces.

Brooks Stevens, Inc., the organization founded in 1934 by its namesake, is still an innovative and successful industrial design firm with offices in Allenton, Wisconsin, and St. Paul, Minnesota, and with dozens of well-known clients including Kohler, Harley-Davidson, General Electric, Rubbermaid, and Herman Miller. In 2007, the firm merged with Ingenium Product Development, bringing together over 70 experts. Recent products included the 100 Series

John Deere lawn tractor, the Remington Code Shaver, and Milwaukee's Tilt-Lok circular saw.

As of 2005, with about 10 employees, Henry Dreyfuss Associates, founded in 1929, continued to operate, with major corporate clients, and was located in Ann Arbor, Michigan, with William Crookes as president.

Nessen Lighting, founded by Walter Von Nessen in 1927 as Nessen Studios, still designs and manufactures historical and contemporary lamps in Mamaroneck, New York.

The firm founded by Dave Chapman in 1936, Dave Chapman, Inc., had been reorganized in 1970 as Goldsmith Yamasaki Specht (GYS). In 1996, First Design, a design firm in Kansas City, Kansas, purchased GYS, Inc., which had survived with Paul Specht as the only remaining partner. A partner of First Design, David McCormick, became president of GYS and retained the Chicago Office until 1999, when it was sold to Sundberg-Ferar.

Deskey Integrated Branding (DIB), currently a firm specializing in branding strategy with headquarters in Cincinnati, Ohio, was founded in 1978 as Deskey, Inc., a division of Donald Deskey Associates of New York, which had been founded in the mid–1940s by Donald Deskey. The companies were combined around 1990 into Deskey Associates, Inc. The New York office was closed in September of 2001. Family trusts of famous designers such as Norman Bel Geddes and Raymond Loewy still merchandize images and books about their namesakes, but they are no longer active design organizations.

Ford and Earl Associates, founded in 1964 by Harley Earl and Walter B. Ford, is still a design firm in Troy, Michigan, but now specializes in architecture and interior design.

This is a pretty good survival record of design firms founded by designers of the Machine Age. We also have their classic, iconic designs, their books, and perhaps most of all, their legacy of today's creative, innovative, and successful industrial design profession, which continues to shape our physical environment of tools, machines, furnishings and equipment in a manner appropriate to our use, needs, and desires. We will continue to be inspired by the courage of the first machine designers to envision the modern world when there was none, and to implement it into reality despite the Great Depression of the 1930s. It was an extraordinary accomplishment that is unlikely to ever be repeated. The guidebook of the 1939 World's Fair probably said it better than anyone could:

> The true poets of the twentieth century are the designers, architects, and the engineers who glimpse some inner vision, create some beautiful figment of the imagination, and then translate it into valid reality for the world to enjoy.[4]

Chapter Notes

Chapter 1

1. Metropolitan Museum of Art, *Exhibition of Contemporary American Industrial Art*, October 12–November 22, 1931.
2. Pulos, *American Design Ethic*, 134.
3. Ibid., 270.
4. Ibid., 134.
5. Ibid., 261.
6. Hughes, *American Genesis*, 329.
7. Gorman, *The Industrial Design Reader*, 104.
8. Gantz, *Design Chronicles*, 47.
9. Heskett, *Industrial Design*, 88.
10. Waltman, *Memoirs of Chick Waltman*.
11. Long, "Kem Weber and the Rise of Modern Design in Southern California."
12. Frankl, *Form and Reform*.

Chapter 2

1. Gorman, *The Industrial Design Reader*, 115.
2. Pulos, *American Design Ethic*, 304.
3. Gorman, *The Industrial Design Reader*, 116.
4. Wilson et al., *The Machine Age in America, 1918–1941*, 51.
5. Pulos, *American Design Ethic*, 304.
6. Ibid., 316.
7. Long, "Kem Weber and the Rise of Modern Design in Southern California."
8. Pulos, *American Design Ethic*, 316.
9. White, "The Machine Age."
10. Calkins, "Beauty, the New Business Tool," 316.
11. Ibid., 324.
12. Patton, *Made in USA*, 235.
13. Gelernter, *1939: The Lost World of the Fair*, 281.
14. Paul Frankl, as quoted in Pulos, *American Design Ethic*, 336.
15. Pulos, *American Design Ethic*, 323.
16. Walter Dorwin Teague, as quoted in Seldes "Profiles: Industrial Classicist—Profile of Walter Dorwin Teague."
17. Walter Dorwin Teague, "Modern Design Needs Modern Merchandising."
18. Gorman, *The Industrial Design Reader*, 124–125.
19. Jo Sinel, as quoted in a 1969 interview by Robert Harper.
20. Jackson, "Jo Sinel."

Chapter 3

1. Szerlip, "Colossal in Scale, Appalling in Complexity."
2. Bach, in catalog of *Contemporary American Design*, 1928.
3. Loewy, *Industrial Design*, 11.
4. Ibid., 10.
5. Pulos, *American Design Ethics*, 323.
6. Ibid., 332.
7. Lesko, "Industrial Design at Carnegie Institute of Technology, 1934–1967," 272.
8. Bel Geddes, *Miracle in the Evening*, 345–346.
9. *Annual of American Design*, 1931.
10. Brooklyn Museum exhibition, American Union of Decorative Artists and Craftsmen, May 2–July 1, 1931.
11. Laine, "Ann Swainson, Master Designer."
12. Metropolitan Museum of Art, Exhibition of Contemporary American Industrial Art, October 12–November 22, 1931.
13. Elmo Calkins, as quoted in Gorman, *The Industrial Design Reader*, 130.
14. Ibid., 131.
15. Philip Johnson, as quoted in Secrest, *Frank Lloyd Wright*, 392.

Chapter 4

1. Bel Geddes, *Horizons*, 24.
2. Loewy, *Industrial Design*, 36.

3. Bel Geddes, *Horizons*, 3.
4. Bel Geddes, *Miracle in the Evening*, 344–345.
5. Ibid., 347.

Chapter 5

1. Oliver Sprague, as quoted in Nelson, "Both Fish and Fowl."
2. Nelson, "Both Fish and Fowl."
3. Bel Geddes, *Miracle in the Evening*, 347.
4. Nelson, "Both Fish and Fowl."
5. Plato, as quoted in Pricola, "Art for Trade's Sake: The Fusion of Commerce and Culture, 1927–1934."
6. Alfred H. Barr, as quoted in Patton, *Made in USA*, 33.
7. Gorman, *The Industrial Design Reader*, 134.
8. Pricola, "Art for Trade's Sake: The Fusion of Commerce and Culture, 1927–1934."
9. Philip Johnson, in a 1991 interview by Sharon Zane.
10. Alfred Barr, as quoted in Patton, *Design in USA*, 236.

Chapter 6

1. Wayt, "Donald Dohner: The Man Who Designed 'Rivets.'"
2. Lesko, "Industrial Design at Carnegie Institute of Technology, 1934–1967."
3. Robert Dirig, e-mails to author.
4. Loewy, *Industrial Design*, 98.
5. Bourke, *The Reminiscences of Robert E. Bourke*.
6. Belle Kogan, in a 1935 interview by Marcy Babbitt, reprinted in Gorman, *The Industrial Design Reader*, 137–139.
7. Kogan, "Making the Most of Materials."

Chapter 7

1. Clark, "When Modernism Was Still Radical: The Design Laboratory and the Cultural Politics of Depression-Era America."
2. Laine, "Ann Swainson: Master Designer."
3. Paul Specht, e-mail to Matranga, July 27, 2013.
4. Bel Geddes, *Miracle in the Evening*, 348.
5. Ibid.
6. Ibid.
7. Lesko, "Industrial Design at Carnegie Institute of Technology, 1934–1967."
8. Pulos, *American Design Adventure*, 196.
9. Ibid., 167.
10. Pulos, *American Design Ethic*, 402.
11. Gruen, nomination of Belle Kogan, January 20, 1994.

Chapter 8

1. Iannelli, "History of Industrial Designers Institute."
2. Clark, "When Modernism Was Still Radical: The Design Laboratory and the Cultural Politics of Depression-Era America."
3. Ockman, "Object Lessons from the Bauhaus."
4. Banham, "Machine Aesthetics."
5. Van Doren, *Industrial Design: A Practical Guide*, 79.
6. New York World's Fair official guidebook.
7. Gelernter, *1939: The Lost World of the Fair*, 60.
8. Ibid., 65–66.
9. Ibid., 19–25.
10. Bel Geddes, *Miracle in the Evening*, 348.
11. Gelernter, *1939: The Lost World of the Fair*, 158–161.
12. Ibid., 45.
13. Ibid., 264–266.
14. Loewy, *Industrial Design*, 108.
15. Gelernter, *1939: The Lost World of the Fair*, 155.
16. Ibid., 113–114.
17. Ibid., 115.
18. Ibid., 116.
19. Ibid., 125.

Chapter 9

1. Clark, "When Modernism Was Still Radical: The Design Laboratory and the Cultural Politics of Depression-Era America."
2. Gelernter, *1939: The Lost World of the Fair*, 352.
3. Loewy, *Never Leave Well Enough Alone*, 162–163.
4. Teague, *Design This Day: The Technique for Order in the Machine Age*, 1940.
5. Ibid.
6. Van Doren, *Industrial Design: A Practical Guide*, 3.
7. Ibid., 16.
8. Bel Geddes, *Miracle in the Evening*, 349.
9. Secrest, *Frank Lloyd Wright*, 392.
10. Eliot Noyes, as quoted in Pulos, *American Design Ethic*, 411.
11. Noyes, "Organic Design in Home Furnishings."
12. Pulos, *American Design Adventure*, 20.
13. Ibid., 21.
14. Bel Geddes, *Miracle in the Evening*, 350.
15. John Tjaarda papers.
16. Durez promotional brochure, ca. 1942, courtesy Vicki Matranga.

Chapter 10

1. Bel Geddes, *Miracle in the Evening*, 350–351.
2. Pulos, *American Design Adventure*, 46–47.

3. Toledo Museum of Art, *The Alliance of Art and Industry: Toledo Designs for a Modern America*, 187.
4. Pulos, *American Design Adventure*, 172–173.
5. Ibid., 198.
6. Ibid., 200.
7. Ibid.
8. Ibid., 175.
9. Walter Dorwin Teague, as quoted in Pulos, *American Design Adventure*, 176.
10. Loewy, *Never Leave Well Enough Alone*, 170.
11. Loewy, "Industrial Design—The Aesthetics of Salesmanship—An American View," 71–72.
12. Toledo Museum of Art, *The Alliance of Art and Industry: Toledo Designs for a Modern America*, 55–58.
13. Bel Geddes, *Miracle in the Evening*, 351.
14. "Patten's Distinctive Designs," *Railroad Heritage*, 13–15.
15. Pile, *Dictionary of 20th Century Design*, 65.

Chapter 11

1. Steinhilber, *Looking Back*.
2. Van Doren, *Industrial Design: A Practical Guide*, 137–138.
3. Bourke, *Reminiscences of Robert E. Bourke*.
4. Nelson, "Business and the Industrial Designer."
5. "He Streamlines the Sales Curve," *Time*.
6. Ibid.
7. Van Doren, "Streamlining: Fad or Function?"
8. Dreyfuss, "The Industrial Designer's Best Friend and Severest Critic."
9. Robert Bourke, interview in *The Reminiscences of Robert E. Bourke*.
10. Pulos, *American Design Adventure*, 110–121.
11. Ibid., 110–121.
12. Vienne, "Good Design MoMA's Campaign against Doodads."
13. Dreyfuss, "The Silent Salesman of Industry," 2.

Chapter 12

1. Müller-Munk, "Industrial Designers Can Face Today's Challenge."
2. Teague, "A Quarter Century of Industrial Design."
3. Pulos, *American Design Adventure*, 120–121.
4. Laine, "Ann Swainson: Master Designer."
5. Dreyfuss, *Designing for People*, title page.
6. Adamson, *Industrial Strength Design*, 129.
7. Gorman, *The Industrial Design Reader*, 131.
8. Bel Geddes, *Miracle in the Evening*.
9. Teague, "The Responsibilities of the Industrial Designer."

Chapter 13

1. John Tjaarda papers.
2. Loewy, "The Classic Criterion: Quality."
3. Adamson, *Industrial Strength Design*, 155–157.
4. "Candid Comments from Design Award Winners," *Merchandising Week*, 20–21.
5. Library of Congress, Manuscript Division, *Raymond Loewy, a Register of His Papers in the Library of Congress*.
6. Teague, *Industrial Designer: The Artist as Engineer*, 172.
7. Larry Sliker, e-mail to author, July 22, 2013.
8. Library of Congress, Manuscript Division, *Raymond Loewy, a Register of His Papers in the Library of Congress*, 1995.
9. Huxtable, "Design Notebook."

Chapter 14

1. Larry Sliker, e-mail to author, July 22, 2013.
2. Hu Jintao, as quoted in Justice, "Going East," 39.
3. Ibid.
4. New York World's Fair guidebook, 1939.

Bibliography

Books

Adamson, Glenn. *Industrial Strength Design*. Cambridge: MIT Press, 2003.
Baker, Eric. *Great Inventions, Good Intentions*. San Francisco: Chronicle Books, 1990.
Bayley, Stephen. *In Good Shape: Style in Industrial Products 1900–1960*. New York: Van Nostrand Reinhold, 1979.
Bel Geddes, Norman. *Horizons*. Boston: Little, Brown, 1932.
_____. *Magic Motorways*. New York: Random House, 1940.
_____. *Miracle in the Evening*. Edited by William Kelley. Garden City, NY: Doubleday, 1960.
Doorden, Dennis P. "Toledo Museum of Art." *The Alliance of Art and Industry: Toledo Designs for a Modern America*. New York: Hudson Hills Press, 2002.
Dreyfuss, Henry. *Designing for People*. New York: Simon & Schuster, 1955.
Frankl, Paul. *Form and Reform*. New York: Harper, 1930.
Gantz, Carroll. *Design Chronicles*. Atglen, PA: Schiffer, 2005.
_____. *The Industrialization of Design*. Jefferson, NC: McFarland, 2011.
Gartman, David. *Auto Opium*. London: Routledge, 1994.
Gelernter, David. *1939: The Lost World of the Fair.* New York: Free Press, 1995.
Gorman, Carla R. *The Industrial Design Reader*. New York: Allworth, 2003.
Heskitt, John. *Industrial Design*. New Yor=k: Thames & Hudson, 1993.
Hughes, Thomas P. *American Genesis*. New York: Viking Penguin, 1989.
Kerr, Ann. *Russel Wright Designs*. Paducah, KY: Collector Books, 1990.
Loewy, Raymond. *Industrial Design*. Woodstock, NY: Overlook Press, 1979.
_____. *Never Leave Well Enough Alone*. New York: Simon & Schuster, 1951.
Meikle, Jeffrey L. *Design in the USA*. Oxford: Oxford University Press, 2005.
Noyes, Eliot. *Organic Design in Home Furnishings*. New York: Museum of Modern Art, 1941.
Papanek, Victor. *Design for the Real World*. New York: Pantheon, 1972.
Patton, Phil. *Made in USA*. New York: Penguin, 1992.
Pile, John. *Dictionary of 20th Century Design*. New York: Da Capo, 1994.
Pulos, Arthur. *The American Design Adventure*. Cambridge: MIT Press, 1988.
_____. *American Design Ethic*. Cambridge: MIT Press, 1983.
Secrest, Meryle. *Frank Lloyd Wright*. New York: HarperCollins, 1992.
Smythe, Michael. *Gifford Jackson: New Zealand Industrial Design Pathfinder*. Auckland, NZ: Creationz Consultants, 2013.
Society of Industrial Designers. *U.S. Industrial Design, 1949–1950*. New York: Studio Publications, 1949.
Teague, Walter Dorwin. *Design This Day: The Technique for Order in the Machine Age*. New York: Harcourt Brace, 1940.
_____. *Industrial Designer: The Artist as Engineer*. Lancaster, PA: Armstrong World Industries, 1998.
Van Doren, Harold. *Industrial Design: A Practical Guide*. New York: McGraw-Hill, 1940.
Wilson, Richard Guy, et al. *The Machine Age in America, 1918–1941*. New York: Brooklyn Museum in association with Harry Abrams, 1986.
Wright, Frank Lloyd. *An Organic Architecture: the Architecture of Democracy*. London: Lund, Humphries, 1939.

Articles

Banham, Reyner. "Machine Aesthetic." *Architectural Review* (London) (April 1955): 225–228, as reprinted in Bayley, *In Good Shape*, 79–81.
Barnes, J.F., and J.O. Reinecke. "Consumers Know

Best." *Electrical Manufacturing* (October 1938). Courtesy Vicki Matranga.
Bel Geddes, Norman. "Ten Years from Now." *Ladies Home Journal* (1930).
Calkins, Ernest Elmo. "Beauty, the New Business Tool." *Atlantic Monthly* (August 1927).
"Candid Comments from Design Award Winners." *Merchandising Week* (July 11, 1966).
Clark, Shannon. "When Modernism Was Still Radical: The Design Laboratory and the Cultural Politics of Depression-Era America." *American Studies* 50, nos. 3/4 (Fall/Winter 2009): 35–61. https://journals.ku.edu/index.php/amerstud/article/view/4176 (accessed June 14, 2013).
Freedgood, Seymour. "Old Business, This Industrial Design." *Fortune* (February 1959).
"A Group of Designs by Gustav Jensen." *Coronet* (March 1940). http://www.oldmagazinearticles.com/Industrial-Designer-Gustav-Jensen_pdf.
Gruber, John. "American Beauty." *Trains* (September 2001).
"He [Raymond Loewy] Streamlines the Sales Curve." *Time* (October 31, 1949): 68–72.
Heller, Steven. "Gustav Jensen Lost?" http://www.printmag.com/design-education/gustav-jensen-lost (accessed June 10, 2013).
Huxtable, Ada Louise. "Design Notebook." *New York Times* (May 24, 1979).
Jackson, Gifford. "Jo Sinel," published by the New Zealand Industrial Design Council's magazine, *Designscape* (February 1977).
Justice, Lorraine. "Going East." *IDSA Innovation* (Summer 2007): 39.
Kogan, Belle. "Making the Most of Materials." *Electrical Manufacturing* (October 1934).
Lesko, Jim. "Industrial Design at Carnegie Institute of Technology, 1934–1967." *Journal of Design History* 10, no. 3 (1997): 269–291.
Loewy, Raymond. "Industrial Design: The Aesthetics of Salesmanship—an American View." *London Times* (November 19, 1945), quoted in Bayley, *In Good Shape*, 71–72.
Long, Christopher. "Kem Weber and the Rise of Modern Design in Southern California." *Antiques* (May 2009). http://www.themagazineantiques.com/articles/kem-weber-southern-california (accessed June 1, 2013).
Müller-Munk, Peter. "Industrial Designers Can Face Today's Challenge." *Electrical Manufacturing* (June 1951).
Nelson, George. "Both Fish and Fowl." *Fortune* (February 1934).
_____. "Business and the Industrial Designer." *Fortune* (July 1949).
Noyes, Eliot. "Organic Design in Home Furnishings." Museum of Modern Art, 1941. http://www.modernism101.com/noyes_organic_design.php (accessed June 21, 2013) and quoted in Pulos, *American Design Ethic*, 411.
Ockman, Joan. "Object Lessons from the Bauhaus." *Art in America* (December 1, 2009). http://www.artinamericamagazine.com/features/object-lessons-from-the-bauhaus/1 (accessed June 14, 2013).
"Patten's Distinctive Designs." *Railroad Heritage* no. 3 (2001).
Porter, Glenn. "Troubled Marriage: Raymond Loewy and the Pennsylvania Railroad." *Invention and Technology* 11, no. 4 (Spring 1996).
Pricola, Jennifer. "Art for Trade's Sake: The Fusion of Commerce and Culture, 1927–1934." American Studies, University of Virginia, 2003. http://xroads.virginia.edu/~ma03/pricola/art/machine.html (accessed June 13, 2013).
Rice, Helen M. "Belle Kogan Remembers." *IDSA Innovation* (Spring 1994).
Seldes, Gilbert. "Profiles: Industrial Classicist—Profile of Walter Dorwin Teague." *New Yorker* (December 15, 1934). http://en.wikipedia.org/wiki/Walter_Dorwin_Teague#cite_ref-newyorker_7-11 (accessed June 13, 2013).
Szerlip, B. Alexandra. "Colossal in Scale, Appalling in Complexity." *Believer* (May 2012). http://www.believermag.com/issues/201205/?read=article_szerlip (accessed June 3, 2013).
Teague, Walter Dorwin. "Modern Design Needs Modern Merchandising." *Forbes* (February 1, 1928).
_____. "A Quarter Century of Industrial Design." *Art & Industry* (London) (October 1951).
Van Doren, Harold. "Streamlining: Fad or Function?" *Design* (October 1949).
Vienne, Veronique. "Good Design: MoMA's Campaign against Doodads." *Communication Arts* (February 1997). http://www.veroniquevienne.com/index.php?rub=article&cat=8&id=22 (accessed June 21, 2013).
Wayt, Hampton C. "Donald Dohner: The Man who Designed 'Rivets.'" *Classic Trains* (Summer 2009): 31–35.
White, E.B. "Machine Age," The Talk of the Town, *New Yorker* (May 21, 1927): 13.

Interviews

Bourke, Robert. *The Reminiscences of Robert E. Bourke* in an interview in 1985, Automotive Design Oral History, Accession 1673, 51 pages, Benson Ford Research Center, The Henry Ford.
Johnson, Philip. Interview by Sharon Zane, 1991. http://www.moma.org/learn/resources/archives/archives_highlights_10_1991 (accessed June 14, 2013).
Kogan, Belle. Interview by Marcy Babbitt, "As a Woman Sees Design," *Modern Plastics* 13, no. 4 (December 1935): 16–17, 49, 51; as reprinted in Gorman *The Industrial Design Reader*, 137–139.
Sinel, Jo. Interview by Robert Harper, librarian, California College of Arts and Crafts, beginning June 4, 1969.

Lectures

Dreyfuss, Henry. "The Industrial Designer's Best Friend and Severest Critic." Address to the American Society of Engineering Educators, Stanford University, California, December 28, 1949.

_____. "The Silent Salesman of Industry." Dinner address at the Industrial Design Conference, 81st annual meeting of the Canadian Manufacturing Association, Toronto, Canada, May 28, 1952, reprinted in *Industrial Canada*, July 1952, 2.

Loewy, Raymond. "The Classic Criterion: Quality." Opening address at the 18th annual conference of IDSA, New York City, November 1, 1962.

Teague, Walter Dorwin. "The Responsibilities of the Industrial Designer." Speech at a dinner of the Detroit Chapter of the Industrial Designers Institute, June 9, 1959.

Miscellaneous Documents

American Management Association. *General Management Series*, no. 99 (1929).

Annual of American Design, published by the American Union of Decorative Artists and Craftsmen (AUDAC), 1931. http://www.modernism101.com/annual_american.php (accessed June 14, 2013).

Bach, Richard. Catalog the exhibition *Contemporary American Design*, Metropolitan Museum of Art, 1928. http://libmma.contentdm.oclc.org/cdm/singleitem/collection/p15324coll10/id/1173/rec/20 (accessed March 15, 2013).

Bourke, Robert E. *Reminiscences of Robert E. Bourke*. http://www.autolife.umd.umich.edu/Design/Bourke_interview.htm (accessed August 10, 2013).

Brooklyn Museum. *American Union of Decorative Artists and Craftsmen* exhibition, May 2–July 1, 1931. http://www.brooklynmuseum.org/opencollection/exhibitions/1439/American_Union_of_Decorative_Artists_and_Craftsmen (accessed March 14, 2013).

Dirig, Robert, archivist at the Art Center College of Design, Pasadena, CA. E-mails of June 12 and 14, 2013, to the author regarding Kem Weber teaching at the Art Center.

Durez promotional brochure, ca. 1942. Courtesy Vicki Matranga.

Gruen, Robert. Nomination of Belle Kogan for IDSA Personal Recognition Award, January 20, 1994.

Iannelli, Alfonso. "History of Industrial Designers Institute." Document, January 20, 1958.

Laine, Christian K. "Ann Swainson: Master Designer." Paper, ca. 1986.

Latham, Richard. Script for taped interview, 1969.

Lehigh University, brochure for 1949 Product Design Seminar in cooperation with the Society of Industrial Designers.

Library of Congress, Manuscript Division. *Raymond Loewy, a Register of His Papers in the Library of Congress*. Prepared by Donna Ellis with the assistance of George Combs, Patricia Craig, Lisa Madison, and Susie Moody, 1995. http://lcweb2.loc.gov/service/mss/eadxmlmss/eadpdfmss/2003/ms003009.pdf (accessed August 10, 2013).

Metropolitan Museum of Art Libraries. 12th Exhibition of Contemporary American Industrial Art, October 12–November 22, 1931. http://libmma.contentdm.oclc.org/cdm/ref/collection/p15324coll10/id/1202 (accessed March 14, 2013).

New York World's Fair official guidebook, 1939.

Sliker, Larry, employee of HDA, 1970–1984. E-mail of July 22, 2013, to author.

Specht, Paul. E-mail of July 27, 2013, to Vicki Matranga.

Steinhilber, Budd. *Looking Back*, informal autobiography, 1995.

John Tjaarda papers, 1916–1962, Henry Ford Museum, accession 1714, published March, 2011. http://catalog.dalnet.lib.mi.us/ipac20/ipac.jsp?session=13M047477G80H.1 2640&menu=search&aspect=subtab303&npp=10&ipp=20&spp=20&profile=henryford&ri=&index=.GW&term=Tjaarda&x=15&y=7&aspect=subta b303 (accessed July 15, 2013).

Waltman, Chauncey. *Memoirs of Chick Waltman*, autobiography. http://debmurray.tripod.com/brown/brnbioref-7.htm (accessed June 12, 2013).

Index

Numbers in ***bold italics*** indicate pages with photographs.

www.ingramcontent.com/pod-product-compliance
Ingram Content Group UK Ltd.
Pitfield, Milton Keynes, MK11 3LW, UK
UKHW051320220526
5938IPUK00022B/148

9 780786 476862